KINEMATICS
AND DYNAMICS
OF MACHINES

McGRAW-HILL SERIES IN MECHANICAL ENGINEERING

KARL H. VESPER, *University of Washington*
Consulting Editor

BARRON | *Cryogenic Systems*
CSANADY | *Theory of Turbomachines*
ECKERT | *Introduction to Heat and Mass Transfer*
ECKERT AND DRAKE | *Analysis of Heat and Mass Transfer*
ECKERT AND DRAKE | *Heat and Mass Transfer*
GRÖBER, ERK, AND GRIGULL | *Fundamentals of Heat Transfer*
HAM, CRANE, AND ROGERS | *Mechanics of Machinery*
HARTENBERG AND DENAVIT | *Kinematic Synthesis of Linkages*
HINZE | *Turbulence*
JACOBSEN AND AYRE | *Engineering Vibrations*
JUVINALL | *Engineering Considerations of Stress, Strain, and Strength*
KAYS | *Convective Heat and Mass Transfer*
KAYS AND LONDON | *Compact Heat Exchanges*
LICHTY | *Combustion Engine Process*
MARTIN | *Kinematics and Dynamics of Machines*
PHELAN | *Dynamics of Machinery*
PHELAN | *Fundamentals of Mechanical Design*
RAVEN | *Automatic Control Engineering*
SCHENCK | *Theories of Engineering Experimentation*
SCHLICHTING | *Boundary-Layer Theory*
SHIGLEY | *Dynamic Analysis of Machines*
SHIGLEY | *Kinematic Analysis of Mechanisms*
SHIGLEY | *Mechanical Engineering Design*
SHIGLEY | *Simulation of Mechanical Systems*
SHIGLEY | *Theory of Machines*
STOECKER | *Refrigeration and Air Conditioning*
SUTTON AND SHERMAN | *Engineering Magnetohydrodynamics*
WILCOCK AND BOOSER | *Bearing Design and Application*

KINEMATICS AND DYNAMICS OF MACHINES

GEORGE H. MARTIN

Associate Professor of Mechanical Engineering
Michigan State University

REVISED PRINTING

McGRAW-HILL
BOOK COMPANY
New York
St. Louis
San Francisco
Auckland
Düsseldorf
Johannesburg
Kuala Lumpur
London
Mexico
Montreal
New Delhi
Panama
Paris
São Paulo
Singapore
Sydney
Tokyo
Toronto

KINEMATICS AND DYNAMICS OF MACHINES

Library of Congress Catalog Card Number 69-12261

07-040637-5

10 11 12 13 KPKP 7 9 8 7 6 5

CONTENTS

PREFACE TO THE REVISED PRINTING

In this revised printing no changes have been made in the subject matter or in the problems for assigned work except for Problem 1 at the end of Chapter 1. However, many refinements have been made in the explanatory material in all but two of the chapters to improve clarity. Also, some definitions have been reworded to increase accuracy. Refinements have been made in many of the figures in the text and four are completely new. In Chapter 12 the presentation and discussion of standard interchangeable gear tooth forms has been updated.

Kinematics of machines is the study of the relative motion of machine parts and is one of the first considerations of the designer in the design of a machine. Dynamics of machines treats with the forces acting on the parts of a machine and the motions resulting from these forces. A dynamic analysis is necessary to ensure that balance is provided for rotating and reciprocating parts and that all members are adequate from the standpoint of strength.

This book is intended as an undergraduate text and contains sufficient material for a full year's work. Part 1 of the text is concerned with kinematics of machines and Part 2 with dynamics of machines. The book may be used for either one of these subjects or both. If the student has already completed a course in kinematics of machines using another text, this book should be readily acceptable for a following course in dynamics of machines. Conventional notation has been used throughout the text. If the time allotted for kinematics and dynamics of machines in the undergraduate curriculum is not sufficient for including all of the topics presented in the text, the material in Chapters 9, 14, 15, and 22 can be presented in a graduate course.

As prerequisites to Part 1, Kinematics of Machines, a course in college physics and mathematics through calculus are sufficient. For Part 2, Dynamics of Machines, kinematics of machines and mechanics courses in statics and dynamics are prerequisites.

Emphasis has been placed on presentation of fundamental principles. Special constructions have been omitted in order to include more basic theory. The major concern has been to present the principles in as simple a manner as possible and to write a book which is easy for the student to read. The text contains numerous examples that are worked out, illustrating application of the theory.

In Part 1, in the chapter on accelerations, equivalent linkages are discussed extensively because of their importance in simplifying the acceleration analysis of direct-contact mechanisms. To aid the student in visualizing the direction of the Coriolis acceleration, it is derived for a special case, but the presentation is then extended to the most general case of plane motion.

The Hartmann construction and Euler-Savary equation as methods for obtaining the radius of curvature of the path of relative motion are given particular emphasis. In the past the author has found that students have often experienced considerable difficulty in understanding and applying these methods, and thus he has endeavored to present them in a simple manner with rules set down for applying them to any problem.

In Chapter 9 the mathematical analysis of velocities and accelerations in mechanisms is treated. The limitations of analysis by trigonometry are pointed out, and most of the chapter is devoted to analysis by complex numbers, including analysis of complex linkages.

Planetary gear trains are discussed in Chapter 10, including planetaries with multiple inputs. The tabular method of analysis and the principle of superposition are used here to aid the student.

In Chapter 14 the synthesis of mechanisms by graphical and mathematical methods is presented, and Chapter 15 is an introduction to analog computer mechanisms.

In Chapter 20 the method of rotating vectors for investigating engine balance has been included because it simplifies the analysis. Gyroscopic effects are treated in Chapter 21, and here the method of angular momentum has been used because it makes it easier to visualize the direction of the gyroscopic forces. Chapter 22, Critical Speeds of Shafts, includes analysis of a shaft with any number of disks and analysis of stepped shafts.

The author believes that problem work is of great importance to students in engineering courses because it aids motivation, understanding, and retention.   At the end of each chapter the text contains a large number of problems of varying difficulty and length, which can all be worked on $8\frac{1}{2} \times 11$ inch paper.   Scales have been specified for all graphical work so that the solutions will come out a reasonable size in order to simplify grading.   There are ample problems so that no additional problem set is necessary.   However, there are several excellent sets of problem plates on the market which may be used along with the text if desired for either the kinematics or the dynamics portion.   Because of the conventional notation used throughout the text, these plates are readily usable.

Material taken directly from other sources is acknowledged in the book, and the author wishes to express appreciation to the manufacturers who generously presented photographs.

GEORGE H. MARTIN

ONE

KINEMATICS
OF MACHINES

FUNDAMENTAL
CONCEPTS

1-1   KINEMATICS

Kinematics of machines is a study of the relative motion of machine parts. Displacement, velocity, and acceleration are considered.

1-2   DYNAMICS

Dynamics of machines treats with the forces acting on the parts of a machine and the motions resulting from these forces.

1-3   MACHINE

A machine is a device for transforming or transferring energy.   It is sometimes defined as consisting of a number of fixed and moving bodies interposed between the source of power and the work to be done for the purpose of adapting one to the other.   The electric motor transforms electrical energy

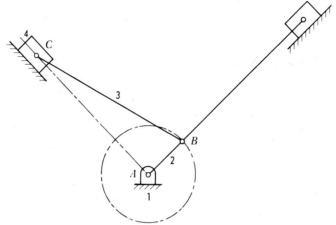

Fig. 1-1

into mechanical energy while its counterpart the electric generator transforms mechanical energy into electrical energy.   In a gasoline engine each
piston connecting rod and the crankshaft act as a machine for transferring
energy.   The input mechanical energy is the product of the force on the
piston and the distance it travels.   This energy is transferred to the crankshaft, where it appears as output mechanical energy and is the product of
the torque and the angle of shaft rotation.

1-4   KINEMATIC DIAGRAM

In studying the motions of machine parts, it is customary to draw the parts
in skeleton form so that only those dimensions which affect their motions
are considered.   The drawing in Fig. 1-1 represents the main elements in
the diesel engine shown in Fig. 1-2.   The stationary members, consisting
of the crankshaft bearing and cylinder wall, are crosshatched and labeled
link 1.   The crank and crankshaft are link 2, connecting rod link 3, and piston
or slider link 4.   *Link* is the name given to any body which has motion relative to another.   Since the bearing and cylinder wall have no motion relative
to one another, they are considered as a single link.   That part of a machine
which is stationary and which supports the moving members is called the
frame and is designated link 1.

In Fig. 1-1 consider the position of the connecting rod for a given
angular position of the crank.   The angular position, velocity, and accelera-

tion of the rod depend only on the lengths of the crank and connecting rod and are in no way affected by the width or thickness of the rod. Thus only the lengths of members 2 and 3 are of importance in a kinematic analysis. The drawing in Fig. 1-1 is known as a *kinematic diagram* and is a scale drawing representing the machine so that only the dimensions which affect its motions are recorded.

All materials have some elasticity. A *rigid link* is one whose deformations are so small that they can be neglected in determining the motions of the various other links in a machine. Bodies 2 and 3 in Fig. 1-1 are considered rigid links. A belt or chain, as illustrated in Fig. 1-3, is a *flexible link*.

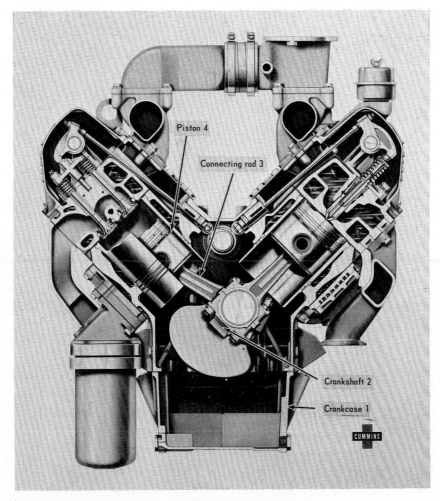

Fig. 1-2. Diesel engine. (*Cummins Engine Company, Inc.*)

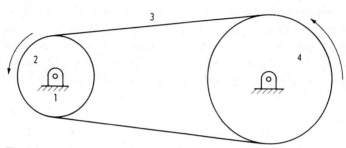

Fig. 1-3

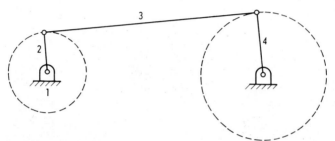

Fig. 1-4

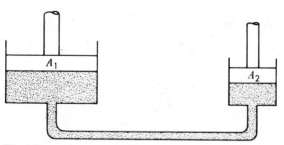

Fig. 1-5

However, if it is always in tension, it may be replaced by a rigid link, as shown in Fig. 1-4, in order to analyze the instantaneous motions of bodies 2 and 4. Similarly, the fluid in the hydraulic press shown in Fig. 1-5 is a flexible link. If the piston areas are A_1 and A_2, then, assuming an incompressible fluid, points A and B of the equivalent rigid link shown in Fig. 1-6 have the same motions as the pistons, provided d_2/d_1 is made equal to A_1/A_2.

1-5 MECHANISM

A *kinematic chain* is a system of links, that is rigid bodies, which are either joined together or are in contact with one another in a manner that permits them to move relative to one another. If one of the links is fixed and the movement of any other link to a new position will cause each of the other links to move to definite predictable positions, the system is a *constrained kinematic chain*. If one of the links is held fixed and the movement of any other link to a new position will not cause each of the other links to move to definite predictable positions, then the system is an *unconstrained kinematic chain*. A *mechanism* or *linkage* is a constrained kinematic chain. When link 1 in Fig. 1-1 is held fixed, the piston and connecting rod each have a definite position for each position of the crank. Thus, the linkage is a constrained kinematic chain and is therefore a mechanism. Suppose, however, we have an arrangement of links as shown in Fig. 1-7. Note that if link 1 is fixed, then with link 2 in the position shown, links 3, 4, and 5 will not have definite predictable positions, but could assume many positions some of which are shown with dashed lines. Thus, this is an unconstrained kinematic chain and is therefore not a mechanism. Further, let us consider Fig. 1-8, in which three bars are pinned

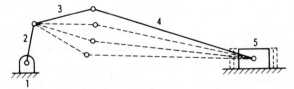

Fig. 1-6

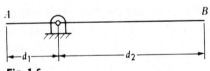

Fig. 1-7

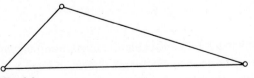

Fig. 1-8

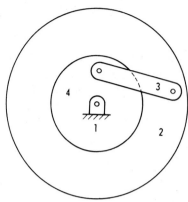

Fig. 1-9

together as shown. An arrangement of this type does not constitute a kinematic chain because there can be no motion of the members relative to one another. Such an assemblage is not a mechanism but a *structure* or *truss*.

A machine is a mechanism which transmits forces. The mechanism in Fig. 1-1 becomes a machine when a force is applied to one of the pistons and is transmitted through the connecting rod and crank to produce rotation of the crankshaft. An electric motor is a machine, but one might question whether it is a mechanism. Actually it is a four-link mechanism which is equivalent to the one shown in Fig. 1-9. In Fig. 1-9 disks 2 and 4 rotate about a common axis and are connected by a connecting bar, which is link 3. In the electric motor the rotating polarity of the field coils is analogous to disk 2, the driver in Fig. 1-9. The magnetic field functions as the connecting bar 3, and the armature is equivalent to the driven disk 4.

Though all machines are mechanisms, not all mechanisms are machines. Many instruments, for example, are mechanisms but are not machines because they do no useful work nor do they transform energy. For example, a clock does no work in excess of that required to overcome its own friction.

1-6   INVERSION

By making a different link in a kinematic chain the fixed member, we obtain a different mechanism. The four mechanisms shown in Figs. 1-10 to 1-13 are derived from the slider-crank chain. The mechanism used in gasoline and diesel engines is shown in Fig. 1-10. If instead of link 1 being fixed, link 2 is held fixed, the result is as shown in Fig. 1-11. This mechanism was used in early radial aircraft engines. The crankshaft was stationary and the crankcase and cylinders rotated. The propeller was attached to the crankcase. Another application of this inversion of the slider-crank chain is the Whitworth quick-return mechanism, which is discussed later in this book in Sec. 3-8. In Fig. 1-12 link 3 is held fixed. This mechanism is used in toy oscillating-cylinder steam engines. In Fig. 1-13 link 4 has been made the fixed member. This mechanism is commonly used for pumps.

It is of importance to note that inversion of a mechanism in no way changes the relative motion between its links. For example, in Figs. 1-10 to 1-13 if link 2 rotates $\theta°$ clockwise relative to link 1, link 4 will move to the

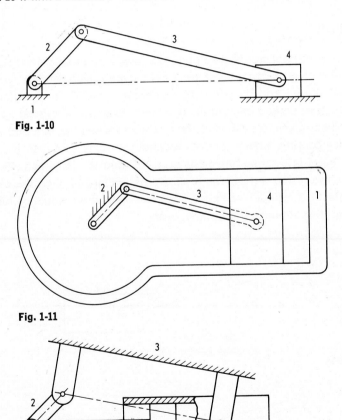

Fig. 1-10

Fig. 1-11

Fig. 1-12

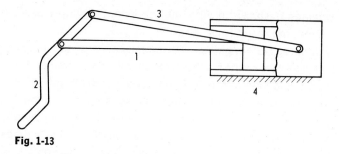

Fig. 1-13

right a definite amount along a straight line on link 1. This is true no matter which link is held fixed.

1-7 PAIRING

Two bodies in contact constitute a pair.   *Lower pairing* exists when two surfaces are in contact.   Examples of lower pairing are a piston and its cylinder wall and a journal and its supporting bearing.   *Higher pairing* refers to contact which exists at a point or along a line.   Examples of higher pairing are a ball bearing, where point contact exists between the ball and race, and a roller bearing, where contact between the roller and race is along a line.   In Fig. 1-14 lower pairing exists at A, B, C, and D.   If instead of being cylindrical, the piston were made a sphere as in Fig. 1-15, then the piston would contact the cylinder wall along a circle.   Thus there would be higher pairing and it would make for greater wear.

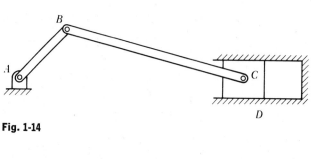

Fig. 1-14

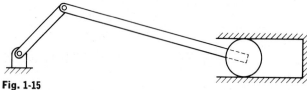

Fig. 1-15

1-8 PLANE MOTION

A body has plane motion if all its points move in planes which are parallel to some reference plane. The reference plane is called the *plane of motion*. Plane motion can be one of three types: translation, rotation, or a combination of translation and rotation.

1-9 TRANSLATION

A body has *translation* if it moves so that all straight lines in the body move to parallel positions. *Rectilinear translation* is a motion wherein all points of the body move in straight-line paths. The piston in Fig. 1-1 has rectilinear translation. A translation in which points in the body move along curved paths is called *curvilinear translation*. In the parallel crank mechanism shown in Fig. 3-2 the connecting link 3 has curvilinear translation.

1-10 ROTATION

In rotation all points in a body remain at fixed distances from a line which is perpendicular to the plane of motion. This line is the axis of rotation, and points in the body describe circular paths about it. The crank in Fig. 1-1 has a motion of rotation if the frame of the engine is fixed.

1-11 TRANSLATION AND ROTATION

Many machine parts have motions which are a combination of rotation and translation. For example in the engine in Fig. 1-16 consider the motion of the connecting rod as it moves from position BC to $B'C'$. These positions are shown again in Fig. 1-17. Here we see that the motion is equivalent to a translation from BC to $B''C'$ followed by a rotation from $B''C'$ to $B'C'$. Another equivalent motion is illustrated in Fig. 1-18. This shows a rotation of the rod about C from position BC to $B'''C$, followed by a translation from $B'''C'$ to $B'C'$. Thus the motion of the connecting rod can be considered as a rotation about some point plus a translation.

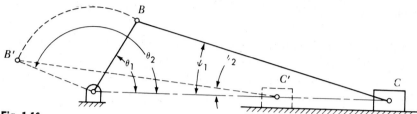

Fig. 1-16

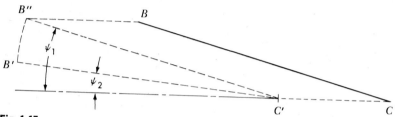

Fig. 1-17

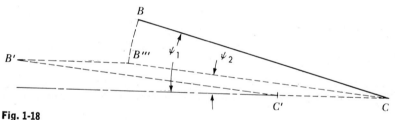

Fig. 1-18

1-12 HELICAL MOTION

A point which rotates about an axis at a fixed distance and at the same time moves parallel to the axis describes a helix. A body has *helical motion* if each point in the body describes a helix. The motion of a nut along a screw is a common example.

1-13 SPHERICAL MOTION

A point has *spherical motion* if it moves in three-dimensional space and remains at a fixed distance from some fixed point. A body has spherical motion if each point in the body has spherical motion. In the ball-and-socket joint in Fig. 1-19, if either the socket or rod is held fixed, the other will move with spherical motion.

1-14 CYCLE, PERIOD, AND PHASE OF MOTION

A mechanism completes a *cycle of motion* when it moves through all its possible configurations and returns to its starting position.   Thus the slider-crank mechanism in Fig. 1-16 completes a cycle of motion as the crank makes one revolution.   The time required for one cycle is the *period.*   The relative positions of the links at any instant during the cycle of motion for the mechanism constitute a phase.   When the crank in Fig. 1-16 is in position θ_1, the mechanism is in one phase of its motion.   When the crank is in position θ_2, the mechanism is in another phase.

1-15 VECTORS

Two types of quantities are treated in mechanics.   *Scalar quantities* are those which have magnitude only.   Examples are distance, area, volume, and time.   *Vector quantities* have magnitude and direction.   Examples are displacement, velocity, acceleration, and force.

A vector quantity can be represented by a straight line with an arrowhead as illustrated in Fig. 1-20. The magnitude of the vector A is represented by its length, which is drawn to any convenient scale.   For example, if we wish to represent a velocity of 20 fps and we let 1 in. on the paper

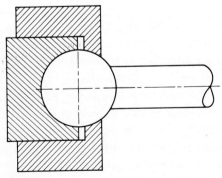

Fig. 1-19

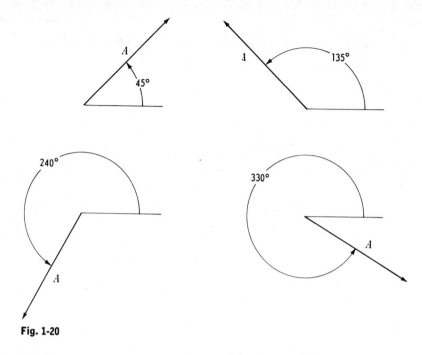

Fig. 1-20

represent 10 fps, then *A* would be drawn 2 in. long. The arrowhead represents the *head* or *terminus* of the vector and the other end is called the *tail*
or *origin*. The direction of a vector may be described by specifying the angle
in degrees which it makes with the horizontal (*x* axis) measuring in the
conventional counterclockwise direction. This is illustrated in Fig. 1-20.

1-16   ADDITION AND SUBTRACTION OF VECTORS

The symbol ++ is commonly used to denote the addition of vector quantities
and the symbol → is used to denote subtraction. The sum of vectors *A*
and *B* is written *A* ++ *B*, and the subtraction of *B* from *A* as *A* → *B*.

Vectors *A* and *B* in Fig. 1-21 can be added by laying them out in the
manner shown in Fig. 1-22 or in the manner shown in Fig. 1-23. Point *O* is
the starting point, called the *pole*, and may be chosen at any convenient
location in the plane of the vectors. From the pole, vectors *A* and *B* are
laid off with the tail of one placed at the head of the other. Their sum is
called the *resultant* and is shown by the dashed line in the figures.

It should be noted that when laying out vectors in order to determine their resultant, their given magnitudes and directions must be maintained but the order in which they are laid out does not affect their resultant. The resultant is always directed outward from the pole and is the closing side of a polygon.

The subtraction of vectors A and B in Fig. 1-21 is accomplished as follows. In order to find the resultant of $A \rightarrow B$ we may write this as $A \lefttwoheadrightarrow (-B)$. That is, we add a minus value of vector B to vector A as shown in Fig. 1-24. Similarly, to find the resultant of $B \rightarrow A$ we may write this as $B \lefttwoheadrightarrow (-A)$. Thus a minus value of vector A is added to vector B as shown in Fig. 1-25. We note that subtraction of a vector consists of adding its negative value.

1-17 COMPOSITION AND RESOLUTION OF VECTORS

Composition refers to the adding together of any number of vectors. The sum is called their *resultant,* and the vectors are called the *components of the resultant.* It is important to note that any given number of vectors has but

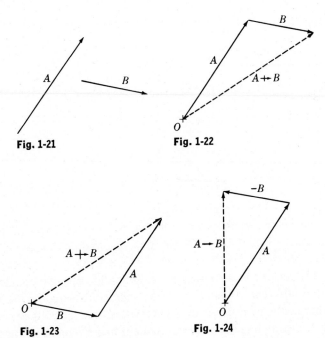

Fig. 1-21   **Fig. 1-22**

Fig. 1-23   **Fig. 1-24**

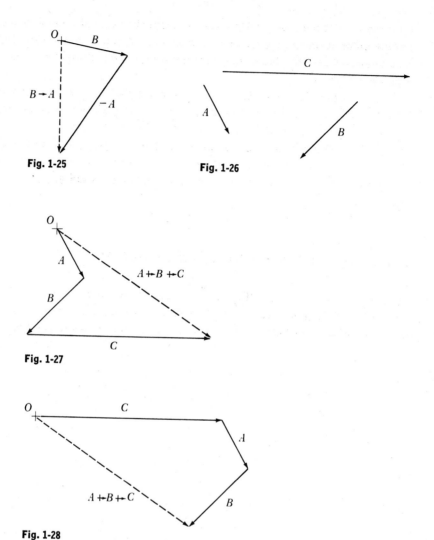

Fig. 1-25

Fig. 1-26

Fig. 1-27

Fig. 1-28

one resultant.   For example, the vectors in Fig. 1-26 can be added in any order, as illustrated by Figs. 1-27 to 1-29, yet their resultant is the same.

Resolution refers to the breaking down of a vector into any number of components.   Any vector can be resolved into an infinite number of sets of components.   It is often convenient to resolve a vector into two components, for example, a horizontal component and a vertical component.   If a vector is resolved into two components, each component has a magnitude and a direction.   When any two of these four quantities are known, the other two can be found.

Suppose we wish to find the two components of vector A, Fig. 1-30, if their directions are known as shown by dashed lines at the left in the figure.   Through the origin and terminus of vector A draw lines parallel to given directions.   The intersection of these two lines determines the magnitudes of components B and C.

In Fig. 1-31 the two components of vector A are to be found when their

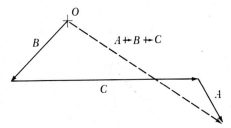

Fig. 1-29

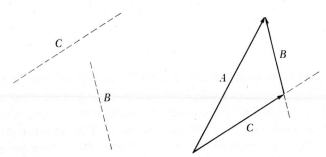

Fig. 1-30

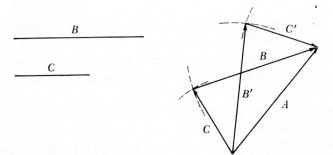

Fig. 1-31

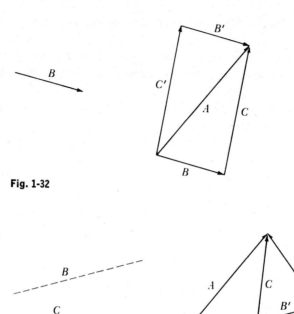

Fig. 1-32

Fig. 1-33

magnitudes B and C, as shown to the left, are known.   From the origin of vector A two arcs are swung, one with a radius equal to B and the other with a radius equal to C.   Similarly from the terminus of vector A we swing two arcs, one with radius B and the other with radius C.   The intersections of the arcs determine the directions of the two components.   There are two possible solutions.   The components are vectors B and C or vectors B' and C'.

In Fig. 1-32 the two components of vector A are to be found; the magnitude and direction of one component, B, is known.   From the origin of vector A this component is laid off.   The closing side of the polygon then determines the other component, C.   Another construction, B' and C', is shown.   This is equivalent to the solution B and C.

Next, let us consider Fig. 1-33.   The two components of vector A are to be found when one is to be in the direction of B and the magnitude of the other is to be C.   A line parallel to B is drawn through the origin of vector A.   An arc of radius C is then drawn from the terminus of vector A.   The intersection of this arc with the line of direction B determines vector C or C'. There are two solutions, BC and $B'C'$. If the arc is tangent to the line of direction B, then there is just one solution.

PROBLEMS

1-1 In Fig. P1-1a the rectangle at point C is used to indicate that BC and CD is one continuous member and that BC and CD are not separate parts pivoted together at C. This symbol is used throughout the text. Which of these drawings represent mechanisms?

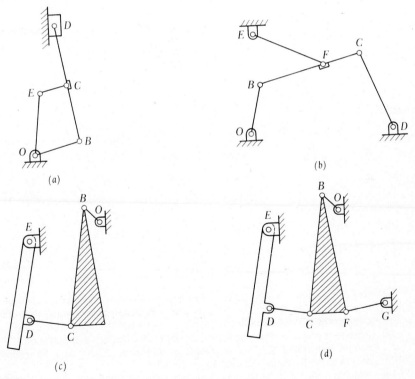

Fig. P1-1

1-2 Given the vectors in Fig. P1-2. Use a scale of 1 in. = 10 units and determine the following vectors: (a) $H = A \nrightarrow B$, (b) $I = A \rightarrow B$, (c) $J = A \rightarrow C \nrightarrow B \rightarrow E$, (d) $K = G \rightarrow F \nrightarrow D \rightarrow C \nrightarrow B$, and (e) $L = -D \nrightarrow E \rightarrow F \rightarrow G$.

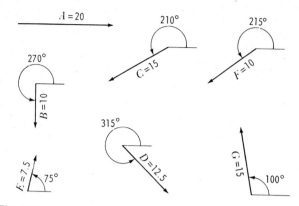

Fig. P1-2

1-3 For each of the vector polygons in Fig. P1-3 write the vector equation giving the resultant R.

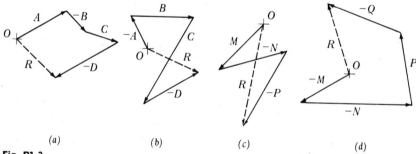

(a)   (b)   (c)   (d)

Fig. P1-3

1-4 Resolve a vector A which has a magnitude of 20 units and a direction of 135° into two vectors B and C. B is to have a direction of 80° and C is to have a direction of 210°. Scale: 1 in. = 10 units. Determine the magnitudes of B and C.

1-5 Resolve a vector T which has a magnitude of 20 units and a direction of 120° into two vectors R and S. The magnitude of R is to be 12 units and the magnitude of S is to be 26 units. Scale: 1 in. = 10 units.

1-6 Resolve a vector A which has a magnitude of 20 units and a direction of 210° into two vectors B and C. C is to be 15 units and is to be directed at 75°. Determine the magnitude of B and its direction in degrees. Scale: 1 in. = 10 units.

1-7 Resolve a vector T which has a magnitude of 24 units and a direction of 345° into two vectors R and S. S is to have a direction of 315°. The magnitude of R is to be 13 units. Scale: 1 in. = 10 units. Indicate on the drawing the magnitude of S.

2

PROPERTIES OF MOTION,
RELATIVE MOTION,
METHODS OF MOTION TRANSMISSION

2-1 INTRODUCTION

The motion of a rigid body can be defined in terms of the motion of one or more of its points. Thus we shall first study the motion of a point.

2-2 PATH OF MOTION AND DISTANCE

The path of a moving point is the locus of its successive positions, and the distance traveled by the point is the length of its path of motion. Distance is a scalar quantity since it has magnitude only.

2-3 LINEAR DISPLACEMENT AND VELOCITY

Displacement of a point is the change of its position and is a vector quantity. In Fig. 2-1 as point P moves along path MN from position B to position C, its linear displacement is the difference in position vectors R_1 and R_2.

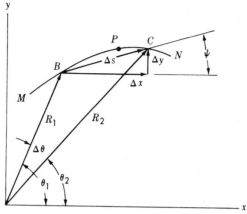

Fig. 2-1

This can be expressed as vector Δs, which is the sum of vectors Δx and Δy. Thus

$$\Delta s = \Delta x + \Delta y \qquad (2\text{-}1)$$

The magnitude of the linear displacement may be expressed in terms of the magnitudes Δx and Δy

$$\Delta s = \sqrt{(\Delta x)^2 + (\Delta y)^2} \qquad (2\text{-}2)$$

and its direction with respect to the x axis is

$$\tan \psi = \frac{\Delta y}{\Delta x} \qquad (2\text{-}3)$$

If the displacement is decreased indefinitely, then vector Δs approaches the tangent to the path at point B. Hence the motion of a point at any instant is in a direction tangent to its path.

Linear velocity is the time rate of change of linear displacement. In Fig. 2-1 point P moves from position B to position C in time Δt. The average velocity during this interval is

$$V_{\text{av}} = \frac{\Delta s}{\Delta t}$$

The instantaneous linear velocity of the point, when it is at position B, is

$$V = \lim_{\Delta t \to 0} \frac{\Delta s}{\Delta t} = \frac{ds}{dt} \qquad (2\text{-}4)$$

and it is directed tangent to the path.

2-4 ANGULAR DISPLACEMENT AND VELOCITY

Consider the body in Fig. 2-2 to be rotating about the fixed axis O and let P be a point fixed in the body. As P moves to P', the angular displacement of line OP or the body is $\Delta\theta$, which occurs in a time Δt. The average angular velocity of the body during this interval is

$$\omega_{av} = \frac{\Delta\theta}{\Delta t}$$

The instantaneous angular velocity of the body for the position OP is

$$\omega = \lim_{\Delta t \to 0} \frac{\Delta\theta}{\Delta t} = \frac{d\theta}{dt} \tag{2-5}$$

In Fig. 2-2 point P has a radius of rotation R equal to length OP. V is the velocity of point P and is tangent to the path PP' and hence perpendicular to radius R. Arc length PP' equals $R\,\Delta\theta$, where $\Delta\theta$ is expressed in radians. The magnitude of the velocity of point P, when in position P, is

$$V = \lim_{\Delta t \to 0} \frac{R\,\Delta\theta}{\Delta t} = R\frac{d\theta}{dt} \tag{2-6}$$

Substitution of Eq. (2-5) into (2-6) gives

$$V = R\omega \tag{2-7}$$

where ω is expressed in radians per unit of time. If ω is expressed in radians per minute and R is expressed in feet, then since a radian is dimensionless, the units for V are feet per minute. Thus

$$V = R\omega$$

$$\frac{ft}{min} = ft \times \frac{radians}{min}$$

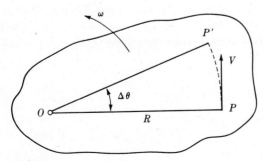

Fig. 2-2

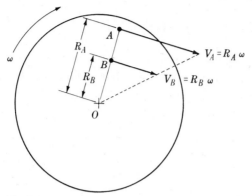

Fig. 2-3

The angular velocity of a machine part is often expressed in revolutions per minute (abbreviated rpm) and is denoted by n. Since each revolution equals 2π rad,

$$\omega = 2\pi n \qquad (2\text{-}8)$$

and

$$V = 2\pi R n \qquad (2\text{-}9)$$

Since the radii of rotation for all points in a rotating body have the same angular velocity ω, we see from Eq. (2-7) that the magnitudes of their linear velocities are directly proportional to their radii. Thus in Fig. 2-3,

$$\frac{V_A}{V_B} = \frac{R_A}{R_B} \qquad (2\text{-}10)$$

2-5   LINEAR AND ANGULAR ACCELERATION

Linear acceleration is the time rate of change of linear velocity. We will first consider the case of a point having rectilinear motion. Hence the velocity can change in magnitude only. Let the initial velocity be V_0 and the velocity after a time interval Δt be V; then during the interval Δt the average acceleration is

$$A = \frac{V - V_0}{\Delta t} = \frac{\Delta V}{\Delta t} \qquad (2\text{-}11)$$

The instantaneous acceleration is

$$A = \lim_{\Delta t \to 0} \frac{\Delta V}{\Delta t} = \frac{dV}{dt}$$

But

$$V = \frac{ds}{dt}$$

Hence

$$A = \frac{d^2s}{dt^2} \tag{2-12}$$

If a point moves with uniformly accelerated motion, A is constant and the average acceleration for any time interval is equal to the value at any instant.   For this type of motion, Eq. (2-11) can be written

$$V = V_0 + At \tag{2-13}$$

where the time interval Δt is expressed simply as t.

If a point moves with constant velocity, its displacement in a time interval t is

$$s = Vt \tag{2-14}$$

For a point having variable velocity, the displacement is the product of the average velocity and the time.   Hence for uniform acceleration, the displacement is

$$s = \tfrac{1}{2}(V_0 + V)t \tag{2-15}$$

where V is the final velocity.   Substituting the value of V from Eq. (2-13) in Eq. (2-15), we obtain

$$s = V_0 t + \tfrac{1}{2}At^2 \tag{2-16}$$

Substituting the value of t from Eq. (2-13) in Eq. (2-15) gives

$$V^2 = V_0^2 + 2As \tag{2-17}$$

Angular acceleration is the time rate of change of angular velocity and is

$$\alpha = \frac{d\omega}{dt} = \frac{d^2\theta}{dt^2} \tag{2-18}$$

For a body having uniform angular acceleration, α is constant.   For motion of this type the same analysis as used for uniform linear acceleration gives Eqs. (2-19) to (2-22), which are similar to Eqs. (2-13), (2-15), (2-16), and (2-17), except that s, V, and A are replaced by θ, ω, and α, respectively.   Thus

$$\omega = \omega_0 + \alpha t \tag{2-19}$$
$$\theta = \tfrac{1}{2}(\omega_0 + \omega)t \tag{2-20}$$
$$\theta = \omega_0 t + \tfrac{1}{2}\alpha t^2 \tag{2-21}$$
$$\omega^2 = \omega_0^2 + 2\alpha\theta \tag{2-22}$$

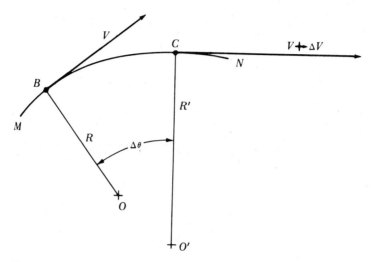

Fig. 2-4

Angular displacement, velocity, and acceleration are either clockwise (cw) or counterclockwise (ccw). Throughout this text counterclockwise quantities will be considered positive (+) and clockwise quantities negative (−).

2-6 NORMAL AND TANGENTIAL ACCELERATION

A point can have acceleration in a direction either normal, tangential, or both to its path of motion. If the point has curvilinear motion, it will have a *normal acceleration* resulting from a change in direction of its linear velocity; if the magnitude of its linear velocity changes, the point will also have a *tangential acceleration*. A point having rectilinear motion has no normal acceleration since its linear velocity does not change direction; it will have a tangential acceleration if the linear velocity changes in magnitude.

In Fig. 2-4 consider a point moving along path MN. Its velocity, when at position B, is V. After time Δt the point is at position C, and its velocity is $V + \Delta V$. R and R' are the radii of curvature at points B and C on the path. The two velocity vectors are shown again in Fig. 2-5 where ΔV, the change in velocity, is the sum of components ΔV^n and ΔV^t. Component ΔV^n results from the change in direction of vector V, and ΔV^t is the change in magnitude of V.

The tangential acceleration A^t of the point when at position B is the time rate of change of the magnitude of its linear velocity; thus

$$A^t = \lim_{\Delta t \to 0} \frac{\Delta V^t}{\Delta t} = \frac{dV^t}{dt} \qquad (2\text{-}23)$$

As Δt approaches zero, point C approaches B, and vector ΔV^t in Fig. 2-5 becomes tangent to the path at B in the limit. Hence A^t is tangent to the path. Substituting the derivative of the expression for V in Eq. (2-7) in Eq. (2-23) gives

$$A^t = R\frac{d\omega}{dt} \qquad (2\text{-}24)$$

Next, substituting the expression for $d\omega/dt$ in Eq. (2-18) in Eq. (2-24) gives

$$A^t = R\alpha \qquad (2\text{-}25)$$

The normal acceleration A^n of the point when at position B is the time rate of change of its velocity in a direction normal to the path. Thus

$$A^n = \lim_{\Delta t \to 0} \frac{\Delta V^n}{\Delta t} = \frac{dV^n}{dt} \qquad (2\text{-}26)$$

In Fig. 2-5 angle $\Delta\theta$ becomes $d\theta$ and the magnitude of ΔV^n becomes equal to the arc length in the limit. Thus

$$dV^n = V\,d\theta \qquad (2\text{-}27)$$

Substituting Eq. (2-27) in Eq. (2-26) gives

$$A^n = V\frac{d\theta}{dt} \qquad (2\text{-}28)$$

Substituting Eqs. (2-5) and (2-7) in Eq. (2-28) gives

$$A^n = V\omega = R\omega^2 = \frac{V^2}{R} \qquad (2\text{-}29)$$

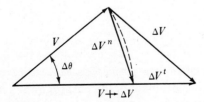

Fig. 2-5

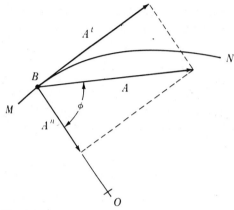

Fig. 2-6

Let us again refer to Fig. 2-5 and Eq. (2-26).   We note that as $\Delta\theta$ approaches zero and ΔV^n becomes dV^n, the latter is directed toward the center of curvature of the path, and thus the direction of A^n is always toward the center of curvature.   A point having curvilinear motion always has a normal component of acceleration.   A point having rectilinear motion has no normal component of acceleration since

$$A^n = \frac{V^2}{R = \infty} = 0$$

The total linear acceleration A of the moving point is the vector sum of A^n and A^t, as shown in Fig. 2-6.   Its magnitude is

$$A = \sqrt{(A^n)^2 + (A^t)^2} \tag{2-30}$$

and its direction is

$$\phi = \tan^{-1}\frac{A^t}{A^n} \tag{2-31}$$

where A^n and A^t are the magnitudes of the normal and tangential components.

2-7 SIMPLE HARMONIC MOTION

A particle having rectilinear translation has *simple harmonic motion* if its acceleration is proportional to the displacement of the particle from a fixed point and is of opposite sign.   The mathematical expression for simple harmonic motion is

$$A = -Kx \qquad (2\text{-}32)$$

where A = acceleration
x = displacement
K = a constant

It is often convenient to represent simple harmonic motion by the projection upon a diameter of a point moving on a circle. In Fig. 2-7 let line OP rotate with constant angular velocity ω and let B be the projection of point P on the x axis. The displacement of point B from point O is

$$x = R \cos \omega t \qquad (2\text{-}33)$$

and its velocity V and acceleration A are

$$V = \frac{dx}{dt} = -R\omega \sin \omega t \qquad (2\text{-}34)$$

$$A = \frac{d^2x}{dt^2} = -R\omega^2 \cos \omega t \qquad (2\text{-}35)$$

A plot of Eqs. (2-33) to (2-35) appears in Fig. 2-8. From an inspection of Eqs. (2-33) and (2-35) we note that

$$A = -\omega^2 x \qquad (2\text{-}36)$$

Since ω is constant, Eq. (2-36) is the same as Eq. (2-32), which is the definition of simple harmonic motion.

A Scotch-yoke mechanism is shown in Fig. 2-9. If link 2 rotates with constant angular velocity, link 4 has simple harmonic motion.

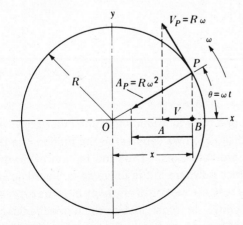

Fig. 2-7

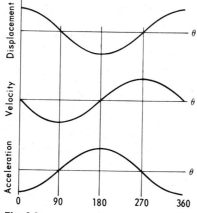

Fig. 2-8

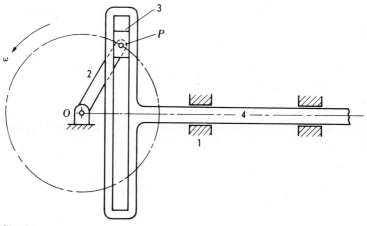

Fig. 2-9

2-8 ABSOLUTE MOTION

Absolute motion is the motion of a body in relation to some other body which is at rest. In our previous discussion we considered the motion of a point in relation to some fixed coordinate axes. Thus the motion of the point was an absolute motion. Since nowhere in the universe is there known to be a body which is at absolute rest, the motion of any body must be expressed in relation to some other body. In most instances in mechanics and kinematics we may regard the earth as fixed. The motion of a body in relation to the earth is then an absolute motion.   When referring to absolute motion, it is common to drop the word *absolute;* that is, if an automobile is traveling at 60 mph, this is its speed relative to the ground.   We do not say that its speed is 60 mph absolute, but merely say that its speed is 60 mph.

2-9   RELATIVE MOTION

A body has motion relative to another body only if there is a difference in their absolute motions.   The displacement of a body M relative to a body N is the absolute displacement of M minus the absolute displacement of N.   Similarly the velocity of body M relative to body N is the absolute velocity of M minus the absolute velocity of N.   Also the acceleration of body M relative to body N is the absolute acceleration of M minus the absolute acceleration of N.   If we consider an automobile moving along a straight path, the absolute displacement of the frame is a translation.   A wheel will have an absolute displacement consisting of a translation which is the same as that of the frame, and in addition a rotation.   Then in accordance with our definition of relative motion, the displacement of the wheel relative to the frame is merely a rotation.

As an illustration of relative motion consider two automobiles A and B in Fig. 2-10 traveling with velocities of 60 mph and 40 mph.   Let V_A and V_B denote their absolute velocities.   Whenever a vector is written with a single letter as a subscript it is understood to be an absolute quantity.   The velocity of A relative to B is written $V_{A/B}$ and is the absolute velocity of A minus the absolute velocity of B.   Thus

$$V_{A/B} = V_A \rightarrow V_B$$

The velocity of A relative to B is the velocity that A would appear to have to an observer in car B if the observer were to imagine that car B were at rest. To the observer, car A would appear to be moving to the left at 20 mph.   This is shown as $V_{A/B}$ in the figure.   The velocity of B relative to A is written

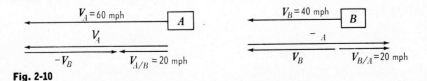

Fig. 2-10

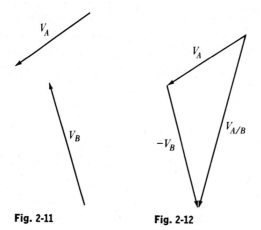

Fig. 2-11   **Fig. 2-12**

$V_{B/A}$ and is the absolute velocity of B minus the absolute velocity of A. Hence

$$V_{B/A} = V_B \rightarrow V_A$$

The velocity of B relative to A is the velocity car B would appear to have to an observer in car A if the observer were to imagine that car A were at rest. To the observer, car B would appear to be moving to the right at 20 mph. This is shown as $V_{B/A}$ in the figure.

Another example of relative motion is illustrated in Fig. 2-11, where V_A and V_B are the velocities of two airplanes. The velocity of A relative to B is the absolute velocity of A minus the absolute velocity of B. Hence

$$V_{A/B} = V_A \rightarrow V_B$$
$$= V_A +\!\!\!+ (-V_B)$$

as shown in Fig. 2-12. Similarly the velocity of B relative to A is the absolute velocity of B minus the absolute velocity of A. Thus

$$V_{B/A} = V_B \rightarrow V_A$$
$$= V_B +\!\!\!+ (-V_A)$$

as shown in Fig. 2-13.

In a vector equation, terms can be transposed, provided their signs are changed. For example,

$$V_{A/B} = V_A \rightarrow V_B$$
$$-V_A = -V_{A/B} \rightarrow V_B$$
$$V_B = V_A \rightarrow V_{A/B}$$

Further, in a vector equation if the subscripts are reversed on a vector, the sign of the vector must be changed.   For example, if we reverse the subscripts on $V_{A/B}$ in the last equations,

$$-V_{B/A} = V_A \rightarrow V_B$$

$$-V_A = V_{B/A} \rightarrow V_B$$

$$V_B = V_A \nrightarrow V_{B/A} \tag{2-37}$$

These equations can be verified by drawing the vector diagrams.

Equation (2-37) is known as the relative-velocity equation; its right-hand side can be shown to be equal to V_B as follows.   By definition,

$$V_{B/A} = V_B \rightarrow V_A$$

Hence

$$V_B = V_A \nrightarrow V_{B/A}$$

may be written

$$V_B = V_A \nrightarrow (V_B \rightarrow V_A)$$

$$= V_A \nrightarrow V_B \nrightarrow (-V_A)$$

$$= V_B$$

Thus Eq. (2-37) states that knowing the velocity of a point A, we can find the velocity of a point B by adding to V_A the velocity of point B relative to A.   In a later chapter we will make extensive use of this idea in finding velocities of points on a mechanism when the velocity of some point on the mechanism is known.

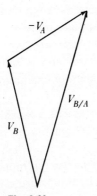

Fig. 2-13

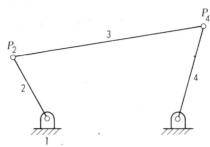

Fig. 2-14

Since linear displacements and linear accelerations are vector quantities, they are handled in the same manner as linear velocities.

If a body 2 and a body 3 have motion in a plane or parallel planes, then
their relative angular motions are defined as the difference in their absolute
angular motions. Thus

$$\theta_{3/2} = \theta_3 - \theta_2$$
$$\omega_{3/2} = \omega_3 - \omega_2$$
$$\alpha_{3/2} = \alpha_3 - \alpha_2$$

where θ, ω, and α are considered positive if counterclockwise and negative
if clockwise.

2-10 METHODS OF TRANSMITTING MOTION

Since all mechanisms transmit motion, it is of interest to classify them as
to basic types in accordance with the manner in which they transmit motion.

In the mechanisms in Figs. 2-14 to 2-16, link 2 is the *driver*. Link 4 is the
driven member and is called the *follower*. For a certain range of angular
motion of the driver the follower is given a definite angular motion. In Fig.
2-14 motion is transmitted from the driver to the follower through link 3,
which is a rigid bar. Hence link 3 is a rigid connector and is called the
coupler. The connecting rod in a gasoline engine is another example of this
type of connector.

Band 3 connecting links 2 and 4 in Fig. 2-15 represents a flexible connector. Belt and chain drives are examples of mechanisms employing
flexible connectors.

In Chap. 1 it was explained how some mechanisms use hydraulic fluid
or a magnetic field as the intermediate link. In the hydraulic press in Fig.
1-5 the fluid serves as the connecting link between the driver, piston A_2, and

the follower, piston A_1.   Further, with the aid of Fig. 1-9, it was explained how the magnetic field acts as the connecting link in the four-bar mechanism which in effect exists in an electric motor.

The linkage in Fig. 2-16 is called a *direct-contact mechanism* because the driver and follower are in direct contact.   The motion imparted to the follower by the driver will depend on the shape of the outlines on links 2 and 4, as well as on the relative position of the links.   In direct-contact mechanisms the driver is usually called the *cam* and the driven member is called the *follower*.   A pair of contacting teeth on a set of gears is another example of a direct-contact mechanism.

2-11   LINE OF TRANSMISSION

Motion is transmitted from the driver to the follower along the line of transmission.   In Fig. 2-14 motion is transmitted from link 2 to link 4 by link 3. Thus P_2P_4 is the line of transmission.   Similarly in the belt drive in Fig. 2-15

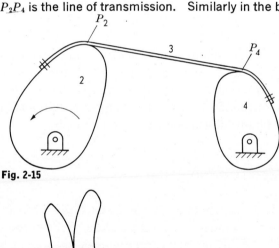

Fig. 2-15

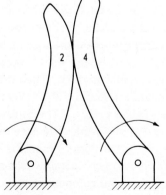

Fig. 2-16

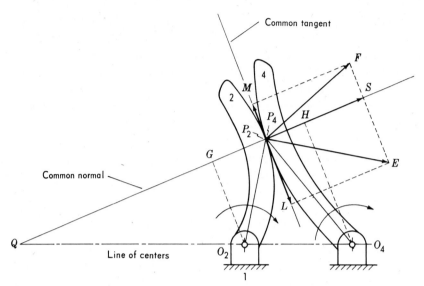

Fig. 2-17

line P_2P_4 is the line of transmission, since motion of the driving pulley is transmitted along this line to the driven pulley. In the direct-contact mechanism in Fig. 2-17 the driving link can transmit motion to the driven link only if the driver has motion in the direction of the common normal. Thus here the line of transmission is the common normal.

2-12   ANGULAR–VELOCITY RATIO

In Fig. 2-17 let P_2 and P_4 be points on bodies 2 and 4, respectively, which are coincident for the instantaneous position of the mechanism shown. Through the contact point the common normal and tangent lines have been drawn. The radii of rotation for P_2 and P_4 are O_2P_2 and O_4P_4. Vector P_2E represents the velocity of P_2 and is perpendicular to O_2P_2. Then the components of P_2E along the normal and tangent are P_2S and P_2L. Vector P_4F represents the velocity of P_4 and is perpendicular to O_4P_4. The components of P_4F along the normal and tangent are P_4S and P_4M. We note that the components of velocities P_2E and P_4F along the normal must be equal; otherwise bodies 2 and 4 would either move out of contact or would deform one another. If velocity P_2E is known, its normal component P_2S can be determined. Since the direction of velocity P_4F is known, its magnitude can then be found by drawing line FS perpendicular to P_4S. This locates point F.

The angular velocities of links 2 and 4 are

$$\omega_2 = \frac{V}{R} = \frac{P_2E}{O_2P_2} \quad \text{and} \quad \omega_4 = \frac{P_4F}{O_4P_4}$$

Hence,

$$\frac{\omega_2}{\omega_4} = \frac{P_2E}{O_2P_2} \frac{O_4P_4}{P_4F} \tag{2-38}$$

Lines O_2G and O_4H are perpendicular to the normal. Then triangle O_2GP_2 is similar to triangle P_2SE, and triangle O_4HP_4 is similar to triangle P_4SF. Hence,

$$\frac{P_2E}{O_2P_2} = \frac{P_2S}{O_2G} \tag{2-39}$$

and

$$\frac{P_4F}{O_4P_4} = \frac{P_4S}{O_4H} \tag{2-40}$$

Dividing (2-39) by (2-40) gives

$$\frac{P_2E}{O_2P_2} \frac{O_4P_4}{P_4F} = \frac{P_2S}{O_2G} \frac{O_4H}{P_4S} = \frac{O_4H}{O_2G} \tag{2-41}$$

Substituting (2-41) into (2-38) gives

$$\frac{\omega_2}{\omega_4} = \frac{O_4H}{O_2G} \tag{2-42}$$

Further, triangle O_2GQ is similar to triangle O_4HQ and

$$\frac{O_4H}{O_2G} = \frac{O_4Q}{O_2Q} \tag{2-43}$$

Substitution of (2-43) into (2-42) gives

$$\frac{\omega_2}{\omega_4} = \frac{O_4Q}{O_2Q} \tag{2-44}$$

In Figs. 2-18 and 2-19 bodies 2 and 4 have identical velocities along the line of transmission P_2P_4. The notation here is similar to that in Fig. 2-17, and the same proof applies. Hence for mechanisms of any of these types the ratio of the angular velocities of the driver and follower is inversely as the lengths of the perpendiculars from their centers of rotation to the line

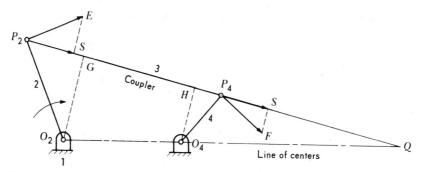

Fig. 2-18

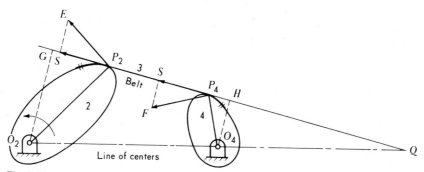

Fig. 2-19

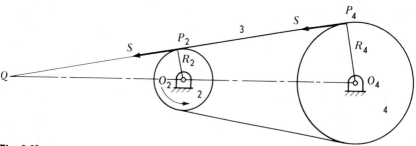

Fig. 2-20

of transmission or inversely as the segments into which the line of transmission divides the line of centers.

Figure 2-20 illustrates the usual form of a belt drive. In this special case of Fig. 2-19, O_2G and O_4H are R_2 and R_4, the radii of the pulleys. Hence for Fig. 2-20, Eq. (2-42) gives

$$\frac{\omega_2}{\omega_4} = \frac{R_4}{R_2}$$

$$(2\text{-}45)$$

2-13 CONSTANT ANGULAR–VELOCITY RATIO

In the preceding section it was shown that

$$\frac{\omega_2}{\omega_4} = \frac{O_4 Q}{O_2 Q}$$

Thus, for the angular-velocity ratio to remain constant, the line of transmission must intersect the line of centers at a fixed point. There are many curves which can be used as the contacting surfaces of the driver and follower in a direct-contact mechanism and which fulfill this condition. Later, in our study of gearing we shall see that mating gear-tooth profiles must satisfy this requirement in order that the angular-velocity ratio of the gears will be constant. From inspection of Fig. 2-20 we note that the condition for constant angular-velocity ratio is fulfilled by a belt drive. In the four-bar linkage in Fig. 2-18 the angular-velocity ratio for driver and follower will be constant only if cranks 2 and 4 are the same length and the coupler length is made equal to length $O_2 O_4$. Then $\omega_2/\omega_4 = 1$.

2-14 SLIDING CONTACT

Sliding exists in a direct-contact mechanism whenever the bodies have relative motion along the tangent through their point of contact. The direct-contact mechanism of Fig. 2-17 is again shown in Fig. 2-21. Vectors $P_2 E$ and $P_4 F$ are the velocities of P_2 and P_4, the points of contact on bodies 2 and 4, respectively. As explained in Sec. 2-12, bodies 2 and 4 have no relative motion along the normal, and hence the normal components of velocities $P_2 E$ and $P_4 F$ must be equal and are represented by vector $P_2 S$. The tangential components of $P_2 E$ and $P_4 F$ are $P_2 L$ and $P_4 M$, respectively. Since these components of velocity along the tangent are not equal in magnitude and direction, bodies 2 and 4 have relative motion in this direction. The difference in their tangential components of velocity is the sliding velocity V_s. Hence

$$V_s = P_2 L \rightarrow P_4 M$$

$$= P_2 L \nrightarrow (-P_4 M)$$

In Fig. 2-21 the velocity with which body 2 slides on 4 is directed from M toward L along the tangent, and its magnitude is represented by the length ML. In the next section it will be shown that sliding exists in a direct-

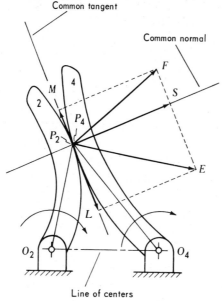

Fig. 2-21

contact mechanism whenever the point of contact lies other than on the line of centers.

2-15 ROLLING CONTACT

In a direct-contact mechanism rolling contact exists only if there is no sliding, and hence the tangential components of velocities P_2E and P_4F in Fig. 2-21 must be equal in magnitude and direction. For this to exist along with the normal components of P_2E and P_4F being equal requires that P_2E and P_4F be equal in magnitude and direction and hence identical. From inspection of Fig. 2-21 we note that velocities P_2E and P_4F can have the same direction only if the radii O_2P_2 and O_4P_4 lie along a common line, namely the line of centers O_2O_4, as shown in Fig. 2-22.

It is important to note that though it is necessary for the point of contact to lie on the line of centers if there is to be rolling, this is not sufficient. There still can be sliding unless the tangential velocities for the bodies are identical. For example, bodies 2 and 4 in Fig. 2-23 will have sliding contact if velocities P_2E and P_4F are not identical. Summarizing, for rolling contact the linear velocities of the bodies at their point of contact must be identical, and this requires that the point of contact lie on the line of centers.

In Fig. 2-22 points P_2 and P_4 are the points of contact, and O_2P_2 and O_4P_4 are the contact radii.   For direct-contact mechanisms we noted earlier that the angular-velocity ratio of driver and follower is inversely as the segments into which the normal through the point of contact divides the line of centers.   Since the point of contact lies on the line of centers in mechanisms having rolling contact, it follows that the angular-velocity ratio of driver and follower is inversely as the contact radii.

2-16   POSITIVE DRIVE

Positive drive exists in a direct-contact mechanism if motion of the driving link compels the follower to move.   In the cam mechanism in Fig. 2-24 link 2 is the driver; let us assume that it is rotating counterclockwise.   The force which 2 exerts on 4 will be directed along the common normal through the point of contact P.   Since this force has a torque arm O_4H about pivot O_4, a counterclockwise movement of link 2 will compel link 4 to rotate clockwise.   Similarly, if link 4 were the driver and were to rotate counterclock-

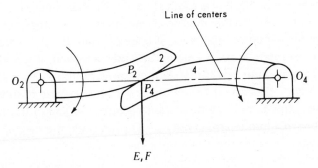

Fig. 2-22

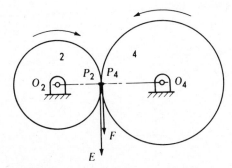

Fig. 2-23

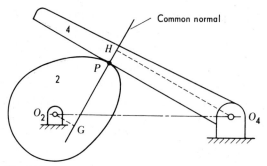

Fig. 2-24

wise, then because of the torque arm O_2G, link 2 would be compelled to
rotate clockwise about its pivot O_2. If the driver, whether it is link 2 or 4,
were to move in a direction so that it would move away from the follower,
then of course the follower would not be compelled to move. Thus for the
positions of the links shown in Fig. 2-24 there is positive drive, provided that
motion of the driver along the normal is toward the follower.

The mechanism of Fig. 2-24 is again shown in Fig. 2-25, where the positions of links 2 and 4 are such that the line of transmission (common normal)
passes through the center of rotation O_2 of body 2. Here if link 2 were the
driver and were to rotate a small amount in either direction, it would not
compel link 4 to move, but would merely move away from it. Thus there
is no positive drive. Likewise if 4 were the driver and we tried to rotate it
counterclockwise, there could be no motion because the line of transmission would not have a torque arm about O_2. This phase of the mechanism
is called *dead center*. Hence there is no positive drive.

Still another direct-contact mechanism is shown in Fig. 2-26. Bodies

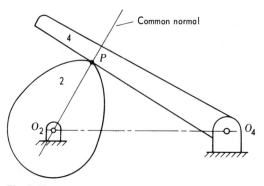

Fig. 2-25

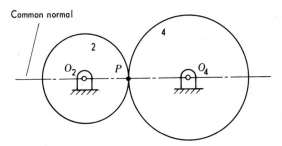

Fig. 2-26

2 and 4 are circular disks and either can be the driver.   The common normal passes through their centers of rotation O_2 and O_4.   Here there can be a rotation of the driver without the follower being compelled to rotate.   Thus there is no positive drive.   Only if there is sufficient friction between bodies 2 and 4 will rotation of one of the bodies cause the other to rotate.   The result is known as a *friction drive*. From the foregoing discussion we note that there can be positive drive in a direct-contact mechanism only if the common normal through the point of contact does not pass through either or both of the centers of rotation.

PROBLEMS

2-1 A steel cylinder 6 in. in diameter is to be machined in a lathe.   The cutting speed is to be 100 fpm.   Determine the speed of rotation in rpm.

2-2 Two points B and C lie on a radial line of a rotating disk.   The points are 2 in. apart. $V_B = 700$ fpm and $V_C = 880$ fpm.   Find the radius of rotation for each of these points.

2-3 The tire of an automobile has an outside diameter of 27 in.   If the rpm of the wheel is 700, determine the speed of the automobile (*a*) in miles per hour, (*b*) in feet per second, and (*c*) the angular speed of the wheel in radians per second.

2-4 An automobile engine has a bore (cylinder diameter) of 3.75 in.   The stroke (distance the piston travels from one extreme position to the other) is 3.5 in. The car is running at 60 mph and the outside diameter of the tires is 27 in.   If the rpm of the engine is four times that of the wheels, find (*a*) the rpm of the wheels, (*b*) the rpm of the engine, (*c*) crankpin velocity (in feet per minute), (*d*) angular velocity of crank (in radians per minute), (*e*) average piston velocity (in feet per minute), and (*f*) distance piston travels per mile of car travel.

2-5 A body moves a distance of 18 in. with a constant velocity of 4 fps.   (*a*) Find the time required in seconds.   (*b*) If the body were to move a distance of 18 in. in 0.2 sec with variable velocity, find the average velocity in feet per second.

2-6 An automobile accelerates from a speed of 20 mph to 60 mph in a distance of 293 ft, which requires 5 sec.   (*a*) If the acceleration is constant, find the acceleration in feet per second squared.   (*b*) Same as part (*a*) except the acceleration is not constant.   What is the average acceleration in feet per second squared?

2-7   A particle starts from rest and accelerates at a constant rate for 4 sec, at the end of which time it has acquired a velocity sufficient to carry it at uniform velocity a distance of 18 ft in 3 sec.   Find the acceleration during the first 4 sec and the velocity at the end of that time.

2-8   An automobile engine accelerates from rest and attains a speed of 2,000 rpm in 5 sec.   Assuming the angular acceleration is constant, find (*a*) the angular acceleration in radians per second squared of the crankshaft and (*b*) the number of revolutions made by the crankshaft in coming up to speed.

2-9   A disk 10 in. in diameter accelerates uniformly from a speed of 1,000 rpm to 2,000 rpm in 20 sec.   Find (*a*) the angular acceleration in radians per second squared and (*b*) the revolutions of the disk during the 20-sec interval.

2-10   In Prob. 2-8 if the stroke of the piston were 3.75 in. (stroke equals twice the crank length), find (*a*) the tangential acceleration of the crankpin in feet per second squared when coming up to speed and (*b*) the normal acceleration in feet per second squared when the speed is 2,000 rpm.

2-11   The rotor of a turbojet engine rotates at 12,000 rpm.   Determine the speed in feet per second and acceleration in feet per second squared of a point on the periphery of the 36-in.-diameter compressor of the rotor.

2-12   For the Scotch-yoke mechanism in Fig. P2-12 $R = 8$ in., $\theta = 60°$, and the crank speed is 200 rpm.   Find the velocity in feet per second and acceleration in feet per second squared of the slider.

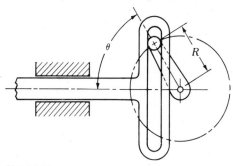

Fig. P2-12

2-13   If the stroke of the slider of the mechanism in Fig. P2-12 were 14 in. and the time for one stroke were 0.125 sec, find (*a*) the crank rpm, (*b*) maximum velocity of the slider in feet per second, and (*c*) maximum acceleration of the slider in feet per second squared.

2-14   Vibration-measuring instruments indicate that a body is vibrating with harmonic motion at a frequency of 420 cycles/min with a maximum acceleration of 29 ips². Determine (*a*) the amplitude of the vibration and (*b*) the maximum velocity.

2-15   An airplane A flies directly north at 400 mph while an airplane B flies directly east at 300 mph.   Find the velocity of A relative to B and the velocity of B relative to A.   In each case write the vector equation and lay out the vectors using a scale of 1 in. = 200 mph.   Assume north in the upward direction on the paper, label all vectors, and determine your answers graphically.

2-16 An airplane flies a straight course due east from city M to city N, 400 miles away. The plane has an airspeed of 180 mph. A crosswind blows due south at 60 mph. In what direction must the plane be headed and how long will the trip take in hours? Write the necessary vector equation using subscripts P for plane and A for air. Lay out the vectors using a scale of 1 in. = 60 mph and scale off results.

2-17 The car in Fig. P2-17 moves to the right with a velocity of 30 mph. Wheels 2 and 4 are 36 and 24 in. in diameter, respectively. Use 1 in. = 20 in. for the drawing. Find V_{O2}, $V_{B/O2}$, V_B, V_C, and $V_{B/C}$ in feet per second. Lay out the vectors, using a scale of 1 in. = 30 fps. Also find ω_2, ω_4, and $\omega_{2/4}$ in radians per second.

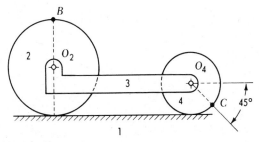

Fig. P2-17

2-18 The disk in Fig. P2-18 has an $\omega = 120$ rpm and $\alpha = 132$ rad/sec^2. Let $OB = 1.5$ in. and $OC = 1$ in. Determine V_B, V_C, A_B^n, A_B^t, A_C^n, and A_C^t. Make a full-size

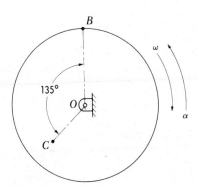

Fig. P2-18

drawing of the disk and show the vectors at points B and C using the following scales: velocity, 1 in. = 1 fps; acceleration, 1 in. = 20 fps^2. Determine graphically $V_{C/B}$, A_B, A_C, and $A_{C/B}$.

2-19 In Fig. P2-19, $\omega = 100$ rpm and $\alpha = 90$ rad/sec^2. Determine V_B, V_C, A_B^n, A_B^t, A_C^n, and A_C^t. Make a full-size drawing of the member and show the vectors at points B and C, using the following scales: velocity, 1 in. = 1 fps; acceleration, 1 in. = 10 fps^2. Determine graphically $V_{B/C}$, A_B, A_C, and $A_{B/C}$.

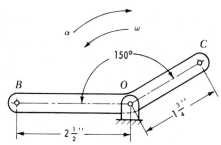

Fig. P2-19

2-20 In Fig. P2-20, $\omega_2 = 120$ rpm ccw.   Determine the angular velocity of link 4 in rpm.

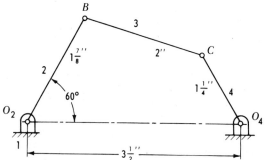

Fig. P2-20

2-21 Same as Prob. 2-20 except find the angular velocity of link 3.   Hint: Use inversion.   Fix link 2 and give $\omega_{1/2}$ the proper value.   Find $\omega_{3/2}$, then $\omega_{3/1} = \omega_{3/2} - \omega_{1/2}$.   Is ω_3 clockwise or counterclockwise?

2-22 In Fig. P2-22, $\omega_4 = 80$ rpm ccw.   Determine the rpm of link 2 and the velocity of sliding in feet per second at the point of contact.   Use a velocity scale of 1 in. = 2.5 fps.

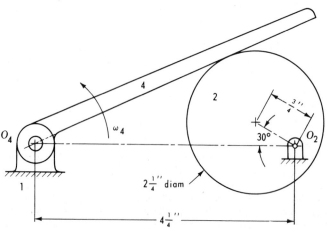

Fig. P2-22

3

LINKAGES

3-1 FOUR–BAR LINKAGE

One of the most useful and most common mechanisms is the four-bar linkage. A four-bar linkage is shown in Fig. 3-1, where link 1 is the frame, links 2 and 4 are the cranks, and 3 is called the coupler. It will be shown later that many mechanisms can conveniently be replaced by a four-bar linkage or a combination of four-bar linkages for the purpose of analyzing their motions.

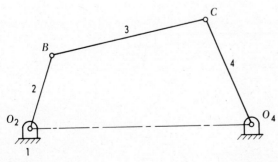

Fig. 3-1

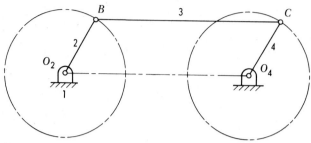

Fig. 3-2

3-2   PARALLEL–CRANK FOUR–BAR LINKAGE

In Fig 3-2 cranks 2 and 4 are of equal length and the coupler 3 is equal in length to the line of centers O_2O_4.   Cranks 2 and 4 always have the same angular velocity.

There are two positions during the cycle when the linkage is not constrained.   These are the positions where the follower, link 4, is colinear with link 3.   At these positions, referred to as *dead points* or *dead center*, the follower could begin to rotate in a direction opposite to that of the driver. Dead points occur in many mechanisms, but usually inertia, springs, or gravity prevent the undesired reversal at the dead point.

3-3   NONPARALLEL EQUAL–CRANK LINKAGE

In Fig. 3-3 cranks 2 and 4 are of equal length and the length of the coupler is equal to the line of centers O_2O_4, but the cranks are nonparallel and rotate in opposite directions.   If crank 2 turns with constant angular velocity,

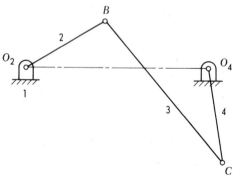

Fig. 3-3

crank 4 will have a varying angular velocity.   In order to ensure that the follower will rotate in the proper sense beyond the dead points, this mechanism can be replaced by a pair of identical elliptical gears.   This is explained later in Chap. 11.

3-4   CRANK AND ROCKER

In Fig. 3-4 crank 2 rotates completely about pivot O_2 and by means of coupler 3 causes crank 4 to oscillate about O_4.   Hence the mechanism transforms motion of rotation into oscillating motion.   In order for this linkage to operate, the following conditions must exist:

$$O_2B + BC + O_4C > O_2O_4$$

$$O_2B + O_2O_4 + O_4C > BC$$

$$O_2B + BC - O_4C < O_2O_4$$

$$BC - O_2B + O_4C > O_2O_4$$

Either 2 or 4 can be the driving crank.   If link 2 drives, the mechanism will always operate.   If 4 is the driver, a flywheel or some other aid will be required to carry the mechanism beyond the dead points B' and B''.   The dead points exist where the line of action BC of the driving force is in line with O_2B.

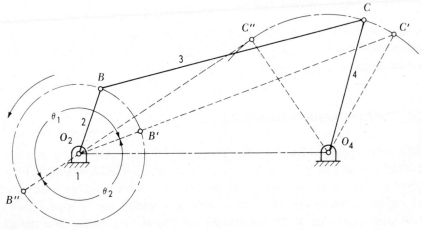

Fig. 3-4

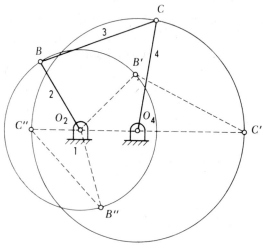

Fig. 3-5

3-5 DRAG LINK

Figure 3-5 shows a four-bar linkage in which the shortest link is fixed. Such a linkage is known as a *drag-link mechanism*. Both 2 and 4 make complete rotations. If one crank rotates at constant speed, the other crank will rotate in the same direction at a varying speed. The proportions of the links must be as follows:

$$BC > O_2O_4 + O_4C - O_2B$$
$$BC < O_4C - O_2O_4 + O_2B$$

These relations can be derived from triangles $O_2B'C'$ and $O_2B''C''$. An application of this mechanism will be discussed under the subject of quick-return mechanisms in Sec. 3.8.

3-6 SLIDER–CRANK MECHANISM

This mechanism is shown in Fig. 3-6. It is a special case of the four-bar linkage of Fig. 3-1. If crank 4 in Fig. 3-1 were made infinite in length, then point C would have rectilinear motion and crank 4 could be replaced by a slider, as shown in Fig. 3-6. The slider-crank mechanism is widely used. Common examples of its application are found in gasoline and diesel engines, where the gas force acts on the piston, link 4. Motion is transmitted through the connecting rod to crank 2. There are two dead-center positions during the cycle, one for each extreme position of the slider. A flywheel mounted on the crankshaft is required to carry the crank beyond these positions. This mechanism is also used in air compressors, where an electric motor or gasoline engine drives the crank and in turn the piston compresses the air.

A modification of the slider-crank linkage of Fig. 3-6 appears in Fig. 3-7 and is known as an *eccentric mechanism*. The crank consists of a circular disk with center B, which is pivoted off-center at O_2 to the frame. The disk rotates inside the ring end of rod 3. The motion of this mechanism is equivalent to that of a slider-crank linkage having a crank length equal to O_2B and a connecting rod of length BC.

3-7 SCOTCH YOKE

The Scotch-yoke mechanism (Fig. 3-8) discussed in Sec. 2-7 is a variation of the slider-crank mechanism. The Scotch yoke is the equivalent of a slider crank having an infinitely long connecting rod. As a result the slider

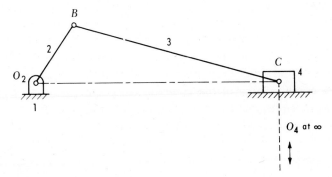

Fig. 3-6

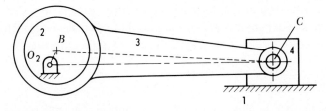

Fig. 3-7

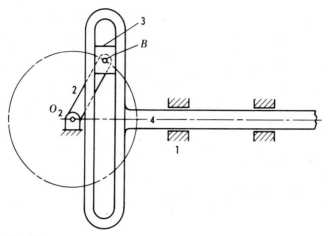

Fig. 3-8

has simple harmonic motion.   The Scotch-yoke mechanism is used in test-
ing machines to simulate vibrations having simple harmonic motion.

3-8   QUICK–RETURN MECHANISMS

Quick-return mechanisms are used in machine tools such as shapers and
power-driven saws for the purpose of giving the reciprocating cutting tool a
slow cutting stroke and a quick-return stroke with a constant angular velocity
of the driving crank.   Some of the common types are discussed below.
The ratio of the time required for the cutting stroke to the time for the return
stroke is called the *time ratio* and is greater than unity.

Crank-shaper

This mechanism employs an inversion of the slider-crank linkage
which is illustrated in Fig. 3-6.   Figure 3-9 shows the arrangement in which
link 2 rotates completely and link 4 oscillates.   If the driver, link 2, rotates
counterclockwise at constant velocity, slider 6 will have a slow stroke to the
left and a fast return stroke to the right.   The time ratio equals θ_1/θ_2.

Whitworth

This mechanism is illustrated in Fig. 3-10 and is obtained by making
the distance O_2O_4 in Fig. 3-9 less than the crank length O_2B.   Both links 2

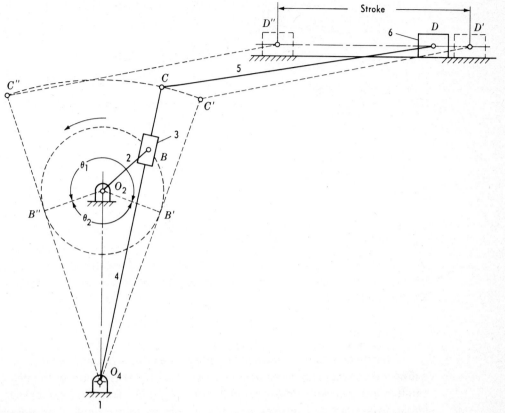

Fig. 3-9

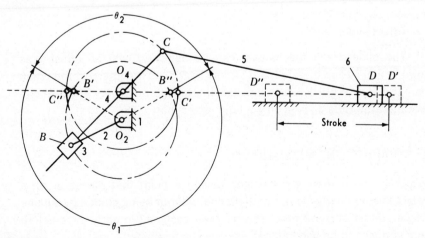

Fig. 3-10

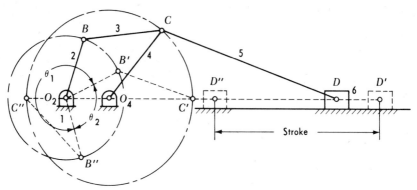

Fig. 3-11

and 4 rotate completely. If the driver, crank 2, rotates counterclockwise with constant angular velocity, slider 6 will move from D' to D'' with a slow motion while 2 rotates through angle θ_1. Then as 2 rotates through the smaller angle θ_2, slider 6 will have a quick-return motion from D'' to D'. The time ratio is θ_1/θ_2.

Drag link

This mechanism is shown in Fig. 3-11 where links 1, 2, 3, and 4 comprise a drag-link mechanism, as explained in Sec. 3-5. If link 2, the driver, rotates counterclockwise with constant angular velocity, then slider 6 makes a slow stroke to the left and returns with a quick stroke to the right. The time ratio is θ_1/θ_2.

Offset slider crank

The slider-crank mechanism can be designed with an offset y as shown in Fig. 3-12 so that the path of the slider does not intersect the crank axis. It is then a quick-return mechanism, though not a very effective one since the time ratio θ_1/θ_2 is only a little larger than 1.

3-9 STRAIGHT–LINE MECHANISMS

Straight-line mechanisms are linkages having a point that moves along a straight line, or nearly along a straight line, without being guided by a plane surface. Most of these mechanisms were designed in early days before plane surfaces to be used as guides could be machined.

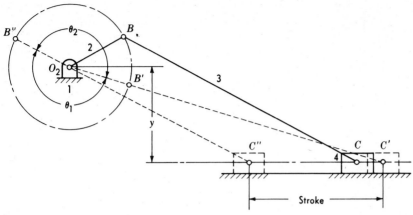

Fig. 3-12

Watt's mechanism (Fig. 3-13) produces approximate straight-line motion. Point P traces a figure-eight-shaped path, a considerable portion of which is approximately a straight line. The lengths must be proportioned so that

$$\frac{BP}{PC} = \frac{CD}{AB}$$

The *Scott-Russell mechanism* (Fig. 3-14) gives exact straight-line motion of point P. Length $AC = BC = CP$. A variation of this mechanism is shown in Fig. 3-15, where the slider is replaced by crank BD. In this linkage, point P has approximate straight-line motion.

Robert's mechanism (Fig. 3-16) produces approximate straight-line motion. Point P moves very nearly along line AB. Length

$$AC = CP = PD = DB$$

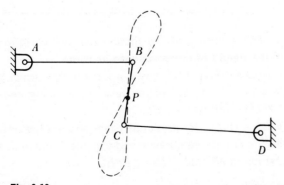

Fig. 3-13

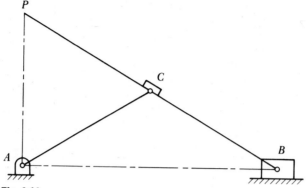

Fig. 3-14

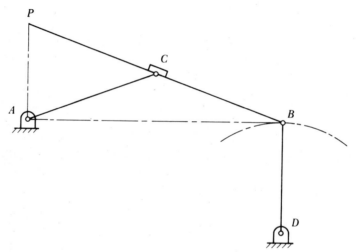

Fig. 3-15

and $CD = AP = PB$. The accuracy of the motion can be increased by increasing the ratio of the height of mechanism to its width.

Tchebysheff's mechanism (Fig. 3-17) gives approximate straight-line motion. Point P, the midpoint of CB, moves very nearly along line CB. Length $AB = CD = 1.25\ AD$, and $AD = 2\ CB$.

Peaucillier's mechanism (Fig. 3-18) produces exact straight-line motion for point P. Peaucillier, a Frenchman, developed this mechanism in 1864. The following relationships must hold: $AB = AE$, $BC = BD$,

$$PC = PD = CE = DE$$

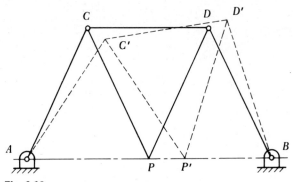

Fig. 3-16

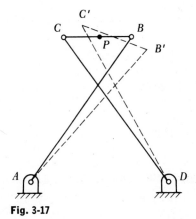

Fig. 3-17

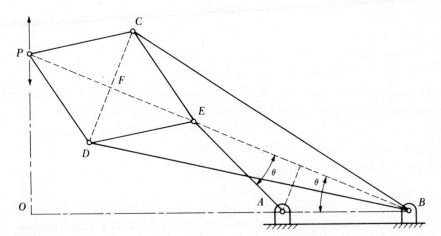

Fig. 3-18

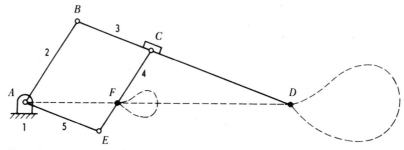

Fig. 3-19

It will be proved that point P moves in a straight line.   Point E will lie on line BP because of symmetry, and CD will bisect PE at F. BFC and BFD are right triangles.   Hence $(BF)^2 = (BC)^2 - (CF)^2$ and

$$(EF)^2 = (CE)^2 - (CF)^2$$

Elimination of $(CF)^2$ from the equations gives

$$(BF)^2 - (EF)^2 = (BC)^2 - (CE)^2$$

$$(BF + EF)(BF - EF) = (BC)^2 - (CE)^2$$

But

$$BF + EF = BP = \frac{BO}{\cos\theta} \quad \text{and} \quad BF - EF = BE = 2AB \cos\theta$$

Then

$$\frac{BO}{\cos\theta} 2AB \cos\theta = (BC)^2 - (CE)^2$$

and

$$BO = \frac{(BC)^2 - (CE)^2}{2AB} = \text{const}$$

Thus point O, the projection of P on a line through AB, is fixed.   It follows that point P moves along PO, which is a straight line perpendicular to AB.

3-10   PARALLEL MECHANISMS

These are linkages which give parallel motion.   The pantograph (Fig. 3-19) is used to enlarge or reduce movements.   Links 2, 3, 4, and 5 form a parallelogram.   Link 3 is extended and contains point D. F is the point of intersection of lines AD and CE. This mechanism finds use in reproducing motions to a different scale. A pen or pencil at F reproduces the motion of a stylus at D except to a reduced scale. The pen and stylus are interchangeable. In order for the motion of F to be parallel to that of D for all positions, it is necessary that the ratio AD/AF be constant. For all positions of D, triangles AEF and DCF are similar, since their three sides are always parallel. Hence

$$\frac{AF}{FD} = \frac{AE}{CD} = \text{const}$$

and

$$\frac{\text{Size of figure at } D}{\text{Size of figure at } F} = \frac{AD}{AF}$$

Pantographs are used for reducing or enlarging drawings and maps. They are also used for guiding cutting tools or cutting torches to duplicate complicated shapes.

Another application of a parallel mechanism is the familiar drafting machine (Fig. 3-20). Parallelograms $ABCD$ and $EFGH$ are coupled by the ring $BECH$. The horizontal and vertical straightedges can be rotated and clamped in any position relative to head FG. By swinging the arms, the straightedges will move to any parallel position on the drawing.

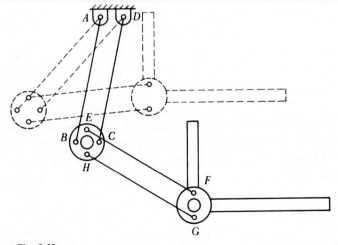

Fig. 3-20

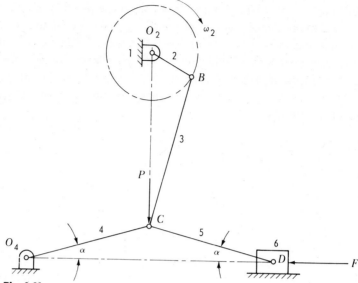

Fig. 3-21

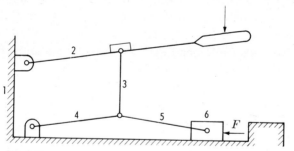

Fig. 3-22

3-11 TOGGLE MECHANISMS

Toggle mechanisms are used whenever a large force acting through a short distance is required. In Fig. 3-21 links 4 and 5 are the same length. P is the vertical component of the force which link 3 exerts on the pin at C. When the angle between BC and O_2C becomes small, a force analysis gives

$$F = \frac{P}{2 \tan \alpha}$$

Thus for a given value of P, as links 4 and 5 approach a colinear position, force F rises rapidly.

Other toggle mechanisms are illustrated in Figs. 3-22 and 3-23. Toggle mechanisms are used in toggle clamps, riveting machines, punch presses, and rock crushers. The kinematic diagram of a rock crusher is shown in Fig. 3-24.

3-12   OLDHAM COUPLING

The Oldham coupling (Fig. 3-25) is a mechanism for connecting two shafts having parallel misalignment. Disk 3 has a tongue on each side. These are at 90° to one another and slide in grooves in members 2 and 4. Since there is no relative rotation between bodies 2, 3, and 4, the coupling transmits a constant-velocity ratio.

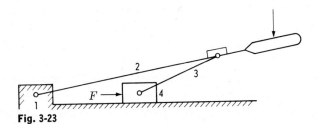

Fig. 3-23

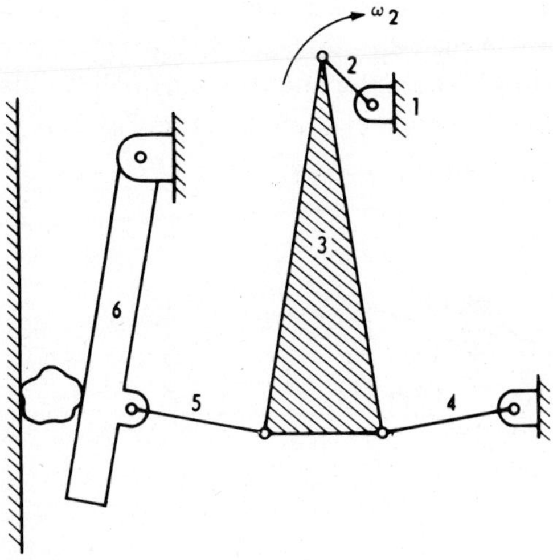

Fig. 3-24

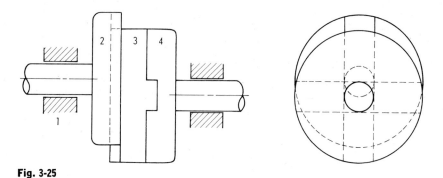

Fig. 3-25

3-13 UNIVERSAL JOINTS

Universal joints are used to connect intersecting shafts. The most common type is the Hooke or Cardan joint (Fig. 3-26). A kinematic diagram is shown in Fig. 3-27a, where the angle between the shafts is δ. For the phase shown the yoke of shaft 2 lies in the vertical plane, and the yoke of shaft 3 is in the horizontal plane. The intermediate member pivots about axes BC and DE. As shaft 3 rotates, point D will move in a circular path of radius R as shown in Fig. 3.27d, which is an end view. As shaft 2 rotates, point B describes a circular path in the plane of projection shown in Fig. 3-27c. Consider now the path of motion of point B in Fig. 3-27b. The projection of this path upon the vertical plane in Fig. 3-27d will not be circular but elliptical, as shown by the dotted line. Let shaft 2 rotate an amount θ_2. Then B moves from B to B' as shown in Fig. 3-27c. In Fig. 3-27d the motion of B along the ellipse DBE is from B to B', where the lines OB and OB' lie in the plane of the paper and θ_3 is angle of rotation of shaft 3. From Fig. 3-27c

$$OF = R \cos \theta_2 \qquad \text{and} \qquad B'F = R \sin \theta_2$$

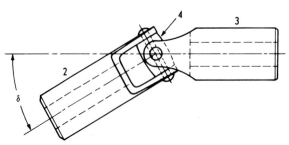

Fig. 3-26

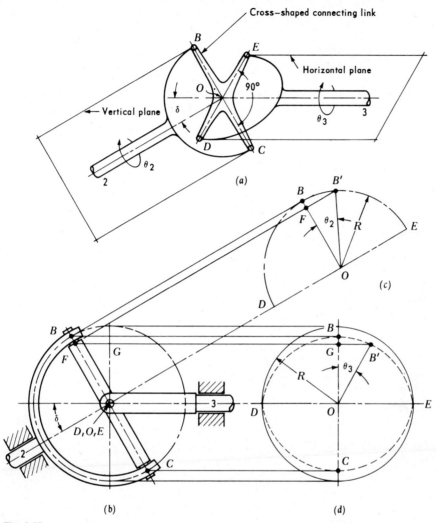

Fig. 3-27

In Fig. 3-27b

$$OG = OF \cos \delta = R \cos \theta_2 \cos \delta$$

Next, in Fig. 3-27d $B'G$ has the same length as $B'F$ in Fig. 3-27c. Thus

$$B'G = R \sin \theta_2$$

and

$$\tan \theta_3 = \frac{B'G}{OG} = \frac{R \sin \theta_2}{R \cos \theta_2 \cos \delta}$$

or

$$\tan \theta_3 = \frac{\tan \theta_2}{\cos \delta} \tag{3-1}$$

δ is usually constant and we will assume it thus. The angular-velocity ratio is obtained by differentiating Eq. (3-1) with respect to time; thus

$$\sec^2 \theta_3 \frac{d\theta_3}{dt} = \frac{\sec^2 \theta_2}{\cos \delta} \frac{d\theta_2}{dt}$$

If we let

$$\omega_3 = \frac{d\theta_3}{dt} \quad \text{and} \quad \omega_2 = \frac{d\theta_2}{dt}$$

then

$$\frac{\omega_2}{\omega_3} = \frac{\sec^2 \theta_3 \cos \delta}{\sec^2 \theta_2} = \frac{\sec^2 \theta_3 \cos \delta}{1 + \tan^2 \theta_2}$$

Substituting Eq. (3-1) into the last equation we can eliminate θ_2. Thus we obtain

$$\frac{\omega_2}{\omega_3} = \frac{\sec^2 \theta_3 \cos \delta}{1 + \tan^2 \theta_3 \cos^2 \delta} = \frac{\cos \delta}{\cos^2 \theta_3 + \sin^2 \theta_3 \cos^2 \delta}$$

Let $\cos^2 \delta = 1 - \sin^2 \delta$. Then

$$\frac{\omega_2}{\omega_3} = \frac{\cos \delta}{1 - \sin^2 \theta_3 \sin^2 \delta} \tag{3-2}$$

For a constant angular velocity ω_3, differentiation of Eq. (3-2) with respect to time gives

$$\alpha_2 = \frac{d\omega_2}{dt} = \frac{d}{dt} \left(\frac{\omega_3 \cos \delta}{1 - \sin^2 \theta_3 \sin^2 \delta} \right)$$

$$= \omega_3 \frac{\cos \delta \sin^2 \delta (2 \sin \theta_3 \cos \theta_3) \, d\theta_3/dt}{(1 - \sin^2 \theta_3 \sin^2 \delta)^2}$$

$$= \omega_3^2 \frac{\cos \delta \sin^2 \delta \sin 2\theta_3}{(1 - \sin^2 \theta_3 \sin^2 \delta)^2} \tag{3-3}$$

From Eq. (3-2) we see that for a constant speed of one shaft, δ can soon become large enough so that the variation in speed of the other shaft is considerable. The accompanying accelerations can then cause vibrations which are intolerable. A solution to this problem can be obtained by using two universal joints arranged so that the second joint compensates for the variation in speed produced by the first. Let Fig. (3-28) represent a drive

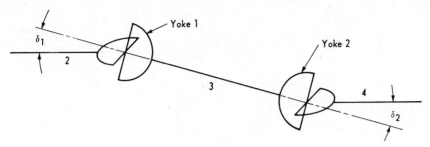

Fig. 3-28

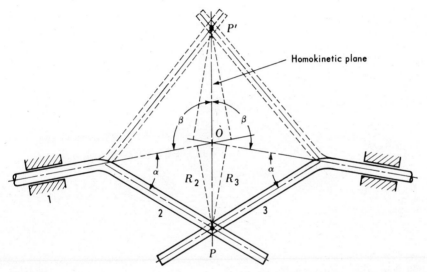

Fig. 3-29

where two universal joints are to be used.   Shafts 2 and 4 need not intersect. In order to have the second universal joint compensate for the variations in speed produced by the first so that $\omega_2/\omega_4 = 1$ at all times, angle δ_1 between shafts 2 and 3 must equal angle δ_2 between shafts 3 and 4, and yoke 1 must be made to lie in the plane of 2 and 3 when yoke 2 lies in the plane of 3 and 4.

Several universal joints have been invented which give a constant-velocity ratio.   The simplest of these (Fig. 3-29) has been used in toys, and the principle of its operation is common to all constant-velocity universal joints.   In the figure the shaft axes which intersect at O lie in the plane of the paper, and for the phase shown point P also lies in this plane.   A plane perpendicular to the paper passing through point P and bisecting the angle

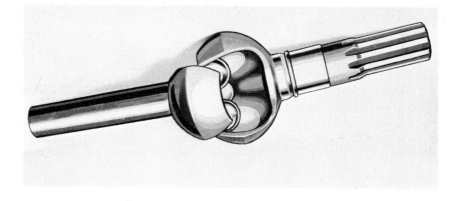

(a)

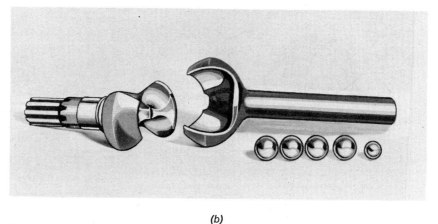

(b)

Fig. 3-30. *(Bendix Aviation Corp.)*

between the shafts is known as the *homokinetic plane.* Point P lies in this plane for all phases, and since the radii R_2 and R_3 will always be equal, the shafts will have equal angular velocities.

The Bendix-Weiss joint is shown in Figs. 3-30 and 3-31. Motion is transmitted from one shaft to the other through four balls which fit between races in the yokes of the shafts. The races are designed so that the center of each ball lies in the homokinetic plane at all times. This is shown in Fig. 3-31. Thus the joint gives a constant angular-velocity ratio. A feature of this joint is that the balls are able to move back and forth in the races allowing end motion without a sliding spline connection. A fifth ball, whose center lies at the intersection of the shaft axes, is used in conjunction with a means for locking the parts in assembly and for carrying end thrust.

3-14 INTERMITTENT–MOTION MECHANISMS

An *intermittent-motion mechanism* is a linkage which converts continuous motion into intermittent motion.   Mechanisms of this type are commonly used on machine tools for indexing a shaft.   *Indexing a shaft* means rotating it through a specified angle with zero velocity at the beginning and the end. For example, the work table of a machine tool is indexed so as to bring a new piece of work into position for machining by the cutters.

Geneva wheel

This mechanism is shown in Fig. 3-32.   Link 2 is the driver and contains a pin which engages slots in the driven link 3.   The slots are positioned so that the pin enters and leaves them tangentially.   Thus an advantage of this mechanism is that it provides indexing without impact loading.   In the particular mechanism shown, the driven member makes one-fourth of a revolution for each revolution of the driver.   However, velocity ratios other than 4:1 may be used.   The locking plate, which is mounted on the driver, prevents the driven member from rotating except during the indexing period.

Ratchets

Ratchets are used to transform motion of rotation or translation into intermittent rotation or translation.   In Fig. 3-33 member 2 is the ratchet

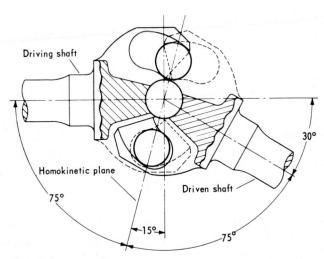

Fig. 3-31

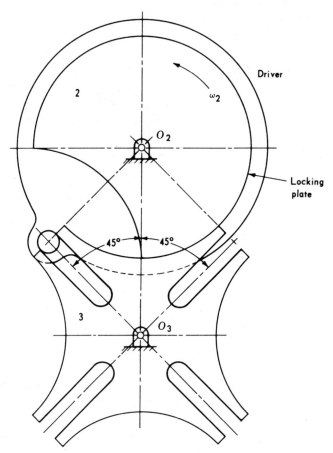

Fig. 3-32

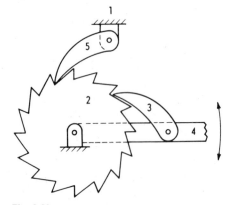

Fig. 3-33

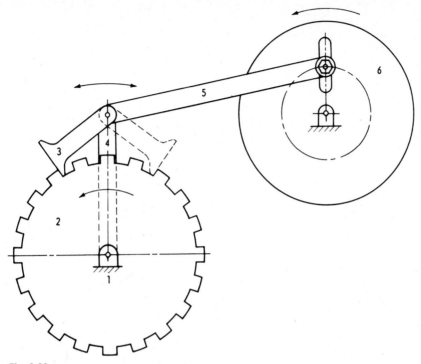

Fig. 3-34

wheel and 3 is the pawl.   As the pawl lever, member 4, is made to oscillate, the ratchet will rotate counterclockwise with an intermittent motion.   A holding pawl, member 5, is often provided to prevent the ratchet from reversing.

Figure 3-34 shows a ratchet-drive mechanism in which the throw of the driving crank, link 6, is adjustable.   For rotation of crank 6 in either direction, link 4 oscillates and member 2 rotates counterclockwise with intermittent motion.   If the pawl 3 is placed in the dotted position, the ratchet will rotate clockwise.

A silent ratchet is shown in Fig. 3-35.   There are no teeth on the ratchet, and the device depends upon the wedging together of smooth surfaces. Link 5 is a holding pawl.

Figure 3-36 shows a ball-type silent ratchet.   The small angle between the flat surface on the inner member and the tangent to the inner surface of the outer member at the point of contact with the ball causes a wedging action when the outer member rotates clockwise relative to the inner member.   Thus member 2 can be the driver if it rotates counterclockwise, or 4

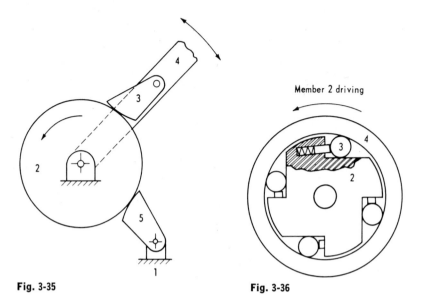

Fig. 3-35   **Fig. 3-36**

Member 2 driving

can be the driver if it rotates clockwise.   This device is also used as an over-running clutch.   When used as a clutch, suppose 2 is the driver and rotates counterclockwise.   If 2 stops, 4 can freewheel.   Similarly, suppose 4 is the driver and rotates clockwise.   If 4 is stopped, 2 can freewheel.   A higher freewheeling speed is possible if 4 is allowed to freewheel rather than 2.

3-15   ELLIPTIC TRAMMEL

The elliptic trammel (Fig. 3-37) is an instrument for drawing ellipses.   Link 3 is pivoted to sliders 2 and 4, which slide in link 1, and point P describes an ellipse.

From the figure

$$x = a \cos \theta$$

$$y = b \sin \theta$$

Then

$$\cos^2 \theta + \sin^2 \theta = \frac{x^2}{a^2} + \frac{y^2}{b^2} = 1 \tag{3-4}$$

which is the equation of an ellipse with center at the origin.   Length a is half the major axis and b is half the minor axis.   When the device is used as

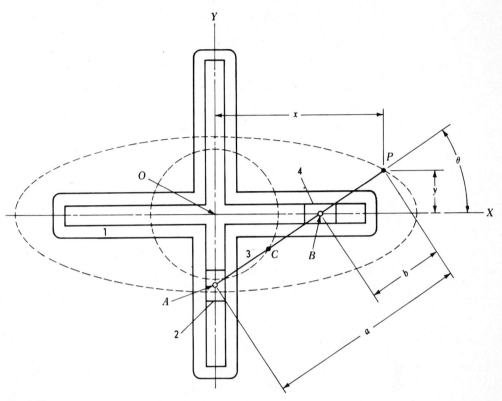

Fig. 3-37

a drawing instrument, a pen or pencil is carried at P and both lengths a and b are adjustable. If P is placed at point C, which is midway between A and B, then a and b are equal and Eq. (3-4) becomes

$$x^2 + y^2 = a^2$$

which is the equation of a circle of radius a.

PROBLEMS

3-1 For the crank-and-rocker mechanism shown in Fig. P3-1, plot point D for each 30° displacement of crank 2. Draw a smooth curve through these points. It will be helpful to draw link 3 containing points B, C, and D on tracing paper. The tracing can then be laid over the drawing. By locating B at each point, C

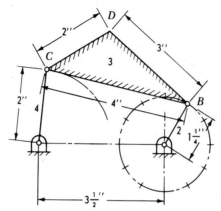

Fig. P3-1

can be located along the circular arc traced by C. A prick can then be made through the tracing to locate D on the drawing.

3-2   In Fig. P3-2 crank 2 is to rotate continuously and 4 is to oscillate.   What are the maximum and minimum values in inches which can be used for the coupler length?

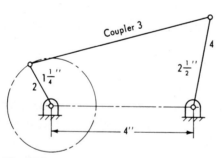

Fig. P3-2

3-3   Design a crank-shaper mechanism (Fig. P3-3) that will give a time ratio of 1.75:1 with a working stroke of 26 in.   Further, path DQ of point D is to be located midway between the highest and lowest points assumed by point C as it moves along the arc of radius O_4C.   The fixed dimensions are given in the figure. That is, compute the required values for O_2B, O_4C, and O_4Q.   Make a drawing of the mechanism using a scale of 1 in. = 10 in. and check these values graphically.   If the crank rotates at a constant speed of 40 rpm, find the average speed in feet per second of slider 6 for the working stroke and also for the return stroke.

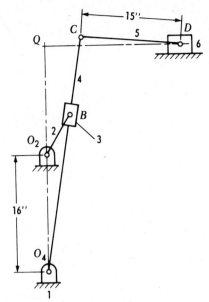

Fig. P3-3

3-4 Design a Whitworth quick-return mechanism similar to the one in Fig. 3-10.   The driving crank is to rotate clockwise at constant speed and the time ratio is to be 2:1.   The slow stroke of slider 6 is to be to the left.   Length O_2O_4 is to be 3 in. and the length of stroke is to be 13½ in.   Further, assume length $CD = 3(O_4C)$. Note: Pivot O_2 may be placed either below or above pivot O_4 as required.   Draw the mechanism to a scale 1 in. = 6 in., showing it in the phase where slider 6 is at the extreme right position and compute the required values for lengths O_2B, O_4C, and CD.

3-5 Design a Watt straight-line mechanism that will give a close approximation to a straight line over a distance of 3 in.   Suggestion: Assume a trial value for $AB = BC = CD = 2$ in. and let D lie 4 in. to the right and 2 in. below A.   Lay out the path of point P for 10° intervals as crank AB moves from its lowest position to its highest position.   Then measure the length of the approximate straight-line portion.   Call it h and indicate its value with a dimension line on the drawing.   The required dimensions will then be equal to the assumed dimensions multiplied by the ratio of the desired length to h.

3-6 From Eq. (3-2) show that for a constant speed of shaft 2,

$$\frac{\omega_{3\,max} - \omega_{3\,min}}{\omega_2} = \sin \delta \tan \delta$$

Then using values of $\delta = 0°, 10°, 20°, 30°, 40°$, and $45°$ plot a curve showing values of $[(\omega_{3\,max} - \omega_{3\,min})/\omega_2] \, 100$ versus δ.

4

INSTANT CENTERS

4-1 INTRODUCTION

In the next two chapters several methods for determining velocities in mechanisms will be presented. One of these requires a knowledge of instant centers. Velocity is important because it affects the time required to perform a given operation, as for example, the machining of a part. Power is the product of force and velocity. Thus for the transmission of a given amount of power, the forces and stresses in the various links of a mechanism can be reduced by altering the velocities through a change in the dimensions of the links. Friction and wear on machine parts are also dependent on velocity. Further, a determination of the velocities in a mechanism is required if an acceleration analysis is to be made.

When determining the velocities in a mechanism we find their values for some instantaneous position of the links. We shall see that any link having plane motion may be considered as rotating at the instant about some point in its plane of motion. Thus the point is a center of rotation for the link and may or may not lie within the link itself. Further, for some links these centers of rotation are stationary, while for others the centers of rotation move. The term *instant center* is used to denote the center of rotation of a body at some instant.

4-2 INSTANT CENTER

An *instant center* is (1) a point in one body about which some other body is rotating either permanently or at the instant; and (2) a point common to two bodies having the same linear velocity in both magnitude and direction in each.   Both parts of this definition are important, as we will make use of them in locating instant centers.

4-3 INSTANT CENTER AT A PIN CONNECTION

In the four-bar linkage shown in Fig. 4-1, each pin connection is an instant center. It is customary to designate these centers using the numbers of the links which are pivoted together at these points.   Thus the point in link 1 about which link 2 rotates is labeled "12" and pronounced "one, two."   If link 2 were held fixed, and link 1 were allowed to rotate, the relative motion for links 1 and 2 would be unchanged, still being a rotation about point 12. Thus instant center 12 may also be regarded as a point in 2 about which link 1 is rotating.   Similarly, instant center 23 (pronounced "two, three") is a point in link 2 about which link 3 is rotating, or it is a point in 3 about which 2 is rotating.   Instant centers 12 and 14 remain fixed in the frame as the mechanism operates, and thus they are called *fixed centers*. Instant centers 23 and 34 are called *moving centers* since they move relative to the frame.

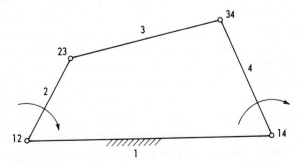

Fig. 4-1

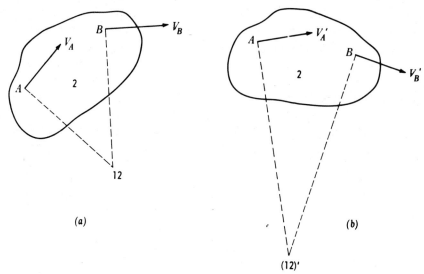

Fig. 4-2

4-4 INSTANT CENTER FOR A BODY WHEN THE VELOCITIES OF TWO POINTS ARE KNOWN IN DIRECTION

Any two bodies having motion relative to one another have an instant center. In Fig. 4-2 the paper is considered the stationary member or body 1. Suppose in Fig. 4-2a points A and B in body 2 have linear velocities whose directions are known. Since the linear velocities of all points in a rotating body are at right angles to their radii of rotation, we can draw in the dashed lines perpendicular to the velocities as shown. Their point of intersection locates instant center 12, the point in body 1 about which body 2 is rotating. Thus when the directions of the linear velocities of two points in a body are known, the instant center can be determined, provided that these points do not lie on the same radial line. As body 2 moves, its position an instant later is shown in Fig. 4-2b. Suppose that the velocities of points A and B have some new values V_A' and V_B'. Then the dashed lines drawn perpendicular to V_A' and V_B' intersect at point (12)' which is the location of the center of rotation at the instant considered. Thus as a body moves, its center of rotation may be a different point at each instant; this explains why it is called an instant center. An instant center is sometimes called a *centro* or a *pole*.

4-5 INSTANT CENTER FOR A SLIDING BODY

In Fig. 4-3 body 2 slides in a circular slot in body 1.   Consequently all points
on the slider move along circular paths whose centers lie at a point in body 1.
Hence point 12 is the instant center for these bodies.

Figure 4-4 shows a slider which has rectilinear motion.   Since all points
on body 2 move along straight-line paths, their radii of rotation will consist of
parallel lines as shown in the figure.   Similar to Fig. 4-2, the center of rotation lies at the intersection of the radial lines.   Recalling that parallel lines
intersect at infinity, it follows that instant center 12 lies at infinity either

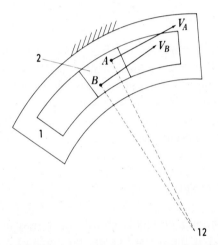

Fig. 4-3

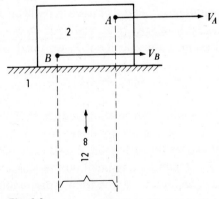

Fig. 4-4

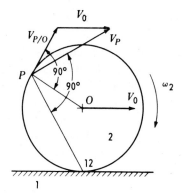

Fig. 4-5

above or below the slider.   Note the manner in which this is indicated in the figure.   Hence rectilinear translation is a special case of rotation wherein the center of rotation lies at infinity and the radii of rotation are of infinite length.   Summarizing, when a body slides with rectilinear motion on another body, their common instant center lies at infinity in either direction along a line which is perpendicular to the direction of sliding.

4-6   INSTANT CENTER FOR A ROLLING BODY

If disk 2 (Fig. 4-5) rolls without slipping on link 1, which may or may not be stationary, the point of contact 12 is the instant center for bodies 1 and 2. That is, 12 is the point in body 1 about which body 2 is rotating at the instant. If body 1 is at rest and disk 2 rotates clockwise as shown, center O of the disk will have a velocity V_O.   The motion of point P relative to O will be one of rotation with radius PO, and $V_{P/O}$, the velocity of P relative to O, will be at 90° to PO.   In order to obtain the absolute velocity of P, we must add V_O to $V_{P/O}$.   Thus

$$V_P = V_{P/O} +\!\!\!+ V_O$$

Next, we recall that the velocity of a point in a rotating body is at 90° to the radius of rotation for the point.   Then a line can be drawn from P perpendicular to V_P.   This line is found to pass through point 12, and length P-12 is the radius of rotation for point P. P can be any point on the disk and by adding V_O and $V_{P/O}$, V_P can always be found.   A line from P perpendicular to V_P will always pass through the instant center 12.

4-7 KENNEDY'S THEOREM

Kennedy's theorem states that any three bodies having plane motion relative to one another have three instant centers, and they lie on a straight line. The proof of this theorem is as follows. Let bodies 1, 2, and 3 in Fig. 4-6 be any three bodies moving relative to one another. For convenience we can assume that one of the bodies is stationary. The stationary member has been called body 1 in the figure. Instant centers 12 and 13 are the points in body 1 about which bodies 2 and 3, respectively, are rotating at the instant. The three bodies need not be connected to one another in any manner. Instant center 23 for bodies 2 and 3 remains to be located. Suppose it were to lie at point P. The only motion body 2 can have relative to 1 at the instant is a rotation about their common instant center 12. Then when considered as a point in 2 the velocity of P must be perpendicular to radius 12-P. Similarly, the only motion body 3 can have relative to 1 at the instant is a rotation about instant center 13. Thus if P is considered as a point in 3, its velocity must be perpendicular to radius 13-P. Next, we must recall that an instant center is a point common to two bodies and has the same linear velocity, in both magnitude and direction in each. Since the directions of the two velocities V_P in the figure do not coincide, point P cannot be the instant center 23. It becomes apparent that their directions can coincide only if instant center 23 lies somewhere along the line 12-13. The exact

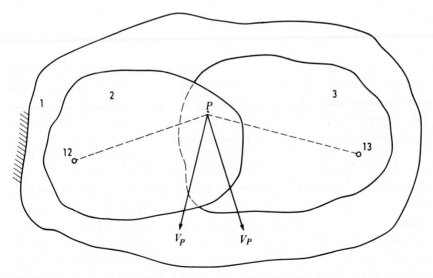

Fig. 4-6

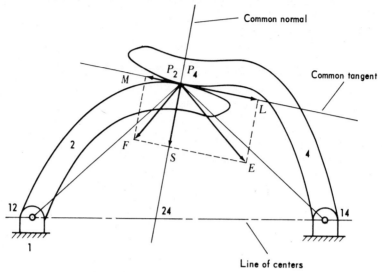

Fig. 4-7

location of 23 along line 12-13 depends on the directions and magnitudes of the angular velocities of 2 and 3 relative to 1.

4-8 INSTANT CENTERS FOR A DIRECT–CONTACT MECHANISM

Sliding contact

In Fig. 4-7 bodies 2 and 4 are in direct contact. P_2 and P_4 are the coincident points at the contact point, and their velocities P_2E and P_4F are perpendicular to 12-P_2 and 14-P_4, respectively. These velocities have normal components P_2S and P_4S which must be equal at all times if the bodies are to remain in contact. The tangential components of velocities P_2E and P_4F are P_2L and P_4M. As explained in Sec. 2-15, if the point of contact does not lie on the line of centers 12-14, these tangential components will not be equal, and sliding exists. Hence the only relative motion which bodies 2 and 4 can have at their point of contact is in the direction of the common tangent, and their center of relative rotation, instant center 24, must then lie along the common normal. However, by Kennedy's theorem instant center 24 must lie along line 12-14. Hence instant center 24 lies at the point of intersection of the common normal and the line of centers 12-14.

Rolling contact

As explained in Sec. 2-15, rolling exists only if the points of contact P_2 and P_4 have velocities which are identical. This requires that the point of contact lie on the line of centers 12-14, as shown in Fig. 4-8. Since an instant center is a point common to two bodies and has the same linear velocity in each, it follows that when bodies 2 and 4 have rolling contact, their common instant center lies at their point of contact.

4-9 NUMBER OF INSTANT CENTERS FOR A MECHANISM

Any two links in a mechanism have motion relative to one another and thus have a common instant center. Hence the number of instant centers for a mechanism is equal to all the possible combinations of two from the total number of links. Let n be the number of links. Then the number of instant centers is

$$N = \frac{n(n-1)}{2} \qquad\qquad (4\text{-}1)$$

4-10 PRIMARY INSTANT CENTERS

All instant centers which can be found merely by inspection are called *primary instant centers*. It is important for the student to be able to recognize their occurrence, because only after all primary instant centers have been located for a mechanism can we locate the remaining instant centers by applying Kennedy's theorem. Primary instant centers can be summarized

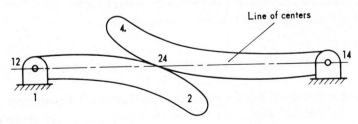

Fig. 4-8

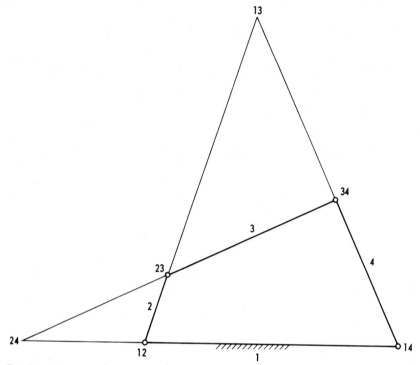

Fig. 4-9

as follows:

1. Instant center for pin-connected links, e.g., instant center 23 in Fig. 4-1.
2. Instant center for a sliding body, e.g., instant center 12 in Figs. 4-3 and 4-4.
3. Instant center for a rolling body, e.g., instant center 12 in Fig. 4-5.
4. Direct-contact mechanisms
   a. If the bodies have sliding contact, their instant center lies where the common normal through the point of contact intersects the line of centers, e.g., instant center 24 in Fig. 4-7.
   b. If the bodies have rolling contact, their instant center lies at the point of contact, e.g., instant center 24 in Fig. 4-8.

4-11 CIRCLE DIAGRAM METHOD FOR LOCATING INSTANT CENTERS

The four-bar linkage in Fig. 4-9 will be used to illustrate the procedure. All the primary instant centers must be located first. These are centers 12, 23, 34, and 14, as shown in the figure. By applying Kennedy's theorem, we can locate the remaining instant centers.   A simple, systematic means for carrying this out is known as the *circle diagram method*.   Points are laid out approximately equally spaced along a circle, as shown in Fig. 4-10.   Each point represents a link in the mechanism.   All the possible straight lines joining these points represent the instant centers.   First, all centers which have already been located are drawn in as solid lines.   Thus since instant centers 12, 23, 34, and 14 have been located in Fig. 4-9, they are drawn in as solid lines in Fig. 4-10.   The instant centers remaining to be located are represented by dotted lines.   In order to locate these centers, we examine the diagram and find any two triangles which a dotted line completes.   For example, we note that line 13 completes triangles 123 and 341; that is, if line 13 were a solid line, these two triangles would be completed (formed by solid lines).   These two triangles can be used to locate instant center 13. Links 1, 2, and 3 have three instant centers, 12, 23, and 13, and the latter are represented by lines 12, 23, and 13 in Fig. 4-10.   By Kennedy's theorem these three instant centers must lie on a straight line.   Hence in Fig. 4-9 instant center 13 is somewhere on a line joining points 12 and 23.   Also links 3, 4, and 1 have three instant centers, 34, 14, and 13, and these are represented by lines 34, 14, and 13 in Fig. 4-10.   Kennedy's theorem states that these three instant centers must lie on a straight line.   Thus in Fig. 4-9 instant center 13 must lie on a line joining points 34 and 14.   Since it was noted earlier that instant center 13 also lies somewhere along line 12-23, it must then be located at the intersection of lines 12-23 and 34-14, as shown in Fig. 4-9.

After an instant center has been located, it is drawn in as a solid line on the circle diagram.   This is illustrated in Fig. 4-11, where line 13 has been made solid.   Next, we observe from Fig. 4-11 that instant center 24 remains

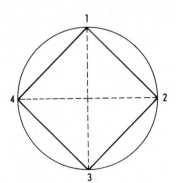

Fig. 4-10

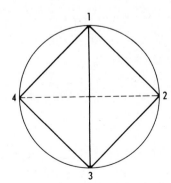

Fig. 4-11

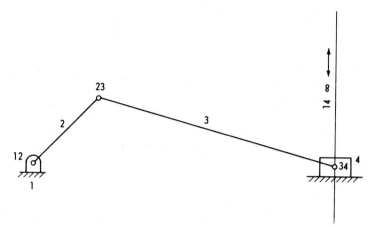

Fig. 4-12

to be located. Since line 24 completes triangle 412, instant center 24 must lie on a line with instant centers 41 and 12 in Fig. 4-9. Also since line 42 in Fig. 4-11 completes triangle 432, instant center 24 must lie on a line with instant centers 34 and 23 in Fig. 4-9. Hence instant center 24 lies where these two lines intersect, as shown in the figure.

When using the circle diagram method, it is important to find all the primary instant centers first; otherwise it may not be possible to find any two triangles which an unlocated instant center completes. Further, after an instant center has been located on the drawing of the mechanism, it should immediately be drawn in as a solid line on the circle diagram. This is necessary when working with mechanisms having more than four links. Otherwise, when locating the remaining instant centers, it may not be possible to find additional pairs of triangles which are composed of solid lines except for having a common side which is dotted.

EXAMPLE 4-1 Locate the instant centers for the slider-crank mechanism in Fig. 4-12.

Solution All the primary instant centers are located first. These are instant centers 12, 23, 34, and 14, as shown in the figure. These instant centers are then drawn in as solid lines on the circle diagram in Fig. 4-13.

Instant center 13 is represented by the dotted line in Fig. 4-13, and we note that it completes triangles 123 and 143. Thus in Fig. 4-12 instant center 13 must lie on a line with instant centers 12 and 23. Also it must lie on a line with instant centers 14 and 34. Hence instant center 13 lies where these lines intersect, as shown in Fig. 4-14.

From the circle diagram which is again shown in Fig. 4-15, we note that the dotted line representing instant center 24 completes triangles 412 and 432. Hence in Fig. 4-14 instant center 24 lies on a line with instant centers 23 and 34 and also on a line with instant centers 12 and 14. Thus instant center 24 lies at the point of intersection of these two lines as shown. Since

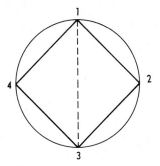

Fig. 4-13

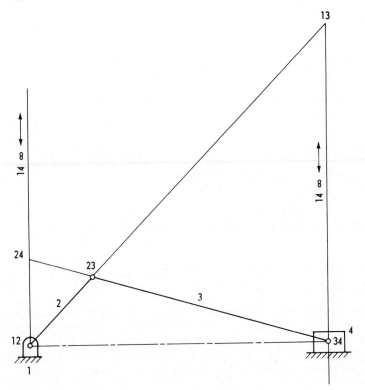

Fig. 4-14

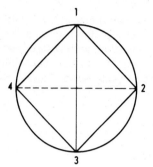

Fig. 4-15

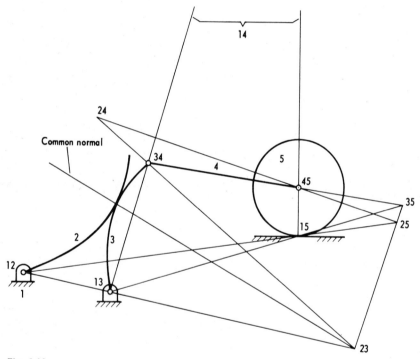

Fig. 4-16

instant center 14 lies at intinity, a line from 12 to 14 must be drawn parallel to line 34-14. This utilizes the concept that parallel lines meet at infinity.

EXAMPLE 4-2 In Fig. 4-16 link 5 is a wheel which rolls on link 1. Locate all instant centers for the mechanism.

Solution The solution is shown in the figure. The number of instant centers = $n(n-1)/2 = 5(5-1)/2 = 10$. All the primary instant centers are located first. These are 12, 13, 34, 45, 15, and 23 and are drawn in as solid lines on the circle diagram in Fig. 4-17. The instant centers remaining to be found are represented by the dotted lines. Since instant center 14 completes triangles 134 and 154, it can be located next. After instant center 14 is located in Fig. 4-16, line 14 is drawn in solid in Fig. 4-17. We then continue using the circle diagram method to locate the remaining instant centers.

EXAMPLE 4-3   Bodies 2 and 3 (Fig. 4-18) rotate about pivots 12 and 13 in the frame. B and C are points in 2 and 3, respectively, and their velocities are as shown. Locate instant center 23.

Solution By Kennedy's theorem instant center 23 lies on line 12-13. Also by the definition of an instant center, 23 is a point common to bodies 2 and 3 and has the same linear velocity in each. As a point in link 2, 23 has a radius of rotation 12-23 and as a point in link 3 its radius of rotation is 13-23. Further, velocities of points in a rotating body are proportional to their radii of rotation. Thus if a line is drawn from 12 and through the terminus of V_B and also one from 13 through the terminus of V_C, the intersection of these lines determines the magnitude of V_{23} as well as the location of instant center 23 along line 12-13.

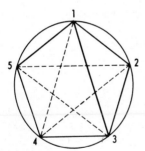

Fig. 4-17

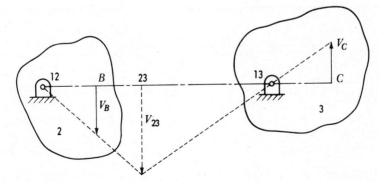

Fig. 4-18

4-12   CENTRODES

We have noted that some instant centers are fixed centers of rotation in the frame, while others are continuously changing position as the mechanism passes through the various phases of its motion.   The path of a moving instant center can be plotted.   A smooth curve through these points is a *centrode*.

In Fig. 4-19 assume disk 2 rolls on body 1.   Instant center 12 is always the point of contact.   The straight line 12-*B* is the centrode of 12 on body 1 and the circle 12-*B′* is the centrode of 12 on body 2.   The centrode which

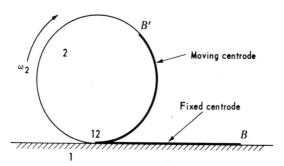

Fig. 4-19

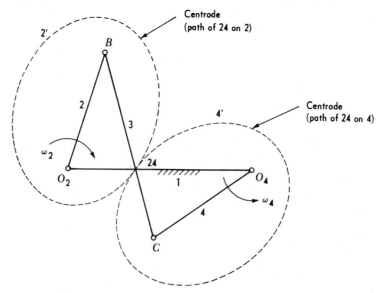

Fig. 4-20

lies on the fixed body is called a *fixed centrode*, and the one which lies in the moving body is a *moving centrode*.

Consider the crossed four-bar linkage in Fig. 4-20, where length $O_2B = O_4C$ and $BC = O_2O_4$. Instant center 24 lies at the intersection of lines BC and O_2O_4, and the centrode which it generates on the frame is line O_2O_4. The centrode which instant center 24 describes on link 2 is an ellipse, and the centrode which it describes on link 4 is also an ellipse. As links 2 and 4 in the four-bar linkage rotate, these ellipses are always in contact at the instant center 24, which moves along the line of centers O_2O_4. Hence the ellipses have rolling contact. If gear teeth are placed on these ellipses, then we have a pair of elliptical gears. Further, if in Fig. 4-20 the original four-bar linkage were replaced by a mechanism consisting of the two rolling ellipses 2' and 4' pivoted to the frame at points O_2 and O_4, then the angular motions of links 2' and 4' would be identical to the motions of links 2 and 4, respectively. Hence we see that if the links of a mechanism are replaced by members whose outlines are made to conform with the centrodes, and if these are made to roll on one another, then we have an equivalent mechanism.

PROBLEMS

4-1 Locate all the instant centers for the mechanism shown in Fig. P4-1.

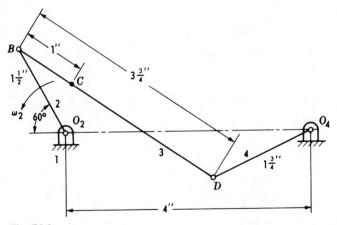

Fig. P4-1

4-2 Locate all the instant centers for the mechanism shown in Fig. P4-2.

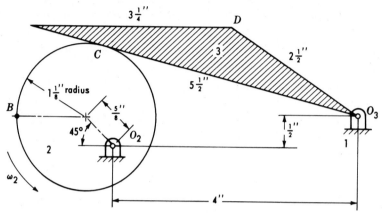

Fig. P4-2

4-3 Locate all the instant centers for the mechanism shown in Fig. P4-3.

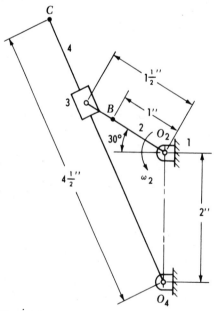

Fig. P4-3

4-4 Locate all the instant centers for the mechanism shown in Fig. P4-4.

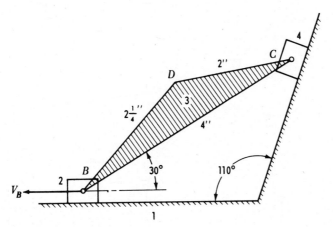

Fig. P4-4

4-5 Figure P4-5 shows a planetary gear train. Gear 3 is integral with the driving shaft and as it rotates causes gear 2 to roll around the inside of stationary gear 1. Each gear 2 rotates freely on a shaft on the carrier 4. The driven shaft is integral with the carrier and rotates at a fraction of the speed of the driving shaft. Locate all instant centers.

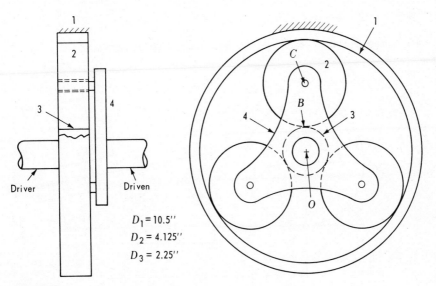

$D_1 = 10.5''$
$D_2 = 4.125''$
$D_3 = 2.25''$

Fig. P4-5

4-6 Locate all instant centers for the mechanism shown in Fig. P4-6.

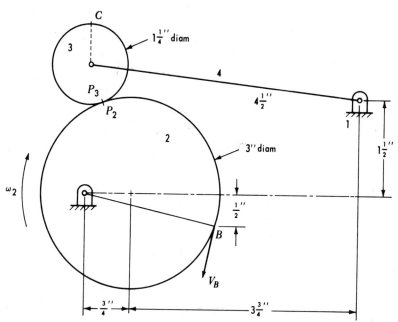

Fig. P4-6

4-7 Locate all instant centers for the mechanism shown in Fig. P4-7.

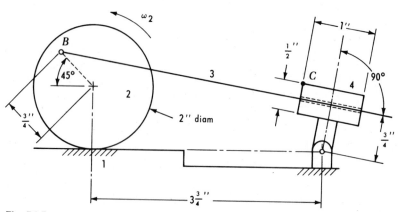

Fig. P4-7

4-8 Locate all the instant centers for the mechanism shown in Fig. P4-8.

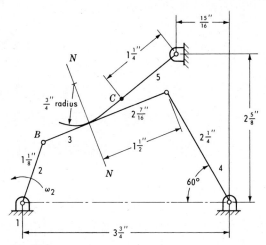

Fig. P4-8

4-9 Locate all the instant centers for the mechanism shown in Fig. P4-9.

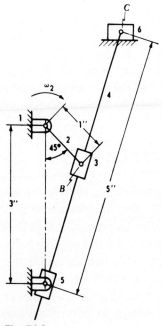

Fig. P4-9

5

VELOCITIES
BY INSTANT CENTERS
AND BY COMPONENTS

5-1 INTRODUCTION

In this chapter two methods for finding linear velocities of points on a mechanism will be presented. The first of these makes use of instant centers; the second consists of resolving velocity vectors into components.

5-2 LINEAR VELOCITIES BY INSTANT CENTERS

When finding linear velocities by the method of instant centers, the following basic principles must be kept in mind: (1) The magnitudes of the linear velocities of points in a rotating body are directly proportional to their radii of rotation.   The radius of rotation of a point is the distance from the point to the instant center in the frame about which the link containing the point is rotating.   (2) The linear velocity of a point is directed perpendicular to the radius of rotation of the point.   (3) An instant center is a point common to two bodies and has the same linear velocity in both magnitude and direction in each.

5-3   VELOCITIES IN A FOUR–BAR LINKAGE

A four-bar linkage will be used as a first example to illustrate two graphical methods for determining linear velocities by use of instant centers.

Rotation-of-radius method

In the mechanism in Fig. 5-1 suppose the linear velocity of point B is known, and the linear velocities of points 23, D, and E are to be found. Points B and 23 lie on link 2, which is rotating about instant center 12 in the frame.   Velocity V_{23} must be directed perpendicular to its radius of rotation 12-23.   Hence its direction is known.   If we draw a line from point 12 through the terminus of V_B, its intersection with the perpendicular at 23 determines the magnitude of V_{23}.   From similar triangles

$$\frac{V_{23}}{12\text{-}23} = \frac{V_B}{12\text{-}B}$$

which satisfies the rule that linear velocities of points in a rotating body are directly proportional to their radii of rotation.   In order to find V_{23}, we considered 23 as a point in link 2.   Next, considering 23 as a point in link 3, the velocity of point D can be found.   Since link 3 is rotating about instant center 13 in the frame, the instant radii of rotation for points 23 and D are lengths 13-23 and 13-D, respectively.   Similar triangles involving these radii

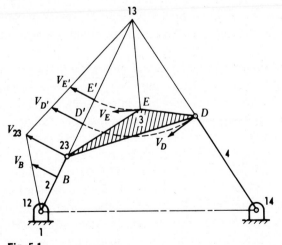

Fig. 5-1

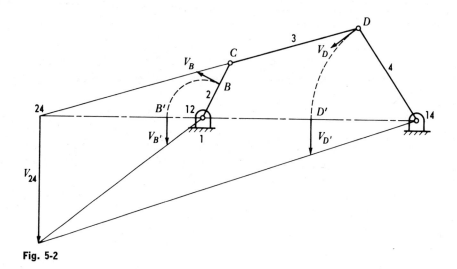

Fig. 5-2

are constructed by rotating radius 13-D about point 13 until it is in line with radius 13-23. Thus 13-D' is the instant radius of point D after it has been rotated, and a perpendicular to 13-D' at point D' indicates the direction for the velocity of point D'. A line drawn from 13 to the terminus of V_{23} will determine the magnitude of $V_{D'}$. Then V_D will be equal in magnitude to $V_{D'}$ but to be in true direction it must be drawn perpendicular to its instant radius 13-D. In a similar manner the velocity of point E is found by rotating its instant radius 13-E in line with instant radius 13-23. Vector $V_{E'}$ represents the velocity of point E when in the rotated position E'. Next, V_E is made equal in magnitude to $V_{E'}$ and must be directed perpendicular to instant radius 13-E.

Another example of the rotation-of-radius method for finding velocities is illustrated in Fig. 5-2. Let us suppose that V_B is known and V_D is to be found. We must first find the velocity of instant center 24, which is a point in both links 2 and 4. Considering 24 as a point in link 2, like all points in 2 it will be rotating about point 12 in the frame, and its instant radius is 12-24. After rotating instant radius 12-B in line with instant radius 12-24, a gauge line from 12 through the terminus of $V_{B'}$ determines the magnitude of V_{24}. Next, instant center 24 is considered as a point in link 4. As points in link 4, 24 and D are rotating about instant center 14 in the frame. Hence their velocities are directly proportional to their instant radii 14-24 and 14-D. A gauge line drawn from 14 to the terminus of V_{24} determines the magnitude

of $V_{D'}$ as shown in the figure. V_D has the same magnitude as $V_{D'}$, but must
be directed perpendicular to instant radius 14-D.

It is to be noted that this method for finding velocities by rotating the
instant radius of a point in line with the instant radius of another can only
be used when the two points are in the same link.   For example, in Fig. 5-1
with V_B known, we wanted to find the velocity of point D which can be
considered as a point in link 3.   It was necessary to first find the velocity of
a point which was common to links 2 and 3, namely point 23.   Such a point
is called a *transfer point*.   Similarly, in Fig. 5-2 when finding V_D, since points
B and D are not on the same link but lie on links 2 and 4 (point D lies on
link 3 or link 4), it was necessary to find the velocity of the point which is
common to links 2 and 4, namely instant center 24.   Thus 24 was used as a
transfer point.

Parallel-line method

The mechanism of Fig. 5-1 is also shown in Fig. 5-3; and again with V_B
known, the velocities for points C, D, and E are to be found.   Vector V_C
is found first in the same manner as illustrated in Fig. 5-1.   Then V_C is
rotated into line 13-C, the instant radius for point C.   From point C' a line
is drawn parallel to line CE.   Where this line intersects line 13-E locates
point E'.   Line $C'E'$ is thus parallel to the base of triangle 13CE.   From a
theorem of geometry which states that a line drawn parallel to the base of a

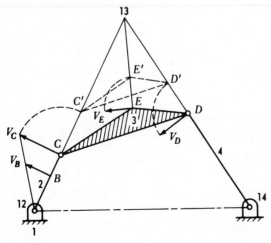

Fig. 5-3

triangle divides the sides of the triangle proportionately, we have

$$\frac{CC'}{C\text{-}13} = \frac{EE'}{E\text{-}13}$$

or

$$\frac{V_C}{C\text{-}13} = \frac{V_E}{E\text{-}13}$$

which satisfies the rule that linear velocities of points in a rotating body are directly proportional to their radii of rotation. In the figure, vector V_E is obtained by rotating length EE' to a position perpendicular to radius 13-E.

In order to find V_D in Fig. 5-3, the procedure just described is repeated. V_E is rotated into its instant radius 13-E. Then from E' line $E'D'$ is drawn parallel to ED. Length DD' will then be the magnitude of V_D. We note that V_D can also be obtained by drawing a line through C' parallel to CD. This also locates point D'.

This method, like the rotation-of-radius method described earlier, can be used to find the velocity of a point only if the velocity of some other point in the same link is known.

5-4   VELOCITIES IN A SLIDER–CRANK MECHANISM

In Fig. 5-4 suppose the angular velocity ω_2 of the crank is known and the velocity of the piston (link 4) is to be found for the crank position shown. The velocity V_{23} is computed first; thus

$$V_{23} = R\omega = (12\text{-}23)\omega_2$$

Vector V_{23} then can be laid off to any convenient scale on the drawing. Points 23 and 34 are points on link 3 which is rotating about instant center 13. Hence their velocities are proportional to their instant radii 13-23 and 13-34. By rotating radius 13-34 in line with radius 13-23, we can find $V_{(34)'}$ as shown. Next, V_{34} must be drawn perpendicular to instant radius 13-34. The magnitude of V_{34} is equal to the magnitude of $V_{(34)'}$. Since 34 is a point on link 4 as well as on link 3, V_{34} is the velocity of the piston as well as the velocity of a point on 3.

5-5   VELOCITIES IN A CAM MECHANISM

In Fig. 5-5 ω_2, the angular velocity of the cam, is assumed known, and the velocity of the follower is to be found for the position of the cam shown. The instant centers are located first. Links 1, 2, and 3 comprise a direct-

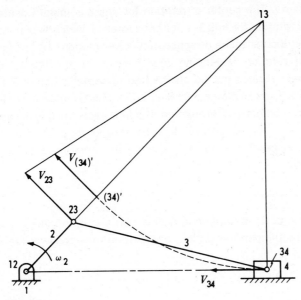

Fig. 5-4

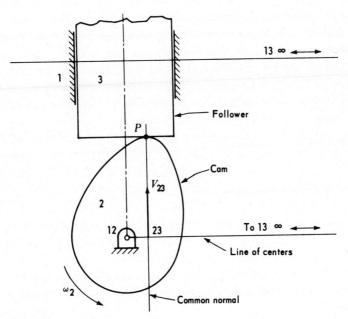

Fig. 5-5

contact mechanism. The center of rotation for link 2 is instant center 12, and the center of rotation for link 3 is instant center 13, which lies at infinity along a line perpendicular to the direction of follower motion. A line joining these centers of rotation is labeled *line of centers* in the figure. Since the point of contact P does not lie on the line of centers, bodies 2 and 3 have sliding contact, as explained in Sec. 2-15. Hence instant center 23 lies where the common normal through point P intersects the line of centers, as proved in Sec. 4-8. Considering 23 as a point on body 2, its velocity V_{23} must be directed perpendicular to 12-23, which is the instant radius for point 23. Its magnitude is

$$V_{23} = R\omega = (12\text{-}23)\omega_2$$

From the definition of an instant center, 23 is also a point on link 3. Since the follower has rectilinear translation, all points on the follower have the same velocity V_{23}.

5-6   VELOCITIES IN A COMPOUND LINKAGE

Mechanisms can be classified as simple mechanisms and compound mechanisms. A simple mechanism consists of three or four links. All other mechanisms, or those consisting of more than four links, are compound mechanisms. Compound mechanisms are usually made up of combinations of simple mechanisms. The mechanism in Fig. 4-16 and shown again in Fig. 5-6 is an example of a compound mechanism. It consists of the simple mechanism composed of links 1, 2, and 3 combined with a second simple mechanism consisting of links 1, 3, 4, and 5.

In Fig. 5-6 suppose the velocity of point B on link 2 is known, and the velocity of point 45 in link 5 is to be found. The analysis here is similar to that used in Fig. 5-2, which was explained in Sec. 5-3. We must first find the velocity of the transfer point (instant center 25) which is a point in both links 2 and 5. As a point in link 2 it has an instant radius 12-25. Hence a gauge line from center 12 through the terminus of $V_B{}'$ determines the magnitude of V_{25}. Next, considering instant center 25 as a point in link 5, it has an instant radius 15-25. Thus a gauge line from 15 to the terminus of V_{25} determines the magnitude of $V_{(45)'}$. V_{45} has the same magnitude as $V_{(45)'}$ but it must be directed perpendicular to line 15-45, which is the instant radius for point 45.

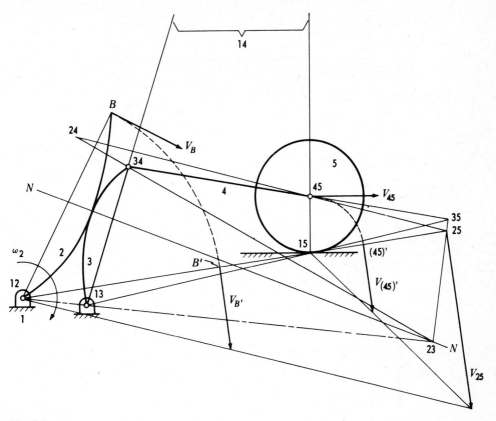

Fig. 5-6

5-7 ANGULAR VELOCITIES

We noted earlier that the angular velocity of a body may be found as follows:

$$\omega = \frac{V}{R}$$

where V is the linear velocity of a point on the body and R is the radius of rotation for the point.   In Fig. 5-1, since 23, the common instant center for bodies 2 and 3, is a point in either link 2 or 3, then

$$\omega_2 = \frac{V}{R} = \frac{V_{23}}{12\text{-}23}$$

and

$$\omega_3 = \frac{V_{23}}{13\text{-}23}$$

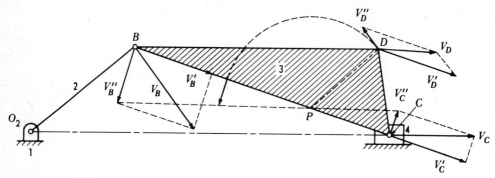

Fig. 5-7

Hence

$$\frac{\omega_2}{\omega_3} = \frac{13\text{-}23}{12\text{-}23}$$

(5-1)

and

$$\omega_3 = \omega_2 \frac{12\text{-}23}{13\text{-}23} \text{ CW}$$

Similarly, in Fig. 5-2, since instant center 24 is a point in either link 2 or link 4,

$$\omega_4 = \omega_2 \frac{12\text{-}24}{14\text{-}24} \text{ CCW}$$

From Eq. (5-1) we can conclude that the angular-velocity ratio for any two links in a mechanism is inversely as the distances from the instant centers in the frame about which the links are rotating to the instant center which is common to the two links.

5-8 VELOCITIES BY COMPONENTS

Velocity analysis of linkages by components consists of resolving velocity vectors into suitable components so that the translation and rotation of the various links can be evaluated.   Suppose in Fig. 5-7, V_B, the velocity of the crankpin, is known, and the velocities of the slider and point D are to be found. V_B' is the component of V_B in the direction of BC and V_B'' is the component of V_B perpendicular to BC.   Since link 3 is a rigid body, V_C', the velocity of C in the direction of BC, will equal V_B'.   The slider must move parallel to its guide in the frame; hence V_C is parallel to the guide. A line from the terminus of V'_C and perpendicular to V'_C determines the magnitude of V_C. V''_C is the component of V_C in the direction perpendicular to BC, and its magnitude is determined by drawing a line from the terminus of V_C perpendicular to the direction of V''_C. The gauge line joining the terminus of V''_B and V''_C locates point P. This point on link 3 has no velocity perpendicular to BC but has a velocity in the direction of BC equal to V'_B. Since V''_D and V''_B are proportional to their distances from P, the magnitude of V''_D is found by rotating line PD into line PB as shown. V'_D has the same magnitude and direction as V'_B. Then V_D is the absolute velocity of D and is the resultant of V'_D and V''_D. The intersection of a line from the terminus of V'_D and parallel to V''_D with a line from the terminus of V''_D and parallel to V'_D determines V_D.

The instantaneous angular velocity of link 3 in Fig. 5-7 may be found as follows:

$$\omega_3 = \frac{V''_B}{PB} = \frac{V''_C}{PC} = \frac{V''_D}{PD} \text{ ccw}$$

In Fig. 5-8 slider 3 is pinned to the end of link 2 and slides on link 4 as link 2 rotates. V_{B_2} is the velocity of a point on link 2 and is known. The velocity of point D is to be determined. V_{B_4} is the velocity of a point on link 4 and is the component of V_{B_2}, which is perpendicular to O_4B_4, the instant radius of B_4. A line from O_4 through the terminus of V_{B_4} determines the magnitude of V_C. Next, V_D is found from V'_D, which is equal to V'_C. Since V'_D is a component of V_D, a line from the terminus of V'_D and perpendicular to V'_D determines the magnitude of V_D.

In the mechanism in Fig. 5-9 link 2 is a cam and link 3 is called the follower. V_B is the velocity of a point on the cam and is known. The velocity of the follower is to be found. The component of V_B in the direction of motion of link 3 is V'_B and is the velocity of a point on the follower. Since the follower has rectilinear translation, all points on link 3 have this velocity. The component of V_B along the face of the follower is V''_B and is the velocity of sliding.

For the linkage in Fig. 5-10, V_B is known and the velocities of points C and D are to be found. V'_B is the component of V_B along line BC. $V'_C(3)$ is made equal to V'_B and is the component of V_C along link 3. A perpendicular from the terminus of $V'_C(3)$ determines the magnitude of V_C, which must be perpendicular to O_4C. Next, from V_C a perpendicular is drawn to link 5 to determine $V'_C(5)$, the component of V_C along link 5. V'_D is made equal to $V'_C(5)$, and a line drawn perpendicular to V'_D determines the magnitude of V_D.

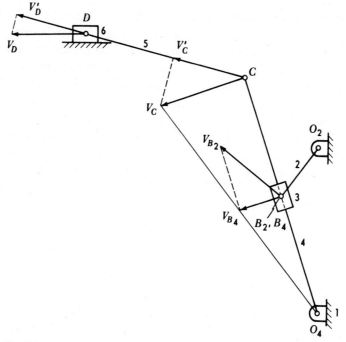

Fig. 5-8

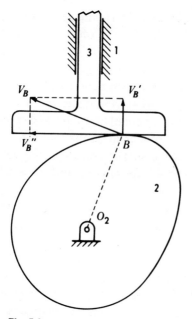

Fig. 5-9

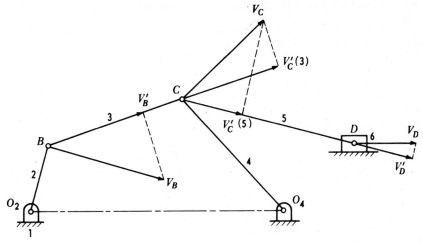

Fig. 5-10

When using the component method for finding velocities, it is important to remember that the absolute velocity of a point must be perpendicular
to its instant radius of rotation and that any components we resolve the
vector into are always smaller than the vector itself.

PROBLEMS

5-1   In Fig. P4-1, let V_B be represented by a vector 1 in. long.   Determine vectors
V_C and V_D by the rotation-of-radius method.

5-2   In Fig. P4-2, let V_B be represented by a vector 2 in. long.   Determine vectors
V_C and V_D by the rotation-of-radius method.   If $\omega_2 = 100$ rpm, determine the
values of V_B and V_C in feet per second.

5-3   In Fig. P4-3, let V_B be represented by a vector 1 in. long.   Determine vector
V_C by the rotation-of-radius method.

5-4   In Fig. P4-4, let V_B be represented by a vector 1 in. long.   Determine vectors
V_C and V_D by the rotation-of-radius method.

5-5   In Fig. P4-5, let V_B be represented by a vector $1\frac{1}{2}$ in. long.   Determine vector
V_C using instant centers.

5-6   In Fig. P4-6, let V_B be represented by a vector 1 in. long.   Determine vector
V_C by the rotation-of-radius method.

5-7   In Fig. P4-7, let V_B be represented by a vector $1\frac{1}{2}$ in. long.   Determine vector
V_C by the rotation-of-radius method.

5-8   In Fig. P4-8, let V_B be represented by a vector $2\frac{1}{2}$ in. long.   Determine vector
V_C by the rotation-of-radius method.

5-9   In Fig. P4-9, let V_B be represented by a vector $1\frac{1}{4}$ in. long.   Determine vector
V_C by the rotation-of-radius method.

5-10 In Fig. P4-1, if $\omega_2 = 100$ rpm, determine ω_3 and ω_4.

5-11 In Fig. P4-2, if $\omega_2 = 100$ rpm, determine ω_3.

5-12 In Fig. P4-3, if $\omega_2 = 150$ rpm, determine ω_4.

5-13 In Fig. P4-4, if $V_B = 20$ fps, determine ω_3 in radians per second.

5-14 In Fig. P4-5, determine the ratio ω_3/ω_4.

5-15 In Fig. P4-6, if $\omega_2 = 150$ rpm, determine ω_3 and ω_4.

5-16 In Fig. P4-7, if $\omega_2 = 120$ rpm, determine ω_4.

5-17 In Fig. P4-8, let $\omega_2 = 75$ rpm. Determine ω_3, ω_4, and ω_5.

5-18 In Fig. P4-9, if $\omega_2 = 75$ rpm, determine ω_3, ω_5, and ω_6.

5-19 In Fig. P4-1, let V_B be represented by a vector 1 in. long. Determine vectors V_C and V_D by the component method.

5-20 In Fig. P4-2, let $\omega_2 = 100$ rpm. Compute the value of V_C in feet per second, the velocity of a point on the cam. Then draw vector V_C, using a scale of 1 in. = 1 fps. Determine vector V_C', the velocity of the coincident point on the follower, and also determine vector V_C'', the velocity of sliding. Scale off their values in feet per second. Using the value found for V_C', compute ω_3 in rpm.

5-21 In Fig. P4-4, let V_B be represented by a vector 1 in. long. Determine vectors V_C and V_D by the component method.

5-22 In Fig. P4-6, let P_2 and P_3 be the coincident points of contact on links 2 and 3, respectively. Let V_{P_2} be represented by a vector $1\frac{1}{8}$ in. long. Locate instant center 13 and then determine vector V_C by the component method.

6

VELOCITIES IN MECHANISMS
BY METHOD OF RELATIVE VELOCITIES

6-1 INTRODUCTION

In the preceding chapter, velocity analysis in linkages by the method of instant centers and the method of components was explained. A third method utilizing the concept of relative velocity, presented in Chap. 2, will now be discussed. This latter method is most important because the relative velocities must be determined if an acceleration analysis is to be made for a linkage.

6-2 LINEAR VELOCITIES

To illustrate the relative-velocity method for finding velocities in a mechanism, let us first consider the slider-crank mechanism in Fig. 6-1. Suppose the angular velocity of the crank $\omega_2 = 15$ rad/sec ccw and it is desired to find the piston velocity V_C.

V_B is directed perpendicular to O_2B. Then we have

$$V_B = (O_2B)\omega_2 = 2.5 \times 15 = 37.5 \text{ ips}$$

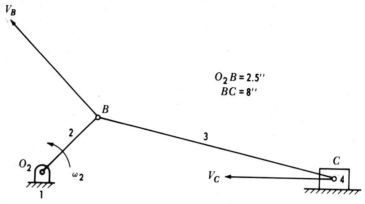

Fig. 6-1

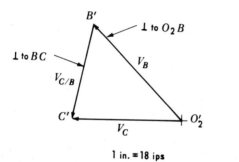

1 in. = 18 ips

Fig. 6-2

Using the relative-velocity equation explained in Sec. 2-9,

$$\overset{-\checkmark}{V_C} = \overset{\checkmark\checkmark}{V_B} +\!\!\!+ \overset{-\checkmark}{V_{C/B}}$$

(6-1)

Each quantity in the equation has both magnitude and direction, and to aid us in keeping in mind which quantities are known and which are unknown, we shall place two marks above each vector. A dash mark will be used to indicate that an item is unknown and a check mark will be used if it is known. We will let the first mark above a vector refer to its magnitude and the second to its direction. The magnitude of V_C is unknown, but its direction is known, because the piston is constrained by the frame to move horizontally. Hence a dash followed by a check are placed above V_C. Since both the magnitude and direction of V_B are known, a check mark for each of these is placed above V_B in the equation. Since link 3 is assumed to be a rigid body, C can have no velocity relative to B along the line CB. Thus if C has any velocity relative to B, it must be in a direction perpendicular to line BC. Then in Eq. (6-1) a dash mark is placed above $V_{C/B}$ to indicate that the magnitude of this vector is unknown, and a check mark signifies that the direction is known. From an inspection of the marks above the equation we see that there are only two unknowns, namely the magnitude of V_C and the magnitude of $V_{C/B}$. A vector equation can be solved only if there are no more than two unknowns.

The velocity polygon is shown in Fig. 6-2. Point O_2' is the pole, and the absolute velocities of all points are laid off from it. The pole represents all points on the mechanism having zero velocity. Primes will be used on the letters in the velocity polygon to represent corresponding points in Fig. 6-1. Thus O_2' represents point O_2 on the mechanism, and since point O_2 has zero velocity, O_2' lies at the pole. $O_2'B'$ and $O_2'C'$ represent the velocities of points B and C, respectively. On the original drawing of Fig. 6-2 a scale of 1 in. = 18 ips was used. The velocity pole O_2' is placed at any convenient location on the paper. Then V_B is laid off from the pole and is directed perpendicular to O_2B. We note from Eq. (6-1) that $V_{C/B}$ is to be added to V_B. Since the direction of $V_{C/B}$ is known to be perpendicular to line BC, the line $B'C'$ is next drawn in this direction. Its length is yet unknown. Next, line O_2C' is drawn and is made parallel to the direction of slider motion. The magnitudes of V_C and $V_{C/B}$ are then revealed from the intersection of lines $O_2'C'$ and $B'C'$. The length of O_2C', when measured from the original drawing, was found to be 1.80 in. Then multiplying by the velocity scale, we obtain

$$V_C = 1.80 \times 18 = 32.4 \text{ ips}$$

As another illustration of the relative-velocity method, let us consider the linkage in Fig. 6-3. The angular velocity of the driving crank $\omega_2 = 20$ rad/sec cw, and the velocity of point D is to be determined.

The velocity of point B is

$$V_B = (O_2B)\omega_2 = 6 \times 20 = 120 \text{ ips}$$

and is represented in Fig. 6-4 by vector $O_2'B'$ drawn from the velocity pole O_2'. By relative velocities

$$\overset{-\,-}{V_D} = \overset{\checkmark\checkmark}{V_B} + \!\!+ \overset{-\checkmark}{V_{D/B}} \tag{6-2}$$

Both the magnitude and direction of V_D are unknown. The magnitude of $V_{D/B}$ is unknown, but its direction is known, i.e., perpendicular to BD. Since Eq. (6-2) contains more than two unknowns, it cannot be solved. V_D

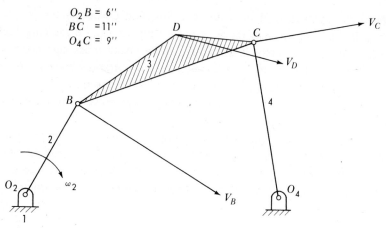

$$O_2 B = 6''$$
$$BC = 11''$$
$$O_4 C = 9''$$

Fig. 6-3

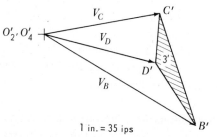

1 in. = 35 ips

Fig. 6-4

can be found, however, by first finding the velocity of C. For V_C we may write

$$V_C = V_B + V_{C/B} \tag{6-3}$$

Since there are only two unknowns in this equation, namely the magnitudes of V_C and $V_{C/B}$, point C' in Fig. 6-4 is readily located as follows. Line $B'C'$ is drawn from B' perpendicular to BC. This is the direction for $V_{C/B}$, which we see from Eq. (6-3) is to be added to V_B. Next, a line is drawn from O'_4 perpendicular to O_4C. This is the direction of V_C. The intersection of these two lines locates point C'. $V_{C/B}$ is then represented by line $B'C'$ and is directed from B' toward C'.

Next, for V_D we have

$$V_D = V_C + V_{D/C}$$

Since this equation contains more than two unknowns, we cannot find V_D by it alone.   However, since the right side of this equation and that of Eq. (6-2) both give V_D, we can equate them.   Then

$$\overset{\vee\vee}{V_B} \overset{-\vee}{\twoheadrightarrow V_{D/B}} = \overset{\vee\vee}{V_C} \overset{-\vee}{\twoheadrightarrow V_{D/C}}$$

This last equation contains only two unknowns and thus can be solved.   A line drawn from B' perpendicular to BD is the direction for $V_{D/B}$.   Its intersection with a line drawn from C' and perpendicular to CD locates point D'. Hence V_D has been found by adding to V_B and V_C the velocities $V_{D/B}$ and $V_{D/C}$, respectively.   Scaling values from the original drawing of Fig. 6-4, it was found that $V_D = 77.8$ ips and $V_C = 79.2$ ips.

Lines drawn from the pole to points on the velocity polygon represent the absolute velocities of the corresponding points on the mechanism.   A line connecting any two points on the velocity polygon represents the relative velocity for the two corresponding points on the mechanism.   In Figs. 6-3 and 6-4 a vector directed from C' toward D' represents the velocity of D relative to C, while a vector directed from D' toward C' is the velocity of C relative to D.

6-3 VELOCITY IMAGE

Each link in a mechanism has an image in the velocity polygon.   In Fig. 6-4 lines $B'C'$, $C'D'$, and $B'D'$ were drawn perpendicular to lines BC, CD, and BD in Fig. 6-3, respectively.   Hence triangle $B'C'D'$ is similar to triangle BCD and is called its image.   Similarly $O_2'B'$ is the image of O_2B, and $O_4'C'$ is the image of O_4C.   The velocity image is a useful concept.   If the velocities for any two points on a link have been found in the velocity polygon, the velocity of a third point on the link can readily be found by drawing the velocity image.   For example, in Fig. 6-4, if points B' and C' have been located, point D' can be located by constructing triangle $B'C'D'$ so that it will be similar to triangle BCD.   This requires that $B'C'$, $B'D'$, and $C'D'$ be perpendicular to BC, BD, and CD, respectively.

6-4 ANGULAR VELOCITIES

The angular velocity of a rigid link is equal to the relative velocity of any two points on the link divided by the distance between the points.   Since the distance between points in a rigid body remains fixed, the only velocity one point can have relative to another on the same link must be perpendicular to a line joining the points. Thus the motion of one point relative to the other is one of rotation, where the radius of rotation is the distance between the points. For example, in Fig. 6-3 the angular velocity of link 3 is

$$\omega_3 = \frac{V}{R}$$

$$= \frac{V_{B/C}}{BC} = \frac{V_{B/D}}{BD} = \frac{V_{C/D}}{CD} \text{ ccw}$$

From Fig. 6-4 $V_{B/C}$ is directed from C' toward B'; thus in Fig. 6-3, B is moving downward relative to C and hence is rotating counterclockwise about C. Therefore ω_3 is counterclockwise. Similarly, the directions of $V_{B/D}$ and $V_{C/D}$, as seen from the velocity polygon, indicate that B is rotating counterclockwise about D and C is rotating counterclockwise about D. Further, from the last equation we note that all straight lines in a body have the same angular velocity.

EXAMPLE 6-1   For the mechanism in Fig. 6-5 suppose $\omega_2 = 5$ rad/sec cw and the velocity of point D is to be found, along with the angular velocity of link 3. V_B is directed perpendicular to O_2B; thus

$$V_B = (O_2B)\,\omega_2$$

$$= 3 \times 5 = 15 \text{ ips}$$

The velocity polygon appears in Fig. 6-6 where $O_2'B'$ represents V_B to scale. A scale of 1 in. = 6 ips was used for the original drawing. By relative velocities

$$\overset{--}{V_D} = \overset{\checkmark\checkmark}{V_B} +\!\!\!+ \overset{-\checkmark}{V_{D/B}} \tag{6-4}$$

The velocity of D is unknown in magnitude and direction and thus dash marks are placed above V_D in the vector equation to signify this, while the check marks above V_B indicate that its magnitude and direction are known. Though the magnitude of $V_{D/B}$ is unknown, its direction is known to be perpendicular to BD. Thus a check mark is used to denote that its direction is known. We note there are three unknowns in Eq. (6-4), the magnitude of V_D, its direction, and the magnitude of $V_{D/B}$. Length $O_2'B'$ represents V_B and is laid off first from the pole O_2'.

Equation (6-4) states that in order to obtain V_D in Fig. 6-6, we are to add vector $V_{D/B}$ to vector V_B. Line $B'D'$ represents $V_{D/B}$ and $O_2'D'$ represents V_D. Since the magnitude of $V_{D/B}$ is unknown, and since the magnitude and direction of V_D are not known, point D' cannot yet be located. However,

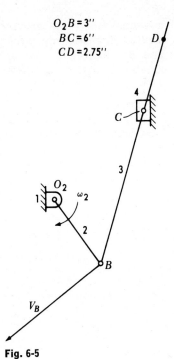

$O_2B = 3''$
$BC = 6''$
$CD = 2.75''$

Fig. 6-5

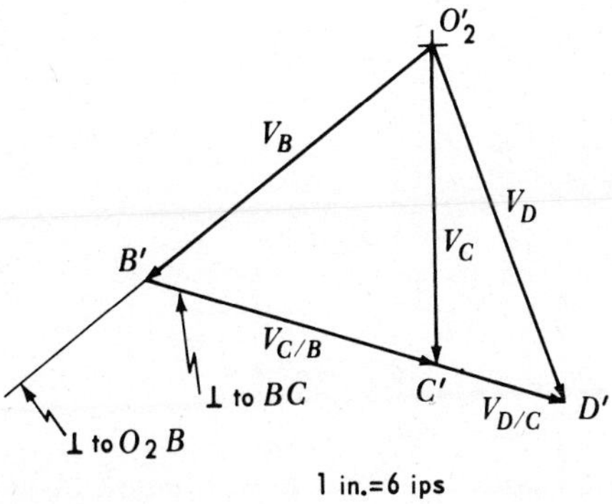

Fig. 6-6

1 in. = 6 ips

the problem can be solved by first determining V_C.   By relative velocities

$$\overset{-\checkmark}{V_C} = \overset{\checkmark\checkmark}{V_B} \nrightarrow \overset{-\checkmark}{V_{C/B}} \tag{6-5}$$

Equation (6-5) states that $V_{C/B}$ is to be added to V_B. $B'C'$ is $V_{C/B}$ and is drawn from point B'.   Its length, however, is not yet known. V_C is an absolute velocity and thus it must emanate from the pole O_2'.   Further, since

C is a point on link 4 as well as on link 3, it is constrained to move parallel to the guide on which link 4 slides. Thus line $O_2'C'$ represents the direction of V_C. The intersection of this line with line $B'C'$ locates C'. We are now able to make use of our original vector equation to obtain V_D; that is,

$$\overset{--}{V_D} = \overset{\vee\vee}{V_B} +\!\!\!+ \overset{\vee\vee}{V_{D/B}}$$

The magnitude of $V_{D/B}$ can be found by proportion. $BC = 6$ in. and $BD = 8.75$ in. Hence

$$BD = \frac{8.75}{6}(BC) = 1.46(BC)$$

Since points B, C, and D lie on the same link, B', C', and D' on the velocity polygon must be the image of BCD on the mechanism, and thus

$$B'D' = 1.46\ B'C'$$

Point D' is then located by drawing $B'D'$ 1.46 times as long as $B'C'$. Vector $O_2'D'$ represents the velocity of point D and $B'D'$ represents $V_{D/B}$. Scaling their values from the polygon, it was found that $V_D = 15.24$ ips and $V_{D/B} = 18$ ips.

The angular velocity of link 3 is

$$\omega_3 = \frac{V_{C/B}}{BC} = \frac{V_{D/B}}{BD}$$

Using the latter,

$$\omega_3 = \frac{18}{8.75} = 2.06 \text{ rad/sec cw}$$

6-5 VELOCITIES OF POINTS ON A ROLLING BODY

Let disk 2 in Fig. 6-7 be rolling on body 1. Then as explained in Sec. 4-6, body 2 is rotating about point P in link 1 at the instant. The center of the disk will have a velocity

$$V_C = R\omega$$

where R is the radius and ω is the angular velocity of the disk. Any other point on the disk, such as Q, will have a velocity relative to C which is

$$V_{Q/C} = (CQ)\omega$$

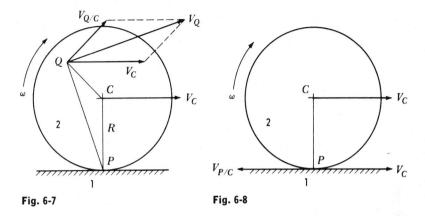

Fig. 6-7   **Fig. 6-8**

Since the motion of Q relative to C is a rotation about C, vector $V_{Q/C}$ must be perpendicular to the radius of rotation CQ. The absolute velocity of Q is then

$$V_Q = V_C \mathbin{+\!\!+} V_{Q/C}$$

as shown in Fig. 6-7. Further, vector V_Q is directed perpendicular to line PQ, which is the instantaneous radius of rotation for point Q.

Next, if instead of point Q we consider point P as the point on the disk, then

$$V_P = V_C \mathbin{+\!\!+} V_{P/C}$$

These vectors are shown in Fig. 6-8, and since $V_{P/C}$ is equal in magnitude to V_C but opposite in sense, point P on the disk has zero absolute velocity. A point P lying in link 1 and coincident with point P of body 2 also has zero velocity, since body 1 is at rest.

EXAMPLE 6-2   A quick-return mechanism is shown in Fig. 6-9. B_2 is a point on link 2 and its velocity V_{B_2} is known. The velocity of D is to be found. The angular velocities of links 4 and 5 are also desired. In Fig. 6-10 vector $O_2'B_2'$ represents V_{B_2}. B_4 is a point on link 4 which is coincident with B_2 at the instant. The magnitude of V_{B_4} is unknown, but its direction is perpendicular to O_4B_4, the radius of rotation for point B_4. Since there can be no motion of B_4 relative to B_2 in a direction perpendicular to link 4, V_{B_4/B_2} must be parallel to the link. Hence from B_2' a line is drawn parallel to O_4C. Its intersection with the line through O_4' that is perpendicular to O_4B_4 locates point B_4'. Vector $O_4'B_4'$ is then V_{B_4}, and $B_2'B_4'$ is V_{B_4/B_2}. Vector $O_4'C'$ is V_C,

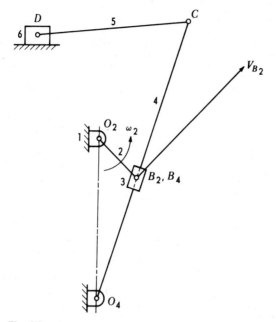

Fig. 6-9

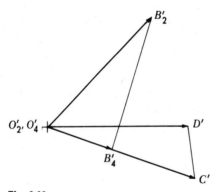

Fig. 6-10

which must be perpendicular to O_4C, the radius of rotation for C.   Length $O_4'C'$ is found by proportion as

$$\frac{O_4'C'}{O_4'B_4'} = \frac{O_4C}{O_4B_4}$$

Thus

$$O_4'C' = \frac{O_4C}{O_4B_4}(O_4'B_4')$$

From point C' a line is drawn perpendicular to CD.   Its intersection with a horizontal line through O_2' locates point D'.   Vector $O_2'D'$ is then V_D and $C'D'$ is $V_{D/C}$.   Next, $\omega_4 = V_C/O_4C$ and $\omega_5 = V_{C/D}/CD$.   From Fig. 6-10 we note that V_C is directed to the right and hence ω_4 is clockwise.   Also, from Fig. 6-10 $V_{C/D}$ is downward and thus ω_5 is clockwise.

EXAMPLE 6-3   In Fig. 6-11 a direct-contact mechanism is shown. P_2 and P_3 are points on links 2 and 3, respectively, which are coincident at the instant.   Since the point of contact lies on the line of centers O_2O_3, bodies 2 and 3 have rolling contact at the instant.   The velocity of B is known, and the velocity of C is to be determined.   The velocity polygon appears in Fig. 6-12, where V_B is laid off as $O_2'B'$ and is directed perpendicular to O_2B. Since V_{P_2} must be perpendicular to the radius of rotation for point P_2, a line is drawn from O_2', perpendicular to O_2P_2.   Next, from B' a line is drawn perpendicular to BP_2.   The intersection of these two lines locates P_2'.   Then $O_2'P_2'$ is V_{P_2} and $B'P_2'$ is $V_{P_2/B}$.   As explained in Sec. 2-15, when two bodies have rolling contact, the velocities of their contacting points are identical.

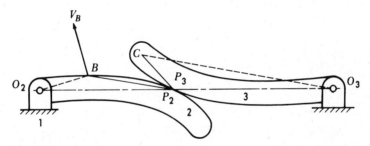

Fig. 6-11

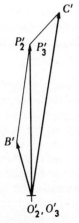

Fig. 6-12

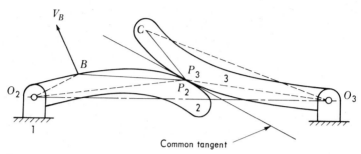

Fig. 6-13

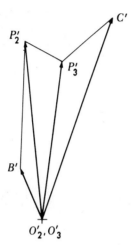

Fig. 6-14

Hence in Fig. 6-12 $O_3'P_3'$ is V_{P_3}. A line drawn from O_3' and perpendicular to radius O_3C is the direction for V_C. Next, from P_3' a line is drawn perpendicular to P_3C. Its intersection with the line drawn from O_3' locates point C'. Then O_3C' is V_C and $P_3'C'$ is V_{C/P_3}.

EXAMPLE 6-4 In Fig. 6-13 the mechanism of Fig. 6-11 is shown again, but in a different phase. It is to be noted that the new point of contact for bodies 2 and 3 is not on the line of centers O_2O_3. Hence the bodies are now in sliding contact, as explained in Sec. 2-14. Again, suppose the velocity of B is known and the velocity of C is to be found. The velocity polygon is shown in Fig. 6-14. Vector $O_2'P_2'$ is perpendicular to O_2P_2 and is determined in the same manner as for Fig. 6-12. Next, since V_{P_3} must be perpendicular to the radius for point P_3, from O_3' line $O_3'P_3'$ is drawn perpendicular to O_3P_3. For sliding contact V_{P_3/P_2} is directed along the tangent. Hence from P_2' a line is drawn parallel to the tangent. Its intersection with the line drawn

from O_3' locates P_3'.   Then $O_3'P_3'$ is V_{P_3} and $P_2'P_3'$ is V_{P_3/P_2}.   Point C' is found
by drawing a line from O_3' perpendicular to O_3C and by drawing a line from
P_3' perpendicular to P_3C.   Then $O_3'C'$ is V_C and $P_3'C'$ is V_{C/P_3}.

EXAMPLE 6-5   In Fig. 6-15 an oscillating-cylinder steam-engine mechanism is shown. V_B is known and V_C is desired. V_B is laid out as $O_2'B'$ in
Fig. 6-16.   By relative velocities

$$\overset{--}{V_C} = \overset{\checkmark\checkmark}{V_B} +\!\!\!+ \overset{-\checkmark}{V_{C/B}}$$

and we note that the magnitude and direction of V_C are unknown.   Also
the magnitude of $V_{C/B}$ is unknown.   Since there are three unknowns, the
vector equation cannot be solved.   In order to find V_C it will be necessary to
find V_{D_3} first.   Point D_3 is a point on an extension of link 3 which is coincident
with pivot O_4 at the instant.   Then

$$\overset{-\checkmark}{V_{D_3}} = \overset{\checkmark\checkmark}{V_B} +\!\!\!+ \overset{-\checkmark}{V_{D_3/B}}$$

Point D_3 can only have velocity in the direction of line BD_3.   Hence the
direction of V_{D_3} is known.   In Fig. 6-16 a line is drawn from B' in a direction

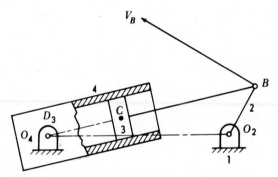

Fig. 6-15

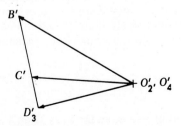

Fig. 6-16

perpendicular to BD_3, and from the pole O'_2, O'_4 a line is drawn parallel to BD_3. The intersection of these lines locates point D'_3. Then $O'_2 D'_3$ is V_{D_3}, and $B'D'_3$ is $V_{D_3/B}$. Next, point C' is located by the proportion

$$\frac{B'C'}{B'D'_3} = \frac{BC}{BD_3}$$

Thus

$$B'C' = \frac{BC}{BD_3}(B'D'_3)$$

Vector $O'_2 C'$ is V_C.

6-6 VELOCITIES IN COMPLEX LINKAGES

Whenever a link in a mechanism does not have a fixed center of rotation, it is called a *floating link*. The coupler in a four-bar linkage is an example. Mechanisms which have two or more floating links are called *complex mechanisms*. An example is the mechanism in Fig. 6-17, in which links 3 and 5 are floating links. When analyzing velocities in a complex linkage by the method of relative velocities, we sometimes encounter too many unknowns in the vector equation to get a solution directly. A trial-and-error method may then be used to obtain the solution. This method is illustrated in the following example.

EXAMPLE 6-6   For the mechanism in Fig. 6-17 suppose V_E is known and V_B is desired. V_C must be found first. Thus

$$V_C = V_E \mathbin{+\mkern-5mu\rightarrow} V_{C/E}$$

In Fig. 6-18, V_E is laid off from the velocity pole O'_2 as $O'_2 E'$. $V_{C/E}$ is directed along a line through E' and perpendicular to CE. $C'*$ denotes that point C' lies somewhere along this perpendicular. Since there are more than two unknowns in the vector equation, we cannot locate point C' with this equation alone. A trial solution may be made to determine the velocities of points on link 3. Consider

$$V_B = V_D \mathbin{+\mkern-5mu\rightarrow} V_{B/D}$$

Any length $O'_2 D'$ as shown in Fig. 6-18 can be assumed for V_D. Then V_B will be $O'_2 B'$ which is perpendicular to $O_2 B$ and $D'B'$ is perpendicular to DB.

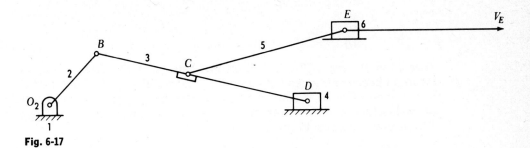

Fig. 6-17

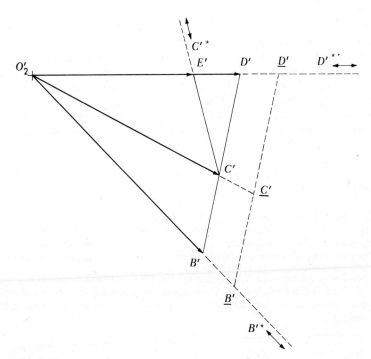

Fig. 6-18

Next, consider

$$V_C = V_D \dplus V_{C/D}$$

where $V_{C/D}$ is $\underline{D}'\underline{C}'$ and is perpendicular to CD.   By proportion

$$\frac{\underline{C'}\underline{D'}}{\underline{B'}\underline{D'}} = \frac{CD}{BD}$$

or

$$\underline{C'}\underline{D'} = \frac{CD}{BD}(\underline{B'}\underline{D'})$$

Then from $\underline{D'}$ length $\underline{C'}\underline{D'}$ is laid off to locate point $\underline{C'}$. This solution is incorrect because $\underline{C'}$ does not lie on the line containing E'. The point of intersection of line C'^* with line $O_2'C'$ locates C'. Next, through C' line of intersection of line C'^* with line $O_2'C'$ locates C'. Next, through C' line $D'B'$ is drawn parallel to $\underline{D'}\underline{B'}$. Then

$$\frac{C'D'}{B'D'} = \frac{CD}{BD}$$

and $B'C'D'$ is the image of BCD and represents the correct solution. Hence we see that the correct solution was obtained by trial.

An alternate solution to this problem which does not require a trial solution is as follows. A velocity polygon can be drawn by first laying off an assumed length $O_2'B'$. Points D', C', and E' can then be located on the polygon, in this order. Length $O_2'E'$ can then be measured. From its value and the known value of V_E, the velocity scale can be computed. Using this scale and the velocity polygon, the velocity of any other point on the mechanism can be determined.

PROBLEMS

The following problems are to be solved by the relative-velocity method. On the velocity polygons label the image of each point labeled on the mechanism. Let the velocity scale be 1 in. = 10 fps, unless otherwise stated. In problems where angular velocities are to be determined, state their magnitudes and directions (i.e., cw or ccw).

6-1   (a) Construct the velocity polygon for Fig. P6-1. Let the length of $V_B = 2$ in.
(b) Place vectors V_B, V_D, and V_E on the drawing of the mechanism and indicate their values in feet per second.   (c) Determine ω_2, ω_3, and ω_4 in radians per second.

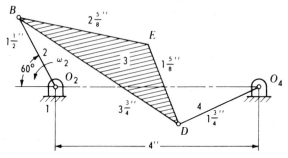

Fig. P6-1

6-2 (a) Construct the velocity polygon for Fig. P4-4.   Let the length of V_B = 2 in.
(b) Determine ω_3 in radians per second.

6-3 (a) Construct the velocity polygon for Fig. P6-3.   Let the velocity scale be 1
in. = 25 ips.   (b) Determine ω_3 in radians per second.

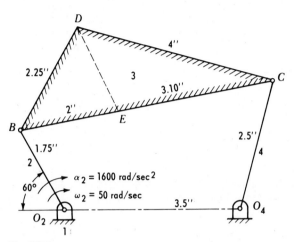

Fig. P6-3

6-4 (a) In Fig. P6-4, link 4 rolls on link 1.   Construct the velocity polygon. ω_2 = 144
rad/sec.   (b) Determine ω_3 and ω_4 in radians per second.

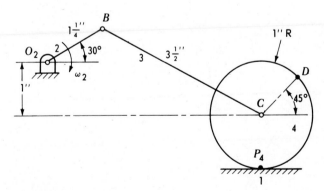

Fig. P6-4

6-5 In Fig. P4-3, at the pin in link 3, label the coincident points P_2 and P_4 on links 2
and 4, respectively. Let ω_2 = 120 rad/sec. (a) Construct the velocity polygon for points O_2, P_2, P_4, and C.   (b) Determine ω_4 in radians per second.

6-6 In Fig. P6-6, body 4 rolls on body 1.   (a) Construct the velocity polygon. $\omega_4 = 18$ rad/sec.   Use a velocity scale of 1 in. = 1 fps.   (b) Determine ω_2 in radians per second.

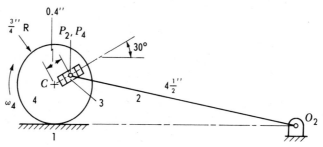

Fig. P6-6

6-7 The cam in Fig. P6-7 rotates at 500 rpm. $O_2 P_2 = 1.15$ in.   Find the velocity of link 4 in feet per second.   Velocity scale: 1 in. = 2 fps.

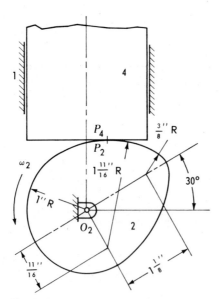

Fig. P6-7

6-8 Construct the velocity polygon for Fig. P6-8. P_2 and P_4 are coincident points on links 2 and 4, respectively. $\omega_2 = 15$ rad/sec.   Use a velocity scale of 1 in. = 1 fps.   Determine the velocity of link 4 in feet per second.

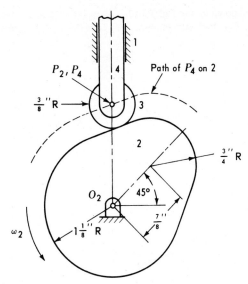

Fig. P6-8

6-9 In Fig. P6-9, P_2 and P_3 are coincident points on links 2 and 3, respectively, and Q_3 and Q_4 are coincident points on links 3 and 4, respectively. $V_{P_2} = 30$ ips. (a) Construct the velocity polygon.   Use a velocity scale of 1 in. = 20 ips.   Determine the linear velocity of link 4 in inches per second.   (b) Determine ω_3 in radians per second.

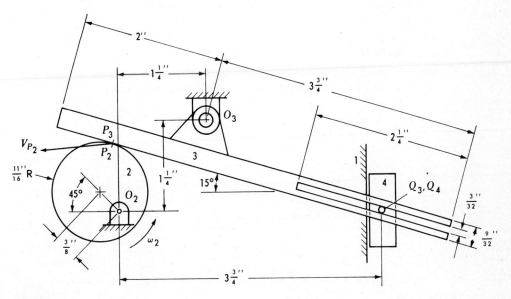

Fig. P6-9

6-10 (*a*) Construct the velocity polygon for Fig. P6-10. $\omega_2 = 144$ rad/sec. Determine the velocity of slider 6 in feet per second. (*b*) Determine ω_3, ω_4, and ω_5 in radians per second.

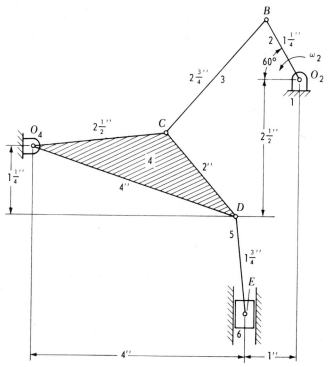

Fig. P6-10

6-11 (*a*) Construct the velocity polygon for Fig. P6-11. Use a velocity scale of 1 in. = 0.5 ips. (*b*) Determine ω_3 and ω_6 in radians per second.

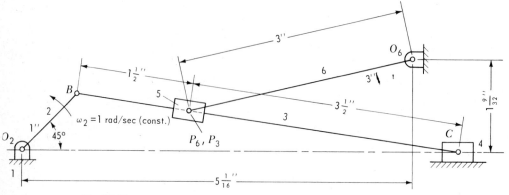

Fig. P6-11

6-12 (*a*) In Fig. P6-12, the velocity of point E is 15 fps. Using the trial solution method, construct the velocity polygon. Determine the velocity of point D in feet per second. (*b*) Determine ω_3 and ω_5 in radians per second.

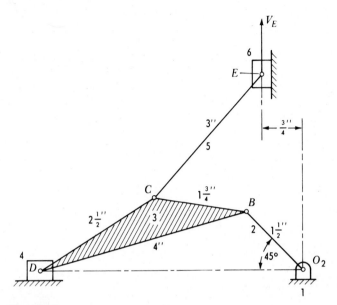

Fig. P6-12

7

ACCELERATIONS
IN MECHANISMS

7-1 INTRODUCTION

In the last two chapters we have seen how the instantaneous linear velocity of any point on a linkage can be determined, as well as the instantaneous angular velocity for any link. In this chapter a determination of the instantaneous linear and angular accelerations will be explained. Acceleration is of interest because of its effect on inertia forces, which in turn influence stresses in the parts of a machine, bearing loads, vibration, and noise.

An acceleration analysis of a mechanism is made by adding relative accelerations. Thus the method is actually similar to the method of relative velocities. When drawing the velocity polygon, the pole was designated O_2'. The pole for the acceleration polygon will be labeled O_2'', and double primes will also be used for points on the polygon. Lines from the pole to points on the acceleration polygon represent the absolute accelerations of the corresponding points on the mechanism, and a line joining any two points in the polygon represents the relative acceleration of the corresponding points on the mechanism.

The following equations, which were developed in Chap. 2 for the acceleration of a point, will be used in solving problems:

$$A^n = \frac{V^2}{R} = R\omega^2 = V\omega \tag{7-1}$$

$$A^t = R\alpha \tag{7-2}$$

$$A = \sqrt{(A^n)^2 + (A^t)^2} \tag{7-3}$$

In addition to the tangential and normal components of acceleration listed here, the Coriolis component of acceleration will be considered later. It may be defined as follows:

$$\text{Coriolis acceleration} = 2V\omega \tag{7-4}$$

7-2 LINEAR ACCELERATION

The slider-crank mechanism in Fig. 7-1 will be used to illustrate the method of finding accelerations for a mechanism.   The crank has a uniform angular velocity of 1,800 rpm.   The acceleration of point C is to be found.   The velocities must be found first; thus

$$V_B = (O_2B)\omega_2 = \frac{2.5}{12} \times \frac{1,800 \times 2\pi}{60} = 39.3 \text{ fps}$$

$$\overset{-\checkmark}{V_C} = \overset{\checkmark\checkmark}{V_B} \overset{-\checkmark}{\nrightarrow} V_{C/B}$$

The velocity polygon is shown in Fig. 7-2.   A scale of 1 in. = 20 fps was used for the original drawing. $V_{C/B}$, when scaled from the polygon, was found to be 34.4 fps.

The acceleration of C can be found from the following equation:

$$A_C = A_B \nrightarrow A_{C/B}$$

which can be written

$$\overset{0}{A_C^n} \nrightarrow \overset{-\checkmark}{A_C^t} = \overset{0}{A_B^n} \nrightarrow \overset{\checkmark\checkmark}{A_B^t} \nrightarrow \overset{\checkmark\checkmark}{A_{C/B}^n} \nrightarrow \overset{-\checkmark}{A_{C/B}^t} \tag{7-5}$$

This equation is solved graphically by drawing the acceleration polygon which is shown in Fig. 7-3.   Point O_2'' is the acceleration pole taken at any convenient location.   A scale of 1 in. = 2,000 fps² was used for the original drawing.   Since the path of motion for point C is a straight line

$$A_C^n = \frac{V^2}{R} = \frac{V_C^2}{\infty} = 0$$

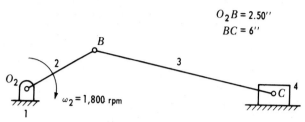

$$O_2B = 2.50''$$
$$BC = 6''$$

Fig. 7-1

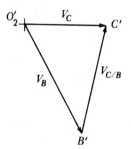

Fig. 7-2

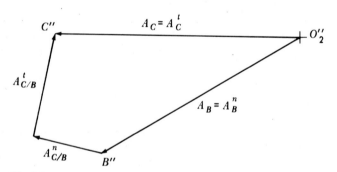

Fig. 7-3

A_C^t is drawn from point O_2'' in the direction of the path of motion of C; its magnitude is unknown. A dash mark is placed above A_C^t in Eq. (7-5) to indicate that its magnitude is unknown. A_C^t is directed tangential to the path of motion for C, and the check mark above this acceleration indicates that its direction is known. The addition of vectors which are on the right side of the equation is performed by laying them out, starting at pole O_2''. Point B moves in a circular path, and thus its normal acceleration is drawn parallel to O_2B.

$$A_B^n = \frac{V_B^2}{O_2B} = \frac{(39.3)^2}{2.5/12} = 7,400 \text{ fps}^2$$

Since ω_2 is constant, $\alpha_2 = 0$ and

$$A_B^t = (O_2B)\alpha_2 = (O_2B)0 = 0$$

$A_{C/B}^n$ and $A_{C/B}^t$ are relative accelerations; to determine their directions, we must consider the path of motion of C relative to B. Point C rotates in a circular path of radius BC about point B, and $A_{C/B}^n$ and $A_{C/B}^t$ are directed normal and tangential to this path, respectively.

$$A_{C/B}^n = \frac{V_{C/B}^2}{BC} = \frac{(34.4)^2}{\%_{12}} = 2,360 \text{ fps}^2$$

$A_{C/B}^n$ is drawn parallel to BC. From the terminus of $A_{C/B}^n$ a line is drawn perpendicular to BC. Its point of intersection with the horizontal line from O_2'' determines the magnitudes of A_C^t and $A_{C/B}^t$.

7-3   ACCELERATION IMAGE

For any mechanism there is an image in the acceleration polygon for each link, just as there is an image for each link in the velocity polygon. Let B and C be two points on a link. Then

$$A_{B/C} = A_{B/C}^n + \!\!\!\!+ A_{B/C}^t$$

The magnitude of the relative acceleration is

$$A_{B/C} = \sqrt{(A_{B/C}^n)^2 + (A_{B/C}^t)^2}$$
$$= \sqrt{[(BC)\omega^2]^2 + [(BC)\alpha]^2}$$
$$= BC\sqrt{\omega^4 + \alpha^2}$$

Since ω and α are properties of the whole link, the last equation indicates that the relative acceleration is proportional to the distance between the points. This provides a convenient means of constructing the acceleration polygon, since the magnitudes of the relative acceleration vectors for all points on a link will be proportional to the distances between the points. Further, this means that points on the acceleration polygon will form an image of the corresponding points on the link. For example, in

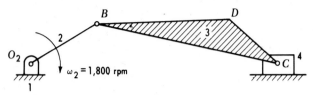

Fig. 7-4

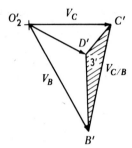

Fig. 7-5

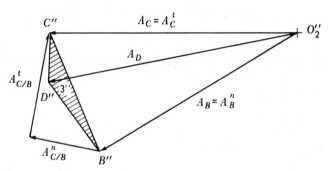

Fig. 7-6

Fig. 7-4 the mechanism of Fig. 7-1 is shown again but with link 3 extended to include point D. The velocity polygon is shown in Fig. 7-5. In Fig. 7-6 the acceleration polygon is shown, and point D'' was located by making $B''C''D''$ the image of BCD in Fig. 7-4; that is,

$$\frac{B''C''}{BC} = \frac{B''D''}{BD} = \frac{C''D''}{CD}$$

where a vector directed from B'' toward C'' represents the acceleration of C relative to B, while a vector from C'' toward B'' represents the acceleration of B relative to C.   When constructing the acceleration image, care must be taken not to flip the image over.   That is, if B, C, and D are in a clockwise order on the link, then B'', C'', and D'' must be in a clockwise order.   Since BC and D are in a counterclockwise order on link 3 in Fig. 7-4, B'', C'', and D'' are in a counterclockwise order.

7-4 ANGULAR ACCELERATION

The angular acceleration of any rigid link in a mechanism is equal to the tangential acceleration of any point on the link relative to any other point on the link divided by the distance between the points.   Since the relative motion for any two points on the link is one of rotation, Eq. (7-2) can be used to compute the angular acceleration.   For example, in Fig. 7-1 the angular acceleration of link 3 is

$$\alpha_3 = \frac{A^t_{C/B}}{BC}$$

The direction of the angular acceleration is determined from an inspection of the acceleration polygon.   In Fig. 7-1 point C rotates about point B. Hence the path of relative motion is a circle of radius BC.   In Fig. 7-3, since $A^t_{C/B}$ is directed upward, we see that point C is accelerating upward in a direction tangent to its path of motion relative to B, and hence α_3 is counterclockwise.   The angular velocity of link 3 is seen from Fig. 7-2 to be counterclockwise.   Therefore the angular velocity of link 3 is increasing.

EXAMPLE 7-1   A four-bar linkage is shown in Fig. 7-7.   The angular velocity and acceleration of link 2 are given, and the acceleration of points C, D, and E are to be determined along with the angular accelerations of links 3 and 4.

Solution The velocity and acceleration polygons appear in Figs. 7-8 and 7-9.   The acceleration of C is found as follows:

$$A_C = A_B \nleftrightarrow A_{C/B}$$

which can be written

$$\overset{\vee\vee}{A^n_C} \nleftrightarrow \overset{-\vee}{A^t_C} = \overset{\vee\vee}{A^n_B} \nleftrightarrow \overset{\vee\vee}{A^t_B} \nleftrightarrow \overset{\vee\vee}{A^n_{C/B}} \nleftrightarrow \overset{-\vee}{A^t_{C/B}} \tag{7-6}$$

The value of A^n_C is computed from the value of V_C found from the velocity polygon and is laid off from the pole O''_2 in Fig. 7-9.   The magnitude of $A^t_C = (O_4C)\alpha_4$ and is unknown because α_4 is unknown.   However, the direction of A_C^t is drawn in as a dashed line from the terminus of A_C^n. Next, the vectors on the right side of Eq. (7-6) are laid off from O_2''. The magnitudes of A_B^n and A_B^t are computed from the given data for the motion of link 2. Their vector sum is A_B, which is represented by $O_2''B''$. $A_{C/B}^n$ is laid off from B'' and is equal to $V_{C/B}^2/BC$ where the value of $V_{C/B}$ is obtained from the velocity polygon. A perpendicular at the terminus of $A_{C/B}^n$ represents the direction of $A_{C/B}^t$. The intersection of this line with the direction line for A_C^t locates point C''. Then $O_4''C''$ is A_C. Point D'' is located by making $B''C''D''$ the image of BCD. Vector $O_2''D''$ is A_D.

Point E'' can be located next by laying off from point D'' vectors $A_{E/D}^t$ and $A_{E/D}^n$. A simpler way of locating E'' is by making $B''E''C''$ the image of BEC. $O_2''E''$ is then A_E.

The values of $A_{C/B}^t$ and A_C^t can be scaled from the acceleration polygon,

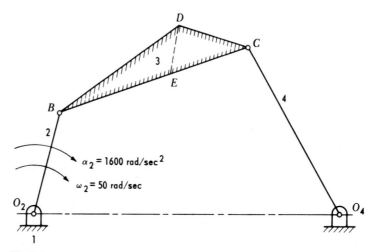

Fig. 7-7

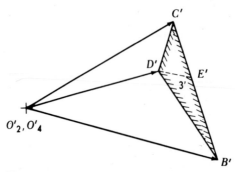

Fig. 7-8

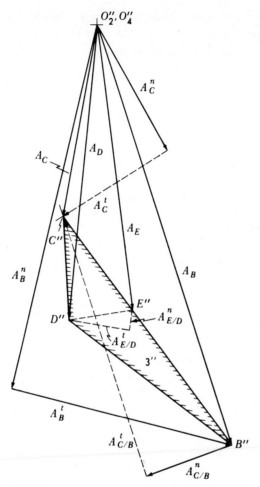

Fig. 7-9

and then α_3 and α_4 can be computed as follows:

$$\alpha_3 = \frac{A^t_{C/B}}{BC} \qquad \alpha_4 = \frac{A^t_C}{O_4C}$$

From the directions of $A^t_{C/B}$ and A^t_C as seen from the acceleration polygon, we note that α_3 and α_4 are both counterclockwise.

   EXAMPLE 7-2   In Chap. 6 it was explained that in some mechanisms a trial solution is required when constructing the velocity polygon.   If a trial solution is required for a link in the velocity analysis, then trial solutions for the same link are required in the acceleration analysis.   In Fig. 7-10 the mechanism of Fig. 6-17 is shown again. Suppose V_E and A_E are known and A_B is desired. The velocity polygon in Fig. 6-18, which required a trial solution, is shown again in Fig. 7-11. In order to find A_B we could try to find A_C first as follows:

$$A_C = A_E \mathbin{+\!\!\!\!\!+} A_{C/E}$$

which can be written

$$A_C = \overset{--}{A_E^n} \mathbin{+\!\!\!\!\!+} \overset{0}{A_E^t} \mathbin{+\!\!\!\!\!+} \overset{\vee\!\!\vee}{A_{C/E}^n} \mathbin{+\!\!\!\!\!+} \overset{-\vee}{A_{C/E}^t}$$

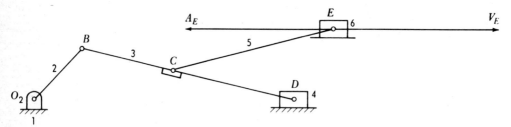

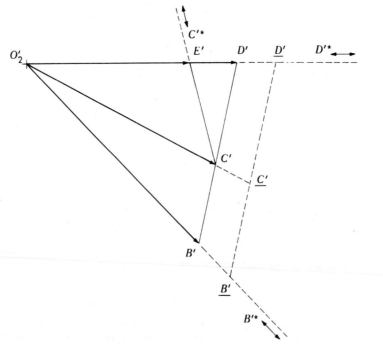

Fig. 7-10

Fig. 7-11

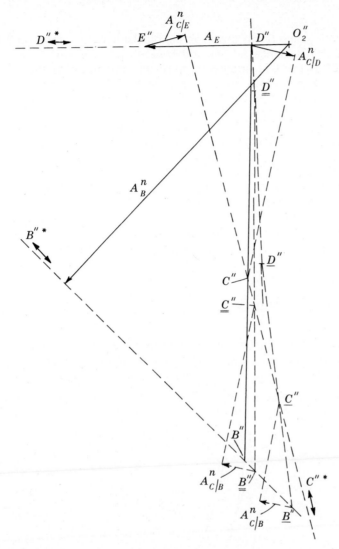

Fig. 7-12

A_C is unknown in magnitude and direction. $A_E^t = A_E$, $A_{C/E}^t = (CE)\alpha_5$, and since α_5 is unknown, the magnitude of $A_{C/E}^t$ is not known. Since there are more than two unknowns in the vector equation, we cannot obtain a solution by this equation alone. However, the acceleration image for link 3 can be found by trial. The acceleration polygon is shown in Fig. 7-12. A_E is laid off from O_2'' as length $O_2''E''$. A_D is known only in direction, as indicated by line $D''*$. The magnitude of $A_{C/E}^n$ can be computed from results obtained from the velocity polygon and is drawn from E'' parallel to CE. From the terminus of $A^n_{C/E}$, line $C''*$ is drawn perpendicular to CE and represents the direction of $A^t_{C/E}$. Hence C'' lies somewhere along this line. Next, the magnitude of A^n_B can be computed from results obtained from the velocity polygon. A^n_B is then drawn from $0''_2$ parallel to O_2B. From the terminus of A^n_B, line $B''*$ is drawn perpendicular to A^n_B and represents the direction of A^t_B. Thus point B'' lies somewhere along line $B''*$.

As shown in Fig. 7-12, a trial point $\underline{B''}$ is chosen somewhere along line $B''*$. To find point $\underline{C''}$, $A^n_{C/B}$ and $A^t_{C/B}$ are added to A_B, which is vector $0''_2 \underline{B''}$. The magnitude of $A^n_{C/B}$ can be computed, and $A^t_{C/B}$ is known in direction. Point $\underline{C''}$ is at the intersection of $A^t_{C/B}$ and line $C''*$. Point $\underline{D''}$ is located by making $B''C''D''$ the image of BCD. A second trial gives the trial image $\underline{B''}\underline{C''}\underline{D''}$. The locus of D'' is thus established to determine the true location of D'' on line $D''*$. Vector $0''_2 D''$ is then A_D. To find the true acceleration image of link 3, $A^n_{C/D}$ and $A^t_{C/D}$ are next added to A_D to determine C''. Point B'' is then determined by the intersection of line $D''C''$ and line $B''*$.

7-5 EQUIVALENT LINKAGES

When an acceleration analysis is to be made for a direct-contact mechanism, the problem can be simplified by replacing the mechanism by an equivalent four-bar linkage. In Fig. 7-13 a direct-contact mechanism consisting of links 1, 2, and 4 is shown. Motion is transmitted from the driving

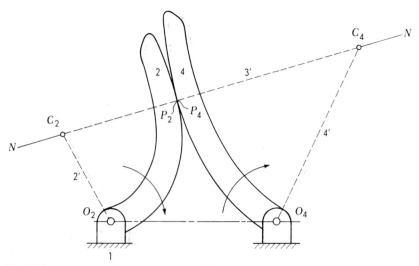

Fig. 7-13

link 2 to the driven link 4 by direct contact; i.e., there is no coupler link 3 as in Fig. 7-7. Cam-and-follower linkages (see Figs. 7-15a, 7-16a, 7-17a, 7-18a, 7-19a, and 7-20a) are examples of direct-contact mechanisms.

In Fig. 7-13 links 1, 2, and 4 will be referred to as the original linkage. An equivalent four-bar linkage is shown dotted and consists of links 1, 2', 3', and 4'. An equivalent four-bar linkage is one whose driving link 2' and driven link 4' have angular velocities and accelerations which are identical at the instant to those of links 2 and 4. Line N-N in Fig. 7-13 is the normal to the contacting profiles. Points C_2 and C_4 are the centers of curvature of

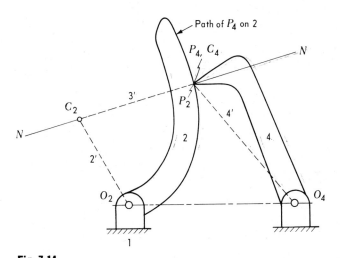

Fig. 7-14

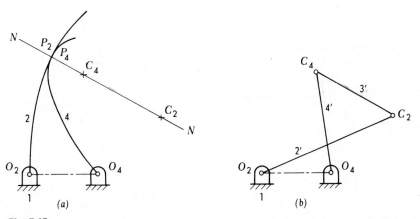

Fig. 7-15

bodies 2 and 4, respectively, for their points of contact P_2 and P_4. If the pin connection for links 2' and 3' is located at point C_2 and the pin connection for links 3' and 4' is located at C_4, then 1, 2', 3', and 4' constitute an equivalent four-bar linkage.[1]

In Fig. 7-14 link 4 is known as **a point follower** because contact always occurs at the same point on 4. The radius of curvature P_4C_4 is then zero; C_4 lies at P_4 and link 4' extends from O_4 to C_4, just as in Fig. 7-13.

In Figs. 7-15 to 7-20 equivalent four-bar linkages are shown for a number of mechanisms. In each case the equivalent linkage was constructed

[1] A proof of this equivalent linkage is given in the Appendix.

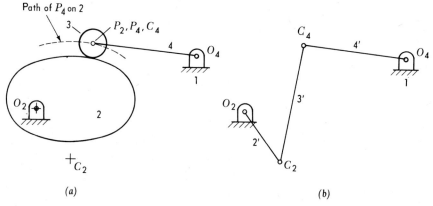

(a)   (b)

Fig. 7-16

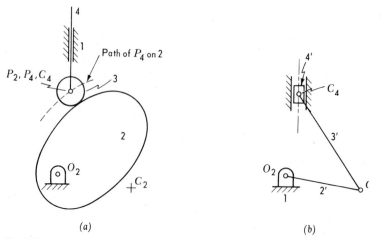

(a)   (b)

Fig. 7-17

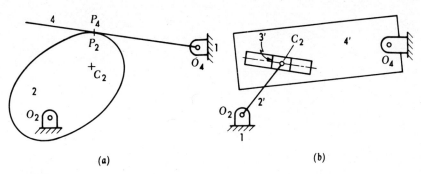

(a)   (b)

Fig. 7-18

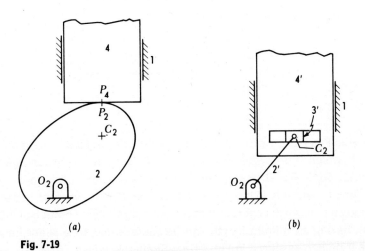

(a)   (b)

Fig. 7-19

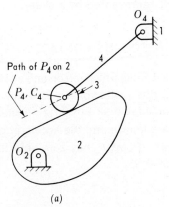

(a)

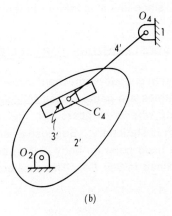

(b)

Fig. 7-20

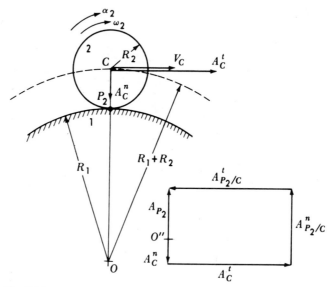

Fig. 7-21

exactly as in Fig. 7-13.   As in Fig. 7-14, the dashed line in Figs. 7-16a and 7-17a
is the path which P_4 describes on body 2, and C_2 is the center of curvature of
the path at point P_2.   Consider Fig. 7-17a.   Since point C_4 has rectilinear
motion, link 4' is infinite in length extending from C_4 to O_4, where O_4 lies at infinity along a line which is perpendicular to the direction of motion of link 4.
Any link having an infinite length can be represented by a slider having
rectilinear motion as shown in Fig. 7-17b.   Similarly, in Figs. 7-18b, 7-19b, and
7-20b, since link 3' is infinite in length, it is represented by a slider.

7-6 ACCELERATIONS FOR MEMBERS IN ROLLING CONTACT

We often encounter mechanisms which have links which roll on one another.
Examples are cams whose followers are fitted with rollers, and linkages
having wheels or gears.   In Fig. 7-21 consider disk 2 rolling on body 1,
which is stationary.   Suppose ω_2 and α_2 are known and the acceleration of
P_2, a fixed point on body 2, is to be found.   Point P_2 has zero velocity as
explained in Sec. 6-5.   The acceleration of P_2 is

$$A_{P_2} = A_C + A_{P_2/C}$$

$$A_{P_2} = A_C^n + A_C^t + A_{P_2/C}^n + A_{P_2/C}^t \tag{7-7}$$

where

$$A_C^n = \frac{V_C^2}{R_1 + R_2} = \frac{(R_2\omega_2)^2}{R_1 + R_2}$$

$$A_C^t = (P_2C)\alpha_2 \tag{7-8}$$

$$A_{P_2/C}^n = (P_2C)\omega_2^2$$

$$A_{P_2/C}^t = (P_2C)\alpha_2$$

The direction and magnitude of A_{P_2} are unknown. The vectors in Eq. (7-7) are laid off from the pole O'' in Fig. 7-21. Since A_C^t and $A_{P_2/C}^t$ are equal and opposite, A_{P_2} is directed along the normal at the point of contact.

Figure 7-22 is similar to Fig. 7-21 except that the surface of body 1 is concave. Equations (7-7) and (7-8) also apply for Fig. 7-22 except that

$$A_C^n = \frac{V_C^2}{R_1 - R_2}$$

From an inspection of the acceleration polygon in Fig. 7-22 we note that since A_C^t and $A_{P_2/C}^t$ are equal and opposite, A_{P_2} is directed along the normal at the point of contact.

If in Figs. 7-21 and 7-22 body 1 were a plane surface, then R_1 would be infinite and

$$A_C^n = \frac{V_C^2}{\infty} = 0$$

Further, if the outlines of either or both bodies are not circular, but have changing curvature along their length, then R_1 and R_2 in Eqs. (7-8) are taken as the radii of curvature at the point of contact.

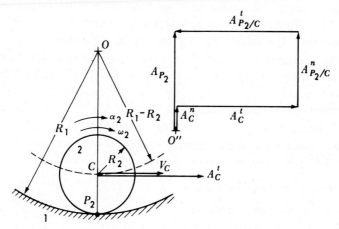

Fig. 7-22

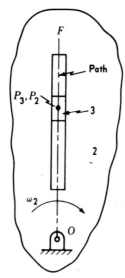

Fig. 7-23

7-7   CORIOLIS ACCELERATION

Whenever a point in one body moves along a path on a second body, and if the path is rotating, then the acceleration of the point in the first body relative to a coincident point in the second body will have a Coriolis component.

In Fig. 7-23 let P_3 be a point on slider 3 which is moving along the path OF in body 2. Let P_2 be a fixed point on the path and let P_3 and P_2 be coincident at the instant. The angular velocity for body 2 is ω_2 and hence also for the path. The path is again shown in Fig. 7-24, where V_{P_3/P_2} is the velocity of P_3 relative to P_2. In a time interval dt, line OF will rotate through an angle $d\theta$ to position OF'. During this time interval P_2 moves from P_2 to P_2'. Point P_3 moves from P_3 to P_3' and this displacement may be considered as the sum of displacements P_2P_2', $P_2'B$, and BP_3'. Displacement P_2P_2' is with constant velocity because ω_2 is constant. Also $P_2'B$ is with constant velocity since V_{P_3/P_2} is constant. However, it will be shown that displacement BP_3' is the result of an acceleration.

$$\text{Arc } BP_3' = \text{arc } CP_3' - \text{arc } P_2P_2'$$

$$= (OC)\, d\theta - (OP_2)\, d\theta$$

$$= (P_2'B)\, d\theta$$

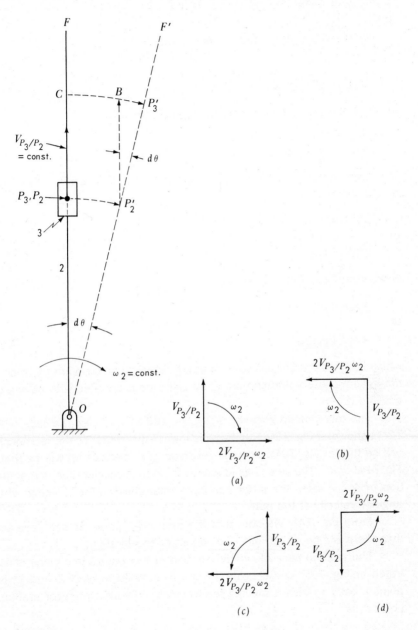

Fig. 7-24

But

$$P_2'B = V_{P_3/P_2}\,dt \qquad \text{and} \qquad d\theta = \omega_2\,dt$$

Thus

$$BP_3' = V_{P_3/P_2}\omega_2(dt)^2 \tag{7-9}$$

In Eq. (7-9) $V_{P_3/P_2}\omega_2$ is an acceleration, i.e., units are

$$\frac{\text{in.}}{\text{sec}} \times \frac{\text{rad}}{\text{sec}} = \frac{\text{in.}}{\text{sec}^2}$$

This acceleration $\times dt \times dt =$ displacement BP_3'.   Further, an infinitesimal displacement

$$ds = \tfrac{1}{2}A(dt)^2$$

or

$$BP_3' = \tfrac{1}{2}A(dt)^2 \tag{7-10}$$

Next, from Eqs. (7-9) and (7-10)

$$V_{P_3/P_2}\omega_2(dt)^2 = \tfrac{1}{2}A(dt)^2$$

or

$$A = 2V_{P_3/P_2}\omega_2 \tag{7-11}$$

which is called the Coriolis component of acceleration for point P_3, after the nineteenth-century French mathematician who is credited with having discovered it.

The relationship between V_{P_3/P_2}, ω_2, and $2V_{P_3/P_2}\omega_2$ for the case of Fig. 7-24 is shown at Fig. 7-24a.   If V_{P_3/P_2} is toward the center 0, the relationship will be that of Fig. 7-24b.   If ω_2 is reversed, the relationship will be that in Fig. 7-24c or d.   The rule is as follows: the Coriolis acceleration is the direction of V_{P_3/P_2} after the latter has been rotated 90° in the direction of the angular velocity of the path.

From Eq. (7-11) we note that if either V_{P_3/P_2} or ω_2, or both, are zero, then there will be no Coriolis component of acceleration.

The general case of relative motion of two bodies in a plane is illustrated in Fig. 7-25.   Here again P_2 is a fixed point in body 2, and P_3 is a point in body 3 which is moving relative to 2.   The absolute acceleration of point P_3 is

$$A_{P_3} = A_{P_2} + A_{P_3/P_2}$$

or

$$A_{P_3}^n + A_{P_3}^t = A_{P_2}^n + A_{P_2}^t + A_{P_3/P_2}^n + A_{P_3/P_2}^t + 2V_{P_3/P_2}\omega_2 \tag{7-12}$$

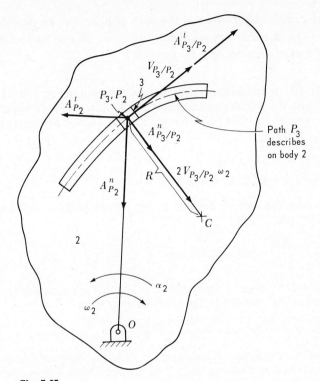

Fig. 7-25

where the Coriolis component $2V_{P_3/P_2}\omega_2$ is part of the acceleration of P_3 relative to P_2. The path which P_2 describes on the frame is a circle of radius OP_2. Hence $A^n_{P_2}$ and $A^t_{P_2}$, which are the normal and tangential components of acceleration of P_2, are directed as shown in the figure. Their values are found as follows:

$$A^n_{P_2} = \frac{V^2_{P_2}}{OP_2} \quad \text{and} \quad A^t_{P_2} = (OP_2)\alpha_2$$

The motion of P_3 relative to body 2 is along the path indicated in the figure. $A^n_{P_3/P_2}$ and $A^t_{P_3/P_2}$ are the normal and tangential components of the acceleration of P_3 relative to P_2, and thus they are normal and tangential, respectively, to this path. $A^n_{P_3/P_2}$ can be computed from

$$A^n_{P_3/P_2} = \frac{V^2_{P_3/P_2}}{R}$$

where R is the radius of curvature of the path at point P_2. In problems, the value of $A^t_{P_4/P_2}$ is either known at the beginning or can be found by laying out the acceleration polygon for Eq. (7-12).

EXAMPLE 7-3   A quick-return mechanism is shown in Fig. 7-26. Link 2 is the driver, and the angular velocity and acceleration of link 4 are to be found. Let P_2 and P_4 be fixed points on links 2 and 4 which are coincident at the instant. Then

$$\omega_2 = 2\pi(9.5) = 59.6 \text{ rad/min}$$

$$V_{P_2} = (O_2P_2)\omega_2 = 0.5(59.6) = 29.8 \text{ fpm}$$

Scaling values from the velocity polygon in Fig. 7-27 we find $V_{P_4} = 14.6$ fpm and $V_{P_4/P_2} = 25.8$ fpm. Then

$$\omega_4 = \frac{V_{P_4}}{O_4P_4} = \frac{14.6}{1.69} = 8.64 \text{ rad/min ccw}$$

Before we can find the angular acceleration of link 4, it is necessary to find the acceleration of P_4. Similar to Eq. (7-12) we may write

$$A_{P_4} = A^n_{P_2} +\!\!\!+ A^t_{P_2} +\!\!\!+ A^n_{P_4/P_2} +\!\!\!+ A^t_{P_4/P_2} +\!\!\!+ 2V_{P_4/P_2}\omega_2$$

In order to solve this equation it is necessary to know the radius of curvature of the path which P_4 describes on body 2. This path is not known. However, the path which P_2 describes on 4 is a straight line along the link. We can use this path if we write the equation for A_{P_2}. Then similar to Eq. (7-12)

$$\overset{\checkmark\checkmark}{A^n_{P_2}} +\!\!\!+ \overset{0}{A^t_{P_2}} = \overset{\checkmark\checkmark}{A^n_{P_4}} +\!\!\!+ \overset{-\checkmark}{A^t_{P_4}} +\!\!\!+ \overset{0}{A^n_{P_2/P_4}} +\!\!\!+ \overset{-\checkmark}{A^t_{P_2/P_4}} +\!\!\!+ \overset{\checkmark\checkmark}{2V_{P_2/P_4}\omega_4} \qquad (7\text{-}13)$$

where

$$A^n_{P_2} = \frac{V^2_{P_2}}{O_2P_2} = \frac{(29.8)^2}{0.5} = 1{,}775 \text{ fpm}^2$$

$$A^t_{P_2} = 0$$

because $\alpha_2 = 0$.

$$A^n_{P_4} = \frac{V^2_{P_4}}{O_4P_4} = \frac{(14.6)^2}{1.69} = 126 \text{ fpm}^2$$

$$A^t_{P_4} = (O_4P_4)\alpha_4$$

but α_4 is unknown.

$$A^n_{P_2/P_4} = \frac{(V^t_{P_2/P_4})^2}{R} = \frac{(25.8)^2}{\infty} = 0$$

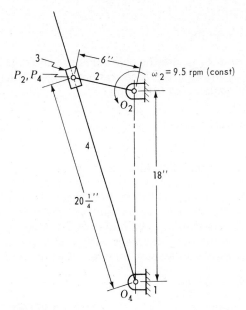

Fig. 7-26

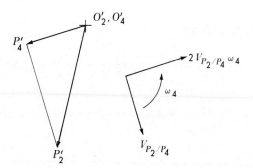

Fig. 7-27

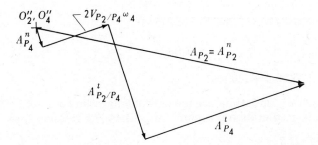

Fig. 7-28

$A^t_{P_2/P_4}$ is unknown, and

$$2V_{P_2/P_4}\omega_4 = 2(25.8)8.64 = 445 \text{ fpm}^2$$

Thus the magnitudes of $A^t_{P_4}$ and $A^t_{P_2/P_4}$ are the only unknowns in Eq. (7-13). Their values may be found by drawing the acceleration polygon which is shown in Fig. 7-28. Then $A^t_{P_4}$ scales 1,088 fpm², and

$$\alpha_4 = \frac{A^t_{P_4}}{O_4P_4} = \frac{1,088}{1.69} = 644 \text{ rad/min}^2 \text{ cw}$$

EXAMPLE 7-4 A cam with oscillating follower is shown in Fig. 7-29, where the angular velocity and acceleration of the cam are indicated. The angular acceleration of link 4 is wanted. Let P_2 and P_4 be points on links 2 and 4 which are coincident at the instant. P_4 is the axis of the roller, and the path which it describes on 2 is shown. The radius of curvature of the path is 5.438 in. and equals the radius of the cam outline plus the roller radius. C is the center of curvature of the cam outline. We have

$$V_{P_2} = (O_2P_2)\omega_2 = (3.28)5 = 16.4 \text{ ips}$$

and the velocity polygon is shown in Fig. 7-30. V_{P_4} scales 8.4 in./sec and V_{P_4/P_2} scales 21 in./sec. Then

$$\omega_4 = \frac{V_{P_4}}{O_4P_4} = \frac{8.4}{7.5} = 1.12 \text{ rad/sec cw}$$

Next, in order to find the angular acceleration of link 4, we must find the acceleration of P_4. Thus

$$\overset{\vee\vee}{A^n_{P_4}} + \overset{-\vee}{A^t_{P_4}} = \overset{\vee\vee}{A^n_{P_2}} + \overset{\vee\vee}{A^t_{P_2}} + \overset{\vee\vee}{A^n_{P_4/P_2}} + \overset{-\vee}{A^t_{P_4/P_2}} + \overset{\vee\vee}{2V_{P_4/P_2}\omega_2} \qquad (7\text{-}14)$$

where

$$A^n_{P_4} = \frac{V^2_{P_4}}{O_4P_4} = \frac{(8.4)^2}{7.5} = 9.42 \text{ ips}^2$$

$$A^n_{P_2} = (O_2P_2)\omega_2^2 = 3.28(5)^2 = 82.0 \text{ ips}^2$$

$$A^t_{P_2} = (O_2P_2)\alpha_2 = 3.28(2.5) = 8.20 \text{ ips}^2$$

$$A^n_{P_4/P_2} = \frac{V^2_{P_4/P_2}}{CP_2} = \frac{(21.0)^2}{5.438} = 81.2 \text{ ips}^2$$

$$2V_{P_4/P_2}\omega_2 = 2(21.0)5 = 210 \text{ ips}^2$$

The magnitudes of $A^t_{P_4}$ and $A^t_{P_4/P_2}$ are the only unknowns in Eq. (7-14). Figure 7-31 shows the acceleration polygon. $A^t_{P_4}$ scales 77.5 ips² and then

$$\alpha_4 = \frac{A^t_{P_4}}{O_4P_4} = \frac{77.5}{7.5} = 10.3 \text{ rad/sec}^2 \text{ cw}$$

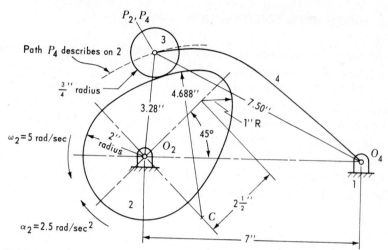

Fig. 7-29

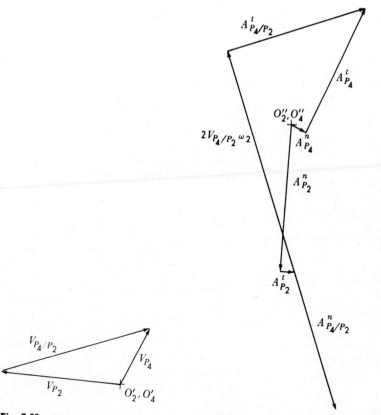

Fig. 7-30   **Fig. 7-31**

7-8 HARTMANN'S CONSTRUCTION

Hartmann's construction is a graphical method for finding the center of curvature for the path of any point of a moving body.

In Fig. 7-32 bodies 2 and 3 have circular outlines having centers C_2 and C_3. Let body 2 be fixed and let body 3 roll on it. P_3 is any point on body 3, and $p\text{-}p$ is the path which it describes on body 2. The location of the center of curvature C for this path is to be found. The procedure is as follows. Since the bodies have rolling contact, their common instant center 23 lies at the point of contact. Point C_3 describes a circular path $b\text{-}b$ about C_2. Vc_3 is any convenient length assumed for the velocity of center C_3 due

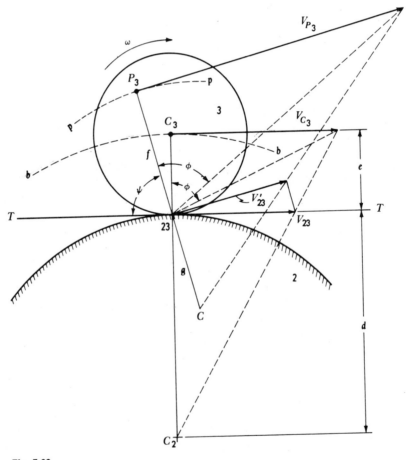

Fig. 7-32

to its rotation about instant center 23 with angular velocity ω. Thus

$$V_{C_3} = e\omega \tag{7-15}$$

The radius of rotation for C_3, as it rotates about instant center 23, is C_3-23 and is called a *ray*. A line joining 23 and the terminus of V_{C_3} is called a *gauge line*. The angle formed by the ray 23-C_3 and the gauge line is called the *gauge angle* ϕ for the moving body. Then

$$V_{C_3} = e \tan \phi \tag{7-16}$$

Instant center 23 is a point common to bodies 2 and 3 having the same linear velocity in each, namely zero. However, as body 3 rolls on 2, a new instant center 23 exists for each point of contact. The velocity with which 23 propagates along the outline of body 2 is V_{23}. Its magnitude is determined by drawing a line from C_2 to the terminus of V_{C_3}. Line 23-P_3 is the ray for point P_3. Next, angle ϕ is laid off from this line in order to draw in the gauge line. V_{P_3} is perpendicular to 23-P_3, and its magnitude is determined by its intersection with the gauge line. V'_{23} is the component of V_{23} in the direction perpendicular to 23-P_3 and is drawn next. Its magnitude is determined by drawing a line perpendicular to V'_{23} from the terminus of V_{23}. Next, a line is drawn through the terminus of V_{P_3} and V'_{23}. Its intersection with line P_3-23 extended locates point C, the center of curvature for the path p-p which P_3 describes on body 2. Then CP_3 is the radius of curvature of the path.

Rules for applying Hartmann's construction are as follows. The path which a point in one body describes on another depends only on the relative motion of the bodies. Hence when using the Hartmann construction, we can hold one of the bodies fixed and consider the motion of the other body relative to it. Call the fixed body 2 and the moving body 3. Then proceed with the construction as explained for Fig. 7-32.

7-9 EULER–SAVARY EQUATION

An analytical expression giving the location for the center of curvature C of the path of P_3 in Fig. 7-32 is easily obtained. In the figure let $f = $ 23-P_3 and $g = C$-23. Then from similar triangles

$$\frac{d+e}{d} = \frac{V_{C_3}}{V_{23}}$$

and by substituting V_{C_2} from Eq. (7-15)

$$\frac{d + e}{d} = \frac{e\omega}{V_{23}}$$

or

$$\frac{d + e}{de} = \frac{\omega}{V_{23}}$$

Hence

$$\frac{1}{d} + \frac{1}{e} = \frac{\omega}{V_{23}} \tag{7-17}$$

Next, the ray for point P_3 is $23\text{-}P_3 = f$ and the velocity of P_3 is

$$V_{P_3} = f\omega \tag{7-18}$$

or

$$V_{P_3} = f \tan \phi$$

Let $T\text{-}T$ be a line through 23 and perpendicular to line $C_2 C_3$, and let ψ be the angle between $P_3\text{-}23$ and a line $T\text{-}T$. Then

$$V'_{23} = V_{23} \sin \psi \tag{7-19}$$

Since $f = 23\text{-}P_3$ and $g = C\text{-}23$, then

$$\frac{f + g}{g} = \frac{V_{P_3}}{V'_{23}} \tag{7-20}$$

Substituting Eqs. (7-18) and (7-19) into (7-20) we obtain

$$\frac{f + g}{g} = \frac{f\omega}{V_{23} \sin \psi}$$

or

$$\frac{f + g}{fg} = \frac{\omega}{V_{23} \sin \psi}$$

and

$$\left(\frac{1}{f} + \frac{1}{g}\right) \sin \psi = \frac{\omega}{V_{23}} \tag{7-21}$$

Combining Eqs. (7-17) and (7-21) we obtain the following, which is known as the Euler-Savary equation:

$$\frac{1}{d} + \frac{1}{e} = \left(\frac{1}{f} + \frac{1}{g}\right) \sin \psi \tag{7-22}$$

where $d = C_2\text{-}23$, $e = 23\text{-}C_3$, $f = 23\text{-}P_3$, $g = C\text{-}23$, and ψ is the angle which line $P_3\text{-}23$ makes with $T\text{-}T$, the perpendicular to $C_2 C_3$. Hence the radius of curvature R of the path described on body 2 by point P_3 is

$$R = CP_3 = f + g \qquad (7\text{-}23)$$

Equations (7-22) and (7-23) have been derived for bodies whose contacting surfaces are circular, but they also apply for bodies whose contact surfaces have varying radii along their length.   Lengths d and e must then be taken as their radii of curvature at the point of contact.   Also, the outline for either body can be flat or concave.   Equations (7-22) and (7-23) hold for either the case where both bodies are convex, or when one is convex and the other is flat or concave, provided the following rule is followed: Each of the lengths in Eqs. (7-22) and (7-23) is to be considered positive if, according to the way it is written below Eq. (7-22), it extends in the same direction that we would move if we went from body 2 to body 3 at the point of contact.

EXAMPLE 7-5   To illustrate the application of Hartmann's construction and also the use of the Euler-Savary equation, let us consider the mechanism in Fig. 7-33.   Suppose that an acceleration analysis were to be made. The simplest procedure would be to draw an equivalent four-bar linkage and analyze its accelerations.   However, if we wish to analyze the given linkage, we must consider the relative acceleration for the coincident points P_2 and P_3, as explained in Sec. 7-7.   In order to evaluate the normal acceleration of P_3 relative to P_2, we must know the radius of curvature of the path which

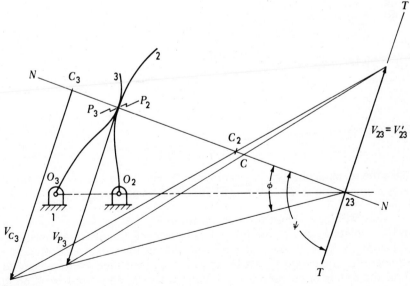

Fig. 7-33

P_3, considered as a fixed point on 3, describes on body 2.   In Fig. 7-33 the radius for this path is P_3C.   Points C_2 and C_3 are the centers of curvature for the outlines of bodies 2 and 3.   Since the point of contact is not on the line of centers O_2O_3, the bodies have sliding contact and instant center 23 lies at the point of intersection of the normal C_2C_3 and the line of centers O_2O_3.

Using Hartmann's construction to locate point C, we proceed in the same manner as explained for Fig. 7-32.   In Fig. 7-32, the centrodes which lie in bodies 2 and 3 are circles.   They pass through instant center 23 and roll on one another.   Similarly in Fig. 7-33 there are two centrodes (curves not shown) passing through 23 which are normal to line C_2C_3.   These centrodes roll on one another.   We hold body 2 fixed and consider the rotation of body 3 about point 23 in link 2. V_{C_3} is drawn first and can be assumed either in the direction shown or in the opposite direction and may be drawn any convenient length.   A line is then drawn from the terminus of V_{C_3} to point 23.   This determines the magnitude of V_{P_3}.   Next, a line from the terminus of V_{C_3} and through C_2 determines the magnitude of V_{23}.   Since V'_{23} is the component of V_{23} in the direction perpendicular to 23-P_3, V'_{23} and V_{23} are identical.   A line from the terminus of V_{P_3} to the terminus of V'_{23} locates point C.

The computation of the path radius CP_3 using the Euler-Savary equation is as follows.   On the original drawing for Fig. 7-33, $d = C_2$-23 $= -2.30$ in.; $e =$ 23-$C_3 = 5.80$ in.; $f =$ 23-$P_3 = 4.80$ in.; $g = C$-23 and is unknown. $\psi = 90°$, $\sin \psi = 1$.   Then by Eq. (7-22)

$$\frac{1}{d} + \frac{1}{e} = \left(\frac{1}{f} + \frac{1}{g}\right) \sin \psi$$

$$\frac{1}{-2.30} + \frac{1}{5.80} = \frac{1}{4.80} + \frac{1}{g}$$

$$-0.435 + 0.1725 = 0.208 + \frac{1}{g}$$

$$\frac{1}{g} = -0.470 \qquad g = -2.125 \text{ in.}$$

By Eq. (7-23)

$$R = CP_3 = f + g = 4.80 - 2.125 = 2.675 \text{ in.}$$

Since the result is positive, line segment CP_3 is directed from body 2 toward body 3 and hence C lies to the right of P_3 in the figure.

EXAMPLE 7-6   The cam mechanism in Fig. 7-34 consists of links 1, 2, and 4.   Points P_2 and P_4 are the points of contact.   Suppose an acceleration analysis were to be made by considering the relative acceleration between the coincident points P_2 and P_4.   It would then be necessary to know the radius of curvature of the path which P_4, as a fixed point on 4, describes on body 2.   Hartmann's construction for locating the center of

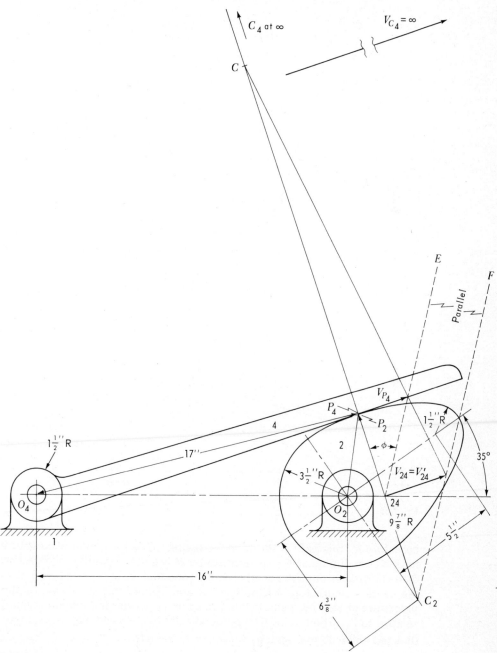

Fig. 7-34

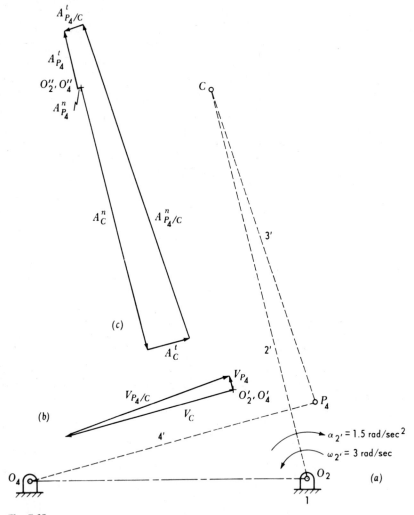

Fig. 7-35

curvature for this path is shown in the figure. C_2 is the center of curvature
for the cam outline at P_2. Instant center 24 lies at the point of intersection
of the normal through the point of contact and the line of centers O_2O_4. C_4 is
the center of curvature for the follower and lies at infinity. Following the
procedure of Sec. 7-8, we hold link 2 fixed and consider the motion of link 4
relative to 2. Then V_{C_4} will be perpendicular to 24-C_4 and may be assumed
directed either to the right, as shown, or to the left.

$$V_{C_4} = \infty$$

because $(24\text{-}C_4) = \infty$.   Line $24\text{-}E$ is then drawn from 24 to the terminus of V_{C_4}.   In this problem line $24\text{-}E$ is drawn at any convenient angle to line $24\text{-}C_4$. Thus it is assumed that $24\text{-}E$, when extended, will meet the terminus of V_{C_4}. Then the angle formed by $24\text{-}E$ and $24\text{-}C_4$ is the gauge angle ϕ. V_{24} is perpendicular to C_2C_4, and its magnitude is determined by line C_2F, which is directed from C_2 to the terminus of V_{C_4}. $24\text{-}P_4$ is the ray for point P_4, and the magnitude of V_{P_4} is determined by the gauge line $24\text{-}E$.   Next, V'_{24} is the component of V_{24} in a direction perpendicular to $24\text{-}P_4$.   Hence V'_{24} is identical to V_{24}.   Then a line drawn through the terminus of V'_{24} and the terminus of V_{P_4} intersects line $P_4\text{-}24$ extended at C.   Point C is the center of curvature of the path which P_4 describes on link 2.

An acceleration analysis of the cam and follower in Fig. 7-34 will be made by considering an equivalent linkage.   The equivalent four-bar linkage shown in Fig. 7-35 and consisting of links 1, 2', 3', and 4' is similar to the one in Fig. 7-14.   In Fig. 7-14 the radius of curvature of the path which P_4 describes on link 2 is P_4C_2; in Fig. 7-34 it is P_4C.   Suppose in Fig. 7-34 $\omega_2 = 3$ rad/sec ccw, $\alpha_2 = 1.5$ rad/sec² cw, and ω_4 and α_4 are to be found. Then considering the equivalent linkage in Fig. 7-35 we have

$$V_C = (O_2C)\omega_{2'} = 22.04(3) = 66.12 \text{ ips}$$

The velocities are then laid off from pole O'_2, O'_4 in Fig. 7-35b. V_{P_4} scales 5.6 ips and $V_{P_4/C}$ scales 66.4 ips.   Then

$$\omega_{4'} = \frac{V_{P_4}}{O_4P_4} = \frac{5.6}{17.04} = 0.328 \text{ rad/sec ccw}$$

For the accelerations,

$$\overset{\checkmark\checkmark}{A^n_{P_4}} \looparrowright \overset{-\checkmark}{A^t_{P_4}} = \overset{\checkmark\checkmark}{A^n_C} \looparrowright \overset{\checkmark\checkmark}{A^t_C} \looparrowright \overset{\checkmark\checkmark}{A^n_{P_4/C}} \looparrowright \overset{-\checkmark}{A^t_{P_4/C}}$$

where the magnitudes of $A^t_{P_4}$ and $A^t_{P_4/C}$ are unknown.   Thus

$$A^n_{P_4} = \frac{V^2_{P_4}}{O_4P_4} = \frac{(5.6)^2}{17.04} = 1.84 \text{ ips}^2$$

$$A^n_C = \frac{V_C{}^2}{O_2C} = \frac{(66.12)^2}{22.04} = 198 \text{ ips}^2$$

$$A^t_C = (O_2C)\alpha_2 = 22.04(1.5) = 33.15 \text{ ips}^2$$

$$A^n_{P_4/C} = \frac{V^2_{P_4/C}}{P_4C} = \frac{(66.4)^2}{18.2} = 242 \text{ ips}^2$$

The acceleration vectors are laid out from pole O''_2, O''_4 in Fig. 7-35c. $A^t_{P_4}$, when scaled from the original drawing, was found to be 42.5 ips².   Then

$$\alpha_{4'} = \frac{A^t_{P_4}}{O_4P_4} = \frac{42.5}{17} = 2.5 \text{ rad/sec}^2 \text{ ccw}$$

PROBLEMS

In the following problems, where accelerations are to be determined, label the image on the acceleration polygon of each point labeled on the mechanism. Where angular accelerations are to be determined, state their magnitude and sense.

7-1 (a) Construct the acceleration polygon for Fig. P6-1. $V_B = 20$ fps, ω_2 is constant. Scales: 1 in. = 10 fps and 1 in. = 1,000 fps². (b) Place vectors A_B, A_D, and A_E on the drawing of the mechanism and indicate their values in feet per second squared. (c) Determine α_3 and α_4 in radians per second squared.

7-2 (a) Construct the acceleration polygon for Fig. P4-4. $V_B = 20$ fps const. Scales: 1 in. = 10 fps and 1 in. = 1,000 fps². Determine the values of A_C and A_D in feet per second squared. (b) Determine α_3 in radians per second squared.

7-3 (a) Construct the acceleration polygon for Fig. P6-3. Scales: 1 in. = 25 ips and 1 in. = 750 ips². (b) Determine α_3 and α_4 in radians per second squared.

7-4 (a) Construct the velocity and acceleration polygons for Fig. P7-4 when the piston is at top dead center and find the velocity in inches per second and acceleration in inches per second squared of the piston. Use a velocity scale of 1 in. = 100 ips and an acceleration scale of 1 in. = 10,000 ips². (b) Same as (a) except when piston is at bottom dead center.

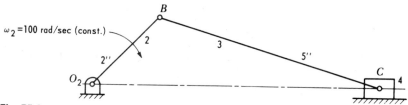

Fig. P7-4

7-5 (a) In Fig. P6-12 the velocity of point E is 15 fps upward as shown, and the acceleration of E is 1,500 fps² downward. Using the method of trial solutions, construct the acceleration polygon. Scales: 1 in. = 10 fps and 1 in. = 500 fps.² Determine the acceleration of point D in feet per second squared. (b) Determine α_3 and α_5 in radians per second squared.

7-6 For the mechanism in Fig. 7-29 draw the equivalent four-bar linkage $O_2CP_4O_4$. Label the equivalent links 1, 2', 3', and 4'. (a) Construct the velocity and acceleration polygons. Use a velocity scale of 1 in. = 5 ips and an acceleration scale of 1 in. = 20 ips². (b) Determine $\omega_{4'}$ and $\alpha_{4'}$ and compare their values with those found in Example 7-4.

7-7 In Fig. P7-7a bodies 2 and 3 are in sliding contact. In Fig. P7-7b links 2 and 3 are shown again. The radius of curvature at the point of contact is $C_2P_2 = 3$ in. for link 2 and $C_3P_3 = 2$ in. for link 3. The motion of 3 relative to 2 can be analyzed by considering 2 as fixed with 3 sliding on 2 at point P. (a) Assume a vector V_{C_1} equal to 4⅞ in. in length and draw it downward to the left from C_3. Using Hartmann's construction, find graphically the location of the center of curvature of the path which P_3 describes on 2.   What is the radius of curvature measured from the drawing?   (b) Using the Euler-Savary equation, compute the radius of curvature.

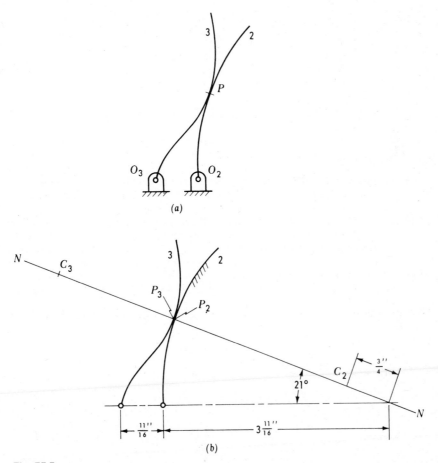

Fig. P7-7

7-8   In Fig. P7-8 bodies 2 and 3 are in sliding contact.   Their radii of curvature at the point of contact are $C_2P_2 = 1$ in. and $C_3P_3 = 2$ in., respectively.   As explained in the preceding problem, the motion of 3 relative to 2 can be analyzed by considering 2 as fixed with 3 sliding on 2 at point P.   (a) Assume a vector $V_{C_3} = 0.75$ in. in length and draw it downward to the left from C_3.   Using Hartmann's construction, find graphically the location of the center of curvature of the path which P_3 describes on 2.   What is its radius of curvature measured from the drawing?   (b) Using the Euler-Savary equation, compute the radius of curvature.

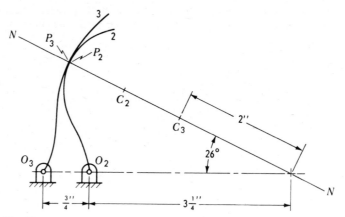

Fig. P7-8

7-9 (a) In Fig. P6-4 link 4 rolls on link 1. Construct the acceleration polygon. $\omega_2 = 144$ rad/sec cw and $\alpha_2 = 1,000$ rad/sec² ccw. Scales: 1 in. = 10 fps and 1 in. = 1,000 fps². (b) Determine α_3 and α_4 in radians per second squared.

7-10 In Fig. P4-3, at the pin in link 3, label the coincident points P_2 and P_4 on links 2 and 4, respectively. Let $\omega_2 = 120$ rad/sec (constant). (a) Construct the acceleration polygon for points O_2, P_2, P_4, and C. Scales: 1 in. = 10 fps and 1 in. = 500 fps². (b) Determine α_4 in radians per second squared.

7-11 Construct the acceleration polygon for Fig. P6-8. $\omega_2 = 15$ rad/sec and is constant. Scales: 1 in. = 1 fps and 1 in. = 20 fps². Determine the acceleration of link 4 in feet per second squared.

7-12 In Fig. P7-12, disk 4 is driven by link 2 sliding in guides on 1 as shown. The

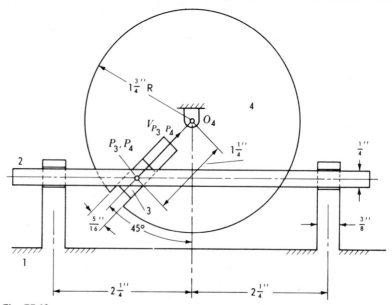

Fig. P7-12

drive is through block 3, which is pivoted on 2 at point P_3. The velocity of link 2 is constant. The instantaneous velocity of slide of block 3 on link 4 is 1.5 ips toward the center of 4. (a) Construct the velocity and acceleration polygons for points O_4, P_3, and P_4. Use a velocity scale of 1 in. = 1 ips and an acceleration scale of 1 in. = 1 ips². (b) Determine α_4.

7-13 In Fig. P6-6, body 4 rolls on body 1. $\omega_4 = 18$ rad/sec and is constant. (a) Construct the acceleration polygon. Scales: 1 in. = 1 fps and 1 in. = 25 fps². (b) Determine α_4 in radians per second squared.

7-14 (a) Construct the acceleration polygon for the mechanism in Fig. P6-11. Scales: 1 in. = 0.5 ips and 1 in. = 0.3 ips². (b) Determine α_3 and α_6 in radians per second squared.

7-15 Suppose the cam in Fig. P6-7 has a constant angular velocity of 300 rpm ccw. (a) Using the Euler-Savary equation find the radius of curvature of the path which point P_4 of body 4 describes on body 2. Label the center of curvature C. (b) Draw an equivalent four-bar linkage using P_4C as the end points of link 3' and let link 4' be a slider. Analyze the equivalent linkage to find the velocity and acceleration of link 4. Scales: 1 in. = 1.5 fps and 1 in. = 30 fps².

7-16 (a) For the cam mechanism in Fig. P7-16 use the Euler-Savary equation to compute the radius of curvature of the path which P_4 describes on body 2. Check your results using Hartmann's construction. For the latter assume a V_{P_4} 1 in. in length and directed to the left. (b) Label the center of curvature

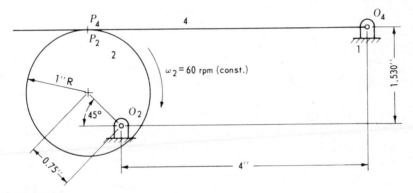

Fig. P7-16

of the path C. Next, draw an equivalent four-bar linkage using point C as one of the pin connections. (c) Analyze the equivalent linkage to obtain ω_4 and α_4 for the given linkage. Scales: 1 in. = 2 ips and 1 in. = 5 ips².

7-17 (a) For Fig. 7-34 use the Euler-Savary equation to compute the radius of curvature for the path which P_4 describes on link 2. Compare your result with that found by the Hartmann construction in Example 7-6. (b) Draw the mechanism one-fourth size and construct the velocity and acceleration polygons for the mechanism in Example 7-6. Analyze the given mechanism and not an equivalent four-bar linkage. In the acceleration analysis utilize the Coriolis component of acceleration for points P_4 and P_2. Also use the radius of curvature of the path found in part (a) of this problem. Use a velocity scale 1 in. = 6 ips and an acceleration scale 1 in. = 10 ips². (c) Determine ω_4 in radians per second and α_4 in radians per second squared.

VELOCITY AND ACCELERATION GRAPHS AND GRAPHICAL DIFFERENTIATION

8-1 INTRODUCTION

In Chaps. 5, 6, and 7 methods for finding the velocities and accelerations for a mechanism when in some particular phase have been discussed. It is often desirable to know how the velocity or acceleration of a given link varies during a whole cycle of operation so that we will know where the velocity or acceleration is a maximum. The velocities and accelerations can be found for a number of phases of the mechanism, and the values plotted so that curves can be drawn. The results are known as velocity and acceleration graphs.

8-2 VELOCITY AND ACCELERATION GRAPHS

A slider-crank mechanism is shown in Fig. 8-1. The linear velocity of the piston was found for 12 equally spaced positions of the crank. For each of these positions of the crankpin B the corresponding position of piston C was located by taking length BC as a radius on a compass and swinging arcs

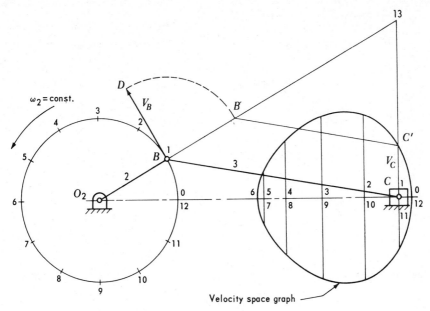

Fig. 8-1

using the points on the crank circle as centers for the compass.   The inter-section of these arcs with the line of piston travel determined the various positions of the piston.   From the known crank rpm, which was assumed constant, V_B was computed in feet per second and drawn perpendicular to O_2B, using a convenient velocity scale.   Next, the parallel-line method, which is explained in Sec. 5-3, was used to find the piston velocity.   The method is illustrated for crank position 1. $BB' = BD$, and CC' represents the magnitude of the piston velocity.   For crank positions 0 through 6 lengths V_c were drawn as positive ordinates perpendicular to the line of piston travel, and for crank positions 6 through 12 they were drawn negative. A smooth curve drawn through the end points forms the velocity-space graph, which is symmetrical about the line of piston travel.   Positive ordi-nates on the graph represent the velocity of the piston during its stroke from right to left, and negative ordinates represent the velocity during the stroke from left to right.   The graph in Fig. 8-1 is called the velocity-space graph because the velocities are plotted at the actual positions of the piston in space at the specified times.

The acceleration-space graph for the piston is shown in Fig. 8-2.   The method of relative accelerations explained in Sec. 7-2 was used to find the

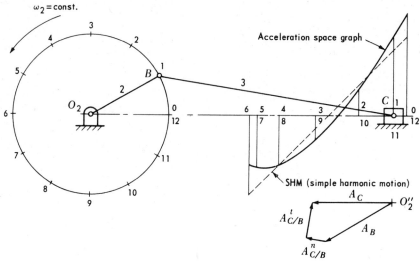

Fig. 8-2

piston acceleration. The acceleration polygon for finding A_C when the crank is in position 1 is shown in the figure. The accelerations of point C are laid off as ordinates from the horizontal line of piston travel. Any convenient scale may be used. Accelerations of the piston in a direction from C toward O_2 were plotted as positive ordinates, and accelerations in a direction from C away from O_2 were plotted negative. The dashed line shows how the acceleration-space graph would appear if the piston were to have simple harmonic motion, which is discussed in Sec. 2-7.

A velocity-time graph and acceleration-time graph are shown in Fig. 8-3. Any convenient scale may be used along the time axis. The length of the diagram is the time required for one revolution of the crank. Since the crank rotates at constant rpm, the time intervals corresponding to the equal spacing on the crank circle are also equal. In Fig. 8-3 the ordinates found in Figs. 8-1 and 8-2 were laid off at the 12 points along the time axis. For the purpose of comparison, the dashed curves show the velocity and acceleration of the piston if it had simple harmonic motion. As the ratio of the length of the connecting rod to the crank is increased, the motion of the piston approaches simple harmonic.

Figure 8-4 shows an oscillating arm quick-return mechanism which was discussed briefly in Sec. 3-8. Crank 2 rotates at constant angular velocity. Slider D is at its extreme positions D' and D'' when crank 2 is perpendicular to arm 4. Hence B' and B'' are the points where link 4 is

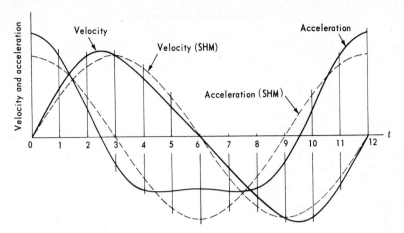

Fig. 8-3

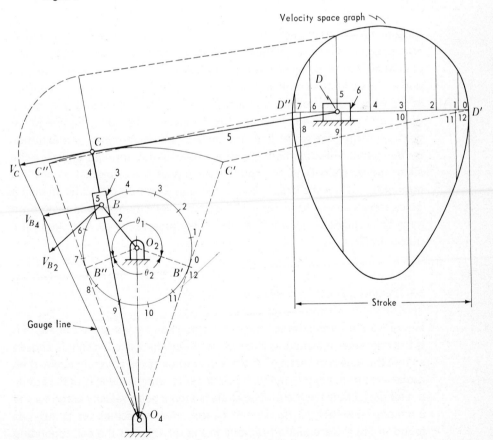

Fig. 8-4

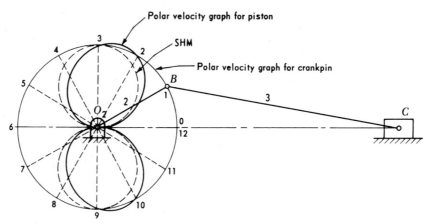

Fig. 8-5

tangent to the crankpin circle. The crankpin circle was divided into 12 equal parts and for each of these positions of B the corresponding positions of D along line $D'D''$ were located and numbered. The velocity construction is shown for the crank in position 5. Any convenient scale can be used in drawing V_{B_2}, the velocity of the crankpin. V_{B_4} is the velocity of a coincident point on 4 and was found by the component method. V_C was found next by drawing the gauge line, and then V_D was determined in magnitude by the parallel-line method. As the crank rotates through angle θ_1, the slider moves from D' to D'', and the velocity of D was plotted as positive ordinates. When the crank rotates through angle θ_2, the slider makes a quick-return stroke from D'' to D', and the velocity of D was plotted as negative ordinates. θ_1/θ_2 is the ratio of the time of the slow stroke to the quick stroke.

8-3 POLAR VELOCITY GRAPHS

Figure 8-5 shows the same slider-crank mechanism as shown in Fig. 8-1. If the crank has constant angular velocity and the velocity scale is chosen so that the length of vector V_B equals the radius of the crankpin circle, then radial lines from point O_2 out to the circle represent the velocity of the crankpin for each crank position. The circle is then a polar velocity graph for the crankpin. Similarly, if from point O_2 the linear velocities for point C as found in Fig. 8-1 are laid off radially in the direction of the corresponding position of the crank, we obtain the polar velocity graph for the piston. The piston velocity for any phase can then be found by drawing the crank in its corresponding position, and then measuring the length of this radial line from O_2 to the curve.   The dashed curves are circles and show how the polar velocity diagram of the piston would appear if the piston had simple harmonic motion.

An oscillating arm four-bar linkage appears in Fig. 8-6.   When crank 2 has constant angular velocity, and the linear velocity of point B is laid off radially from the crank circle O_2B using any convenient scale, the polar velocity graph is a circle as shown.   The parallel-line construction for obtaining the velocity of point C is illustrated and the polar velocity graph for point C is also shown.   The path of C is a circular arc.   For motion of arm 4 to the left, V_C is plotted in the direction outward from the path, and for motion to the right, V_C is plotted inward.

8-4   ANGULAR–VELOCITY AND ACCELERATION GRAPHS

The Geneva mechanism explained earlier in Sec. 3-14 and shown in Fig. 3-32 is illustrated diagramatically in Fig. 8-7a.   Link 2 is the driver and rotates with constant angular velocity.   While link 2 rotates one-fourth revolution, link 3 rotates one-fourth revolution with variable angular velocity.   Then as

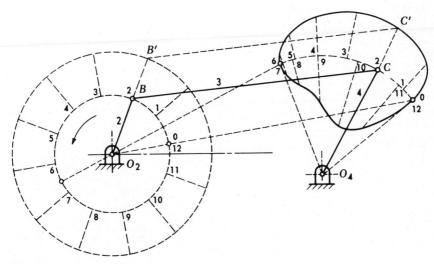

Fig. 8-6

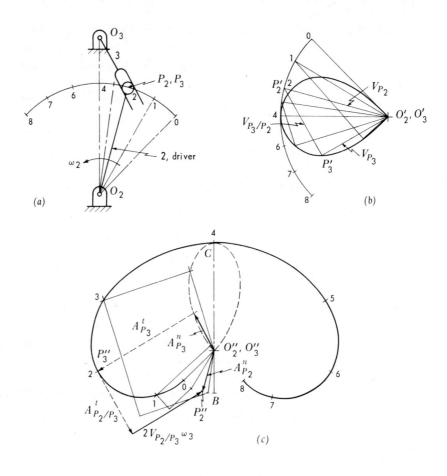

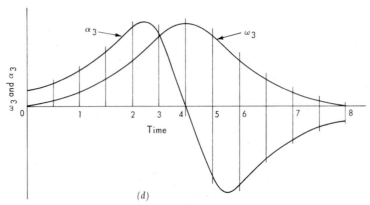

Fig. 8-7

2 rotates the remaining three-fourths revolution, link 3 remains at rest. The angular-velocity and acceleration-time graphs for link 3 are to be constructed.

In Fig. 8-7a let P_2 be a point on 2 and let P_3 be a point on 3 which is coincident with P_2 at the instant.   Points 0, 1, 2, ... , 8 represent successive positions of P_2.   The velocity polygons are shown in Fig. 8-7b, where points 0, 1, 2, . . . , 8 along the circular arc are the corresponding successive positions of P_2'. V_{P_2} can be drawn to any scale.   Here the magnitude of V_{P_2} was made equal to O_2P_2.   The locus of P_3' is the curve.   We note that V_{P_3} is zero when link 2 is in positions 0 and 8, and $V_{P_3} = V_{P_2}$ when link 2 is in position 4.   The magnitudes of V_{P_3} were scaled from Fig. 8-7b and were divided by length O_3P_3 to obtain values of ω_3.   Then the values of ω_3 were plotted as ordinates in Fig. 8-7d.   Any convenient scales may be used for the abscissas and ordinates.

The acceleration polygons are shown in Fig. 8-7c.   For the acceleration vectors we may write

$$\overset{\vee\vee}{A^n_{P_2}} + \overset{0}{A^t_{P_2}} = \overset{\vee\vee}{A^n_{P_3}} + \overset{-\vee}{A^t_{P_3}} + \overset{0}{A^n_{P_2/P_3}} + \overset{-\vee}{A^t_{P_2/P_3}} + \overset{\vee\vee}{2V_{P_2/P_3}\omega_3} \qquad (8\text{-}1)$$

where $A^t_{P_2} = 0$ because link 2 has constant angular velocity, and $A^n_{P_2/P_3} = 0$ because the radius of curvature of the path which P_2 describes on 3 is infinite.   The dash mark above $A^t_{P_3}$ and $A^t_{P_2/P_3}$ indicates that the magnitudes of these vectors are unknown.   With the values obtained from Fig. 8-7b, the magnitudes of the various acceleration components were computed as follows:

$$A^n_{P_2} = \frac{V_{P_2}^2}{O_2P_2} \qquad A_{P_3} = \frac{V_{P_3}^2}{O_3P_3} \qquad \text{Coriolis acceleration} = 2V_{P_2/P_3}\omega_3$$

Next, the acceleration polygons were laid out in Fig. 8-7c from the acceleration pole O_2'', O_3''.   The locations of P_2'' and P_3'' are as indicated when link 2 is in position 2. $A^n_{P_2}$ is parallel to O_2P_2 and $A^n_{P_3}$ is parallel to O_3P_3.   Any convenient scale may be used in laying out the acceleration vectors. $A^t_{P_3}$ and $A^t_{P_2/P_3}$ have been drawn dashed to indicate that their magnitudes are unknown.   Their intersection determines their magnitudes and the location of P_3''.   The locus of P_3'' is the solid curve in the figure, and the locus for the terminus of $A^n_{P_3}$ is the dashed curve.   When link 2 is in its zero position, P_2'' and P_3'' lie at point 0 in Fig. 8-7c. $A^n_{P_3}$ is zero because V_{P_3} is zero, $A^t_{P_2/P_3}$ is zero because $V_{P_2/P_3} = V_{P_2} = \text{const}$, and the Coriolis acceleration is zero because ω_3 is zero.   Hence from Eq. (8-1) we see that $A^n_{P_2} = A^t_{P_3}$.   Further, when link 2 is in position 4, P_2'' lies at B and P_3'' is at C. $A_{P_2}^n$ is $O_3''B$ and $A_{P_3}^n$ is $O_3''C$. $A_{P_3}^t$ becomes zero, the Coriolis component is zero because V_{P_2/P_3} is zero, and length CB represents A_{P_2/P_3}^t. It is to be noted that the diagram is symmetrical about line $O_2''4$.

The magnitudes of $A_{P_3}^t$ were scaled from Fig. 8-7c and divided by length O_3P_3 in order to obtain the values of α_3. Any convenient scale may then be used when plotting these values as ordinates in Fig. 8-7d. From the curve for α_3 we note that link 3 begins and ends its motion with a finite accelera- tion. Just to the left of point 0 along the time axis, α_3 is zero. Thus at position 0 the angular acceleration of link 3 changes suddenly from zero to a positive finite value which results in infinite jerk. Similarly, just to the right of point 8 on the time axis α_3 is zero, and thus at position 8 the angular acceleration of link 3 changes suddenly from a negative finite value to zero resulting in infinite jerk. Accelerations produce inertia forces, and since jerk is the time rate of change of acceleration, a high value of jerk makes for a sudden change in the inertia force resulting in impact loading and high stresses.

The velocity and acceleration curves for the Geneva mechanism are theoretical curves. In an actual mechanism there will be backlash due to manufacturing tolerances. In addition the parts will deform slightly under load. These effects will cause deviations in the curves, particularly at each end of the motion and at the midpoint. Sudden changes in acceleration become increasingly important in high-speed mechanisms. This will be discussed further when we study cams in a later chapter.

8-5 GRAPHICAL DIFFERENTIATION

Thus far in this chapter we have seen how velocity and acceleration curves (graphs) can be constructed for the motion of some point or link in a mech- anism. The procedure consists of determining the velocity or acceleration for a number of different phases of the mechanism, using the methods presented in Chaps. 5, 6, and 7. Another method for obtaining the velocity and acceleration curves consists of constructing a time-displacement curve (often called the displacement diagram) and then by graphical differentia- tion developing the velocity and acceleration curves.

Figure 8-8 shows the displacement-time curve and the velocity-time curve for a point on a mechanism (complete cycle not shown). Since the velocity is the derivative or slope of the displacement curve, ordinates on

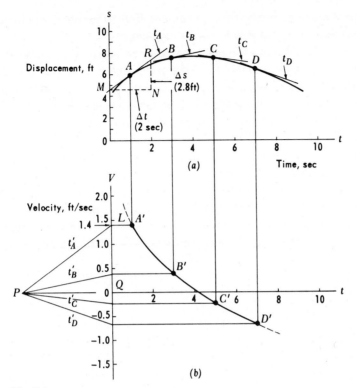

Fig. 8-8

the velocity curve represent the slopes at the corresponding points along the displacement curve. Graphical differentiation consists of obtaining the slopes of various tangents along the first curve and plotting them as ordinates to establish a second curve.

Figure 8-8a shows how the displacement of a point varies with time. Points along the curve are designated A, B, C, etc. These may be taken at equal intervals along the time axis, although they need not be equal. The tangent is drawn to the curve at A and is called t_A. Its slope is $\Delta s/\Delta t$ where Δs and Δt are expressed in the units used along the coordinate axes. Since the velocity equals the slope, the instantaneous velocity at A equals $RN/MN = \Delta s/\Delta t = 2.8/2 = 1.4$ fps, and this can be laid off in Fig. 8-8b as the ordinate for point A' on the velocity curve. The procedure can be repeated for the other points B, C, and D on the displacement curve in order to obtain the corresponding points B', C', and D' along the velocity curve. This method of constructing triangles and measuring their Δs val-

ues is cumbersome, and a more convenient method for obtaining the ordinates on the velocity curve will now be explained.

In Fig. 8-8b, let point P (called the pole) be taken on the time axis at
any convenient distance to the left of the origin, and from it draw t'_A parallel
to t_A. This locates point L, and QL represents to scale the velocity at A
which is found as explained earlier and equals 2.8/2 = 1.4 fps. Thus point
L is marked as 1.4 along the V axis, and the scale for velocities is thereby
established. Then 1.0 fps on the V axis will be laid off from the origin
equal to length QL measured in inches divided by 1.4. Next, lines t'_B, t'_C,
and t'_D are drawn from P parallel to lines t_B, t_C, and t_D, respectively. It is
evident that the velocity at the various points A, B, C, and D will be proportional to the slopes of the tangents t_A, t_B, t_C, and t_D at these points. The
ordinates determined by the intersection of lines t'_A, t'_B, t'_C, and t'_D with the V
axis are proportional to the slopes of these lines, and hence these ordinates

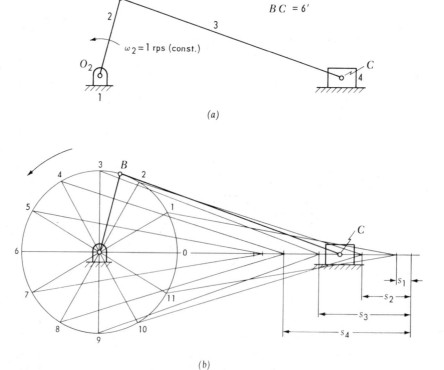

$O_2 B = 2'$
$BC = 6'$

$\omega_2 = 1$ rps (const.)

(a)

(b)

Fig. 8-9

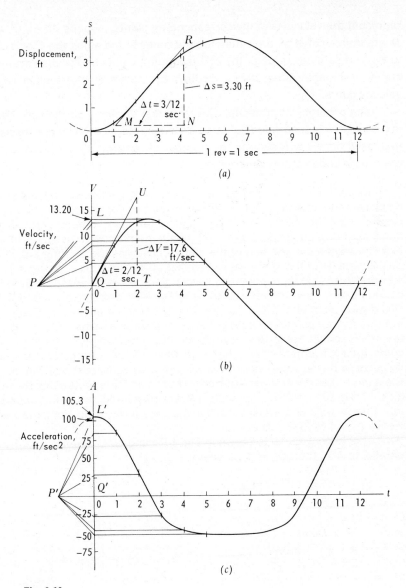

Fig. 8-10

represent the velocities at the corresponding points.   Points A', B', C', and D' are then located by projecting the ordinates found along the V axis over to the vertical lines through the corresponding points on the displacement curve.   A smooth curve drawn through the points of intersection is the velocity curve.

The accuracy of graphical differentiation depends on the care taken in drawing the tangents and the number of increments used for the abscissa. Increased accuracy results as the number of increments is increased and their magnitudes are made smaller.

EXAMPLE 8-1   Suppose for the slider-crank mechanism in Fig. 8-9a the displacement, velocity, and acceleration curves for the slider are to be drawn.   First the mechanism is drawn for a number of different crank positions, as shown on Fig. 8-9b, where 30° crank intervals were used.   With the extreme right-hand position of the slider chosen as the starting point, the displacements of the slider were then laid off in Fig. 8-10a.   Triangle MNR was drawn so that its hypotenuse was tangent to the steepest point on the curve.   Later, when the velocity curve in Fig. 8-10b was constructed, line PL was drawn parallel to MR.   By drawing MR tangent at the steepest point in Fig. 8-10a, we can then choose P at an appropriate distance to the left of the origin so that the largest ordinate on the velocity curve, denoted L, will come out a convenient size.   That is, we do not want the height of the velocity curve to be too small, nor do we want it to be so large that it will fall off the paper.   Length MN was arbitrarily chosen as three units along the time axis.   Thus $\Delta t = \frac{3}{12}$ sec.   Length NR was measured and found to be 1.65 in.   On the original drawing a scale of 1 in. = 2 ft was used along the s axis. Hence $\Delta s = 1.65(2) = 3.30$ ft.

The velocity curve is shown in Fig. 8-10b and since line PL was drawn parallel to MR, then $QL = V_{max}$, where

$$V_{max} = \frac{\Delta s}{\Delta t} = \frac{3.30}{\frac{3}{12}} = 13.20 \text{ fps}$$

On the original drawing, QL was found to measure 1.44 in.   The velocity scale is then 1.44 in. = 13.20 fps or 1 in. = 13.20/1.44 = 9.13 fps.   Next, from pole P lines are drawn parallel to the tangent at each time interval along the displacement curve.   From their points of intersection with the V axis we project across to find the points of intersection with the vertical lines through the corresponding points along the time axis.   A smooth curve through these points is the velocity graph.   Since the displacement curve is symmetrical about its midpoint, the velocity and acceleration curves are also symmetrical about this point.

In Fig. 8-10c the acceleration curve is shown where line $P'L'$ was drawn parallel to QU, which is the line of steepest slope on the velocity curve.   The location of pole P' was chosen so that $Q'L'$ was a convenient length.   In

Fig. 8-10b the maximum slope is at Q and represents A_{max}, where

$$A_{max} = \Delta V / \Delta t$$

QT is Δt and was arbitrarily taken $= \frac{2}{12}$ sec. $TU = \Delta V$ and was found to measure 1.92 in. on the original drawing. Since the velocity scale, as found earlier, is 1 in. $= 9.13$ fps, $\Delta V = 1.92(9.13) = 17.6$ fps. Then

$$A_{max} = \frac{\Delta V}{\Delta t} = \frac{17.6}{\frac{2}{12}} = 105.3 \text{ fps}^2$$

Next, length $Q'L'$ on the original drawing was found to measure 1.75 in. Hence the acceleration scale is 1.75 in. $= 105.3$ fps² or 1 in. $= 105.3/1.75 = 60.2$ fps². From pole P', lines were drawn parallel to the tangents at each time interval along the velocity curve. The ordinates, as found along the A axis, were then projected across to find the points of intersection with the vertical lines through the corresponding points along the time axis. A smooth curve through these points gives the acceleration graph.

EXAMPLE 8-2 The oscillating arm four-bar linkage of Fig. 8-6 is shown again in Fig. 8-11. Crank 2 rotates with a constant speed of 4 rps. The displacement, velocity, and acceleration curves for link 4 are to be constructed. First the mechanism is drawn for various positions of crank 2 as shown in Fig. 8-11. Thirty-degree positions were used, and the angular positions of link 4 were measured for each position of link 2. The zero position for link 2 was assumed to be the point where link 4 is at its extreme right position. The positions of crank 4 were measured in degrees and then converted to radians by multiplying by 0.01745. $(360° = 2\pi$ rad, or $1° = 0.01745$ rad.$)$ The measurements were as follows:

CRANK POSITION	θ, deg	θ, rad
0	0	0
1	7.2	0.13
2	21.9	0.38
3	38.3	0.67
4	53.2	0.93
5	64.3	1.12
6	72.0	1.26
7	70.8	1.24
8	65.2	1.14
9	53.2	0.93
10	33.2	0.58
11	10.3	0.18
12	0	0

Convenient time and displacement scales were selected, and the displacements were plotted to obtain the displacement curve shown in Fig. 8-12a. Tangents MR and CD were then drawn at the points of maximum positive and maximum negative slope, respectively. Next, in 8-12b pole P was assumed at a convenient distance to the left of the origin so that, after drawing PL parallel to MR and PH parallel to CD, the height HL of the velocity curve would come out a reasonable size. Then MN was arbitrarily taken as four units ($\frac{1}{12}$ sec) along the time axis and NR was measured and found to be 1.14 rad. Thus

$$\omega_{\max(+)} = \frac{\Delta\theta}{\Delta t} = \frac{1.14}{\frac{1}{12}} = 13.7 \text{ rad/sec}$$

and point L was marked 13.7 as shown in Fig. 8-12b. The scale for ω was thereby established. On the original drawing QL was found to measure 1.30 in. Hence 1.30 in. = 13.7 rad/sec or 1 in. = 10.5 rad/sec. Length BC was arbitrarily taken as two units ($\frac{1}{24}$ sec) along the time axis, and BD was found to measure 0.783 rad. Thus

$$\omega_{\max(-)} = \frac{\Delta\theta}{\Delta t} = \frac{0.783}{\frac{1}{24}} = 18.8 \text{ rad/sec}$$

Tangents were next drawn along the displacement curve at each point along the time axis. Then from point P lines were drawn parallel to these in order to determine the ordinates at the corresponding points along the time axis in Fig. 8-12b. A smooth curve connecting the ordinates then gave the velocity curve.

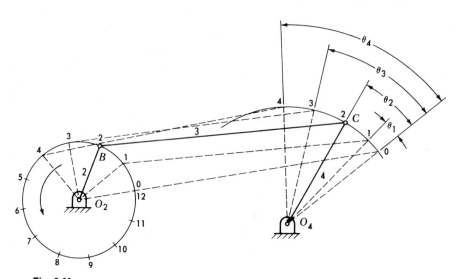

Fig. 8-11

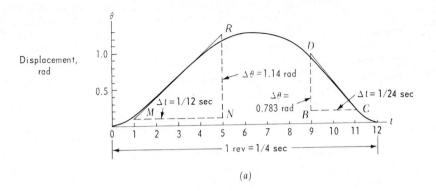

(a)

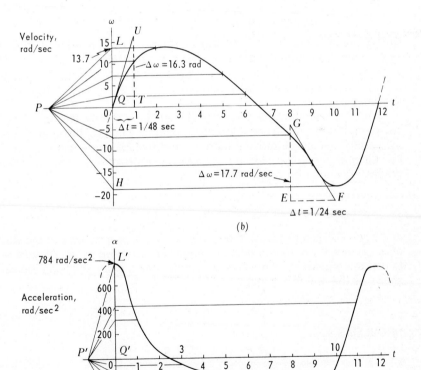

(b)

(c)

Fig. 8-12

The acceleration curve is shown in Fig. 8-12c, and was constructed by first drawing tangents QU and FG at the points of maximum positive and a negative slope in Fig. 8-12b. Then pole P' was assumed at a convenient distance to the left of the origin so that after drawing $P'L'$ parallel to QU and $P'H'$ parallel to FG, the height $H'L'$ of the acceleration curve would be reasonable. In Fig. 8-12b length QT was assumed as one unit ($\frac{1}{48}$ sec) along the time axis and TU was found to measure 16.3 rad. Thus

$$\alpha_{max(+)} = \frac{\Delta\omega}{\Delta t} = \frac{16.3}{\frac{1}{48}} = 784 \text{ rad/sec}^2$$

and point L' was marked 784 as shown. On the original drawing $Q'L'$ measured 2.08 in.; hence the acceleration scale became 2.08 in. = 784 rad/sec² or 1 in. = 377 rad/sec². From triangle EFG

$$\alpha_{max(-)} = \frac{\Delta\omega}{\Delta t} = \frac{17.7}{\frac{1}{24}} = 426 \text{ rad/sec}^2$$

Next, tangents were drawn to the velocity curve at each point along the time axis. Then from pole P' lines were drawn parallel to each of these in order to determine the ordinates at the corresponding points along the time axis in Fig. 8-12c. A smooth curve through these ordinates then formed the acceleration curve.

In this example the accuracy of both the velocity and acceleration curves can be improved considerably by plotting values of the angular velocity directly instead of deriving them from the displacement curve. In Fig. 8-6 values of the linear velocity of point C can be scaled from the velocity graph. The angular velocity of link 4 is

$$\omega = \frac{V_C}{O_4 C}$$

and since $O_4 C$ is a constant, the values of V_C are proportional to the values of ω. Hence if we were to take the values of V_C, found in Fig. 8-6, and were to plot them along the time axis, we would obtain a curve having the same shape as the one in Fig. 8-12b.

PROBLEMS

8-1   (a) Plot the velocity-space graph for slider 6 of the drag-link quick-return mechanism in Fig. P8-1. Let V_B be represented by a vector 0.5 in. long. $\omega_2 = 60$ rpm (constant). Plot the velocity of point D for the 15 positions of B which are shown. Positions 1 to 7 are 45° apart. The remaining positions are 10° apart. In finding the velocities of point D, use the parallel-line method, as explained in Sec. 5-3 and also illustrated in Fig. 8-1. (b) From the known rpm of link 2 and the given length of vector V_B, compute the velocity scale expressing the velocities in feet per second.

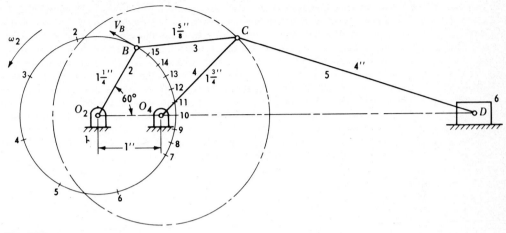

Fig. P8-1

8-2 (*a*) Construct the velocity-space graph for slider 4 in Fig. P8-2. $\omega_2 = 40$ rpm (constant). Let V_B be represented by a vector 1 in. long. Use the velocity-polygon method for determining the slider velocity. For each of the 12 positions place the pole for the polygon at the corresponding position of B. If V_C is directed to the left, plot it upward on the graph, and if it is to the right, plot it downward. Compute the velocity scale expressing velocities in feet per second. (*b*) Construct the acceleration-space graph for slider 4. Use the acceleration polygon method for determining the slider acceleration for each of the 12 crank positions. Use an acceleration scale of 1 in. $= 1$ fps². If A_C is directed to the left, plot it upward on the graph, and if it is to the right, plot it downward.

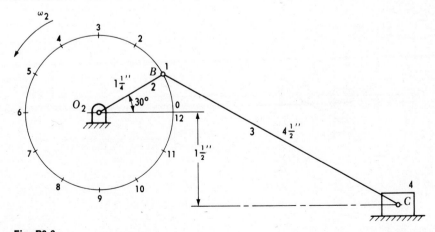

Fig. P8-2

8-3   (*a*) Design a Geneva wheel.   Make a drawing similar to Fig. 3-32 except that the driver is to make five revolutions for each revolution of the driven.   Use a center distance of 3 in.   The driving pin is to enter the slot in the driven member tangentially so that there will be no impact.   (*b*) Fig. P8-3 shows the mechanism schematically.   In Sec. 2-12 it is explained that the angular-velocity ratio $\omega_3/\omega_2 = O_2Q/O_3Q$, where line N-N in Fig. P8-3 is the line of transmission.   Make a separate full-size drawing similar to Fig. P8-3 and divide the angle, which the driver makes from the time the pin enters a slot until the time it leaves, into eight equal divisions.   By measurement determine the ratio $O_2 Q/O_3 Q$ for each of the eight positions.   (*c*) Plot the angular-velocity–time graph for the follower. Make the time axis 4 in. long.   The driver rotates at a constant speed of 120 rpm. Let 1 in. of ordinate represent 10 rad/sec.

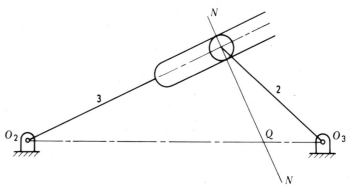

Fig. P8-3

8-4   A Whitworth quick-return mechanism similar to Fig. 3-10 is shown in Fig. P8-4. For the phase shown, slider 6 is at the left end of its stroke.   The corresponding position of point B is labeled position 0.   (*a*) Make a drawing similar to Fig.

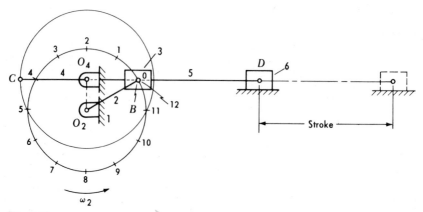

Fig. P8-4

P8-4 and find the location of point D for each of the equally spaced positions of B on circle O_2B.   Number the positions of D to correspond with the numbers on circle O_2B. $O_2O_4 = 0.5$ in., $O_2B = 1$ in., $O_4C = 1.125$ in., and $CD = 4$ in. $\omega_2 = 60$ rpm (constant).   (*b*) Plot the displacement-vs.-time graph for slider 6. Let the time for one revolution of crank 2 be represented by 4 in. along the time axis and plot the 12 displacements half their actual lengths.   (*c*) By graphical differentiation construct the velocity-vs.-time and acceleration-vs.-time graphs. (*d*) Compute the scales for the time, displacement, velocity, and acceleration axes in seconds, inches, inches per second, and inches per second squared, respectively, and place calibration marks along these axes.

9

MATHEMATICAL
ANALYSIS

9-1 INTRODUCTION

We have studied three graphical methods for determining velocities in mechanisms: instant centers, relative velocities, and graphical differentiation. Also we have covered two graphical methods for finding accelerations: relative accelerations and graphical differentiation. In this chapter several mathematical methods for determining velocities and accelerations in mechanisms will be discussed.

In general a mathematical solution is more accurate than the graphical, but it is more time-consuming. Also, errors are easier to make and are less easily detected. However, we shall see that an advantage of mathematical analysis is that the mathematical expressions for velocity and acceleration reveal how the various parameters of a mechanism, such as the lengths and the angular positions of the links, affect the motion. This is valuable in the design of mechanisms to produce desired output motions (kinematic synthesis). The mathematical methods are a powerful tool in synthesis and are more readily understood if one first learns to analyze mathematically the motion of given mechanisms.

9-2 ANALYSIS BY TRIGONOMETRY

Equations specifying the positions of the various links in a mechanism can be written using trigonometry, and these equations can be differentiated to obtain velocities and accelerations.   The slider-crank mechanism (Fig. 9-1) will be used to illustrate the method.   Let $n = L/R$ and $m = d/L$.   From the figure

$$BD = L \sin \phi = R \sin \theta$$

$$\sin \phi = \frac{\sin \theta}{n} \tag{9-1}$$

If we use the trigonometric identity

$$\cos \phi = \sqrt{1 - \sin^2 \phi}$$

Then

$$\cos \phi = \sqrt{1 - \frac{\sin^2 \theta}{n^2}}$$

The binomial expansion is

$$(a + b)^n = a^n + na^{n-1} b + \frac{n(n-1)}{2!} a^{n-2}b^2$$

$$+ \frac{n(n-1)(n-2)}{3!} a^{n-3}b^3 + \cdots$$

and for the term $\sqrt{1 - (\sin^2 \theta/n^2)}$

$$a = 1 \qquad b = \frac{-\sin^2 \theta}{n^2} \qquad n = \tfrac{1}{2}$$

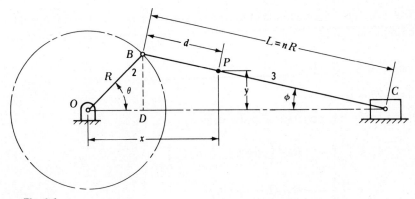

Fig. 9-1

The expansion then becomes

$$\sqrt{1 - \frac{\sin^2 \theta}{n^2}} = 1 - \tfrac{1}{2} \frac{\sin^2 \theta}{n^2} - \tfrac{1}{8} \frac{\sin^4 \theta}{n^4} - \tfrac{1}{16} \frac{\sin^6 \theta}{n^6} \cdots$$

The ratio n is usually at least equal to 4. Using a value $n = 4$ and the maximum value of $\sin \theta = 1$ in the above series, the terms become

$$1 - \tfrac{1}{32} - \tfrac{1}{2048} - \tfrac{1}{65536} \cdots$$

We can drop all but the first two terms. Then

$$\cos \phi = \frac{1}{n} \left(n - \frac{1}{2n} \sin^2 \theta \right)$$

approximately. The displacement of any point P on the connecting rod is

$$x = R \cos \theta + d \cos \phi = R \cos \theta + \frac{d}{n} \left(n - \frac{1}{2n} \sin^2 \theta \right) \qquad (9\text{-}2)$$

$$y = (L - d) \sin \phi = \frac{L - d}{n} \sin \theta \qquad (9\text{-}3)$$

Differentiating these equations with respect to time we get the velocity of point P in the x and y directions

$$V_p^x = \frac{dx}{dt} = -R\omega_2 \left(\sin \theta + \frac{m}{2n} 2 \sin \theta \cos \theta \right)$$

$$= -R\omega_2 \left(\sin \theta + \frac{m}{2n} \sin 2\theta \right) \qquad (9\text{-}4)$$

$$V_p^y = \frac{dy}{dt} = R\omega_2 (1 - m) \cos \theta \qquad (9\text{-}5)$$

The velocity of the piston is obtained by substituting $m = 1$ into the last two equations. Thus

$$V_c^x = -R\omega_2 \left(\sin \theta + \frac{1}{2n} \sin 2\theta \right) \qquad (9\text{-}6)$$

$$V_c^y = 0 \qquad (9\text{-}7)$$

If the crank has constant angular velocity

$$A_p^x = \frac{dV_p^x}{dt} = -R\omega_2^2 \left(\cos \theta + \frac{m}{n} \cos 2\theta \right) \qquad (9\text{-}8)$$

$$A_p^y = \frac{dV_p^y}{dt} = -R\omega_2^2 (1 - m) \sin \theta \qquad (9\text{-}9)$$

The acceleration of the piston is obtained by substituting $m = 1$ into these last two equations. Thus

$$A_c^x = -R\omega_2^2 \left(\cos \theta + \frac{1}{n} \cos 2\theta \right) \tag{9-10}$$

$$A_c^y = 0 \tag{9-11}$$

From Eqs. (9-6) and (9-10) we see that when n is infinite, the motion of the piston becomes simple harmonic, which was discussed in Sec. 2-7.

The angular displacement of the connecting rod is found from Eq. (9-1) to be

$$\phi = \sin^{-1} \left(\frac{\sin \theta}{n} \right) \tag{9-12}$$

The angular velocity of the connecting rod is

$$\omega_3 = \frac{d\phi}{dt} = \frac{\omega_2 \sqrt{(1 - \sin^2 \theta)}}{\sqrt{(n^2 - \sin^2 \theta)}} \tag{9-13}$$

and the angular acceleration is

$$\alpha_3 = \frac{d\omega_3}{dt} = \frac{\omega_2^2 \sin \theta \cos^2 \theta}{(n^2 - \sin^2 \theta)^{3/2}} - \frac{\omega_2^2 \sin \theta}{\sqrt{n^2 - \sin^2 \theta}} \tag{9-14}$$

With the exception of the slider-crank mechanism, in general the analysis of velocities and accelerations in mechanisms by trigonometry becomes very cumbersome. For this reason in the remainder of this chapter a more concise method of analysis, which utilizes complex numbers, is explained.

9-3 VECTORS AS COMPLEX NUMBERS

A convenient method of expressing vectors is to write them as complex numbers. Before going into the next topic, which is the analysis of velocities and accelerations in mechanisms by complex numbers, we shall review some definitions pertaining to the latter.

In Fig. 9-2 the directed-line segment, vector OP, represents a complex number denoted by Z. The quantity Z can be expressed as

$$Z = x + iy \tag{9-15}$$

where x is called the *real component* and y the *imaginary component*. The quantity i in Eq. (9-15) is an imaginary unit defined as follows: $i = \sqrt{-1}$.

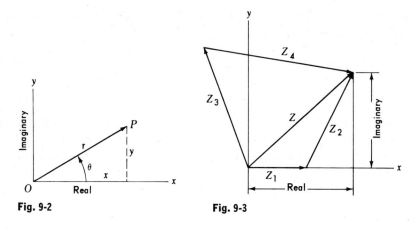

Fig. 9-2   Fig. 9-3

However, x and y are real numbers.   When used in this manner, the cartesian plane is called the *complex plane* or *Z plane*.   Length r is the magnitude or absolute value of the vector and can be expressed as

$$r = \sqrt{[\Re(Z)]^2 + [\mathcal{I}(Z)]^2} \tag{9-16}$$

where $\Re(Z)$ and $\mathcal{I}(Z)$ denote the real and imaginary components, respectively.   Magnitude r is considered positive when directed from O toward P

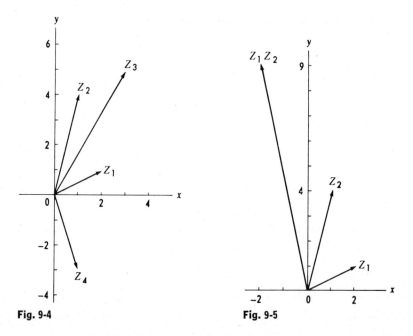

Fig. 9-4   Fig. 9-5

and considered negative when directed from P toward O. The direction of the vector is specified by angle θ, which is measured from the x axis. θ is considered positive when measured counterclockwise as shown and is negative when measured in the clockwise sense. From Fig. 9-2

$$\tan \theta = \frac{y}{x} = \frac{\mathcal{I}(Z)}{\mathcal{R}(Z)}$$

$$\theta = \tan^{-1} \frac{\mathcal{I}(Z)}{\mathcal{R}(Z)} \tag{9-17}$$

It is important to note that a vector can be expressed in a variety of ways as the sum of a number of vectors. For example, in Fig. 9-3

$$Z = Z_1 + Z_2 = Z_3 + Z_4$$

Further, two vectors are equal if and only if the sums of their real parts are equal and the sums of their imaginary parts are equal. Again, in Fig. 9-3 $(Z_1 + Z_2) = (Z_3 + Z_4)$; that is

$$\mathcal{R}(Z_1) + \mathcal{R}(Z_2) = \mathcal{R}(Z_3) + \mathcal{R}(Z_4)$$

and

$$\mathcal{I}(Z_1) + \mathcal{I}(Z_2) = \mathcal{I}(Z_3) + \mathcal{I}(Z_4)$$

Addition, subtraction, and multiplication of complex numbers are performed in the same manner as for real numbers. In Fig. 9-4 let $Z_1 = 2 + i$ and $Z_2 = 1 + i4$. Then $Z_3 = Z_1 + Z_2 = 3 + i5$ and $Z_4 = Z_1 - Z_2 = 1 - i3$. In multiplication the products of powers of the quantity i should be reduced to their briefest form. Thus

$$i^2 = \sqrt{-1} \sqrt{-1} = -1$$
$$i^3 = ii^2 = \sqrt{-1}\,(-1) = -i$$
$$i^4 = i^2 i^2 = -1(-1) = 1$$

If $Z_1 = 2 + i$ and $Z_2 = 1 + i4$, then $Z_1 Z_2 = 2 + i + i8 + i^2 4 = -2 + i9$. The vectors are shown in Fig. 9-5.

From Fig. 9-2 it can be seen that vector $Z = x + iy$ can be expressed in polar coordinates as follows:

$$Z = r \cos \theta + ir \sin \theta$$
$$= r(\cos \theta + i \sin \theta)$$

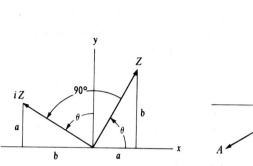

Fig. 9-6   **Fig. 9-7**

From the following MacLaurin series expansions

$$e^{i\theta} = 1 + i\theta - \frac{\theta^2}{2!} - \frac{i\theta^3}{3!} + \frac{\theta^4}{4!} + \frac{i\theta^5}{5!} - \frac{\theta^6}{6!} - \cdots$$

$$\cos\theta = 1 - \frac{\theta^2}{2!} + \frac{\theta^4}{4!} - \frac{\theta^6}{6!} + \cdots$$

$$i\sin\theta = i\theta - \frac{i\theta^3}{3!} + \frac{i\theta^5}{5!} - \frac{i\theta^7}{7!} + \cdots$$

we note that the first series is the sum of the second and third. Thus

$$e^{i\theta} = \cos\theta + i\sin\theta$$

and expressed in exponential form

$$Z = re^{i\theta} \tag{9-18}$$

Multiplication of a vector by i rotates the vector 90° in a positive (counterclockwise) direction. This is illustrated in Fig. 9-6, where $Z = a + ib$. Then $iZ = ia + i^2b = -b + ia$. Thus

$$ire^{i\theta} = re^{i(\theta+90°)} \tag{9-19}$$

Vectors expressed in complex exponential form are differentiated in the usual manner. For example,

$$\frac{d}{dt}e^{i\theta} = e^{i\theta}\frac{d}{dt}i\theta = ie^{i\theta}\frac{d\theta}{dt} = i\omega e^{i\theta} \tag{9-20}$$

where

$$\omega = \frac{d\theta}{dt}$$

If we differentiate again and let $\alpha = \dfrac{d\omega}{dt}$,

$$\frac{d}{dt}(i\omega e^{i\theta}) = i\left(\omega \frac{d}{dt} e^{i\theta} + e^{i\theta} \frac{d\omega}{dt}\right)$$

$$= i[\omega(i\omega e^{i\theta}) + e^{i\theta}\alpha]$$

$$= -\omega^2 e^{i\theta} + i\alpha e^{i\theta} \qquad\qquad (9\text{-}21)$$

The vector $-\omega^2 e^{i\theta}$ in Eq. (9-21) is illustrated as OA in Fig. 9-7. The quantity $e^{i\theta}$ indicates the vector is oriented at angle θ with the x axis. The magnitude of the vector is ω^2, and the minus sign indicates the vector is directed from O toward A rather than from A toward O. The second vector $i\alpha e^{i\theta}$ is represented by OB. The magnitude is α. If the expression were not preceded by an i, this vector would be oriented at an angle θ with the real axis, as indicated by the exponent $i\theta$. However, since multiplication of a vector by i rotates the vector 90° ccw, the vector becomes OB.

9-4 ANALYSIS BY COMPLEX NUMBERS

In this section some general equations are developed for obtaining the angular velocity and acceleration of the various links in a mechanism in terms of the angular velocity and acceleration of the driving link. In addition, it is shown how the linear velocity and acceleration of any point on the linkage can be computed.

Four-bar linkage

For the four-bar linkage in Fig. 9-8 let a, b, c, and d denote the lengths of links 1, 2, 3, and 4, respectively. The frame is regarded as link 1. Angles θ_2, θ_3, and θ_4 denote the angular positions of links 2, 3, and 4, respectively, and are considered positive when measured counterclockwise as shown. The length of the diagonal from A to D is denoted by s and the angle which it makes with line OD is indicated as β. Link 2 is considered as the driver and its angular position θ_2 is assumed known.

Angles θ_3 and θ_4 can be found as follows. Consider triangle OAD. Thus

$$s = \sqrt{a^2 + b^2 - 2ab\cos\theta_2} \qquad\qquad (9\text{-}22)$$

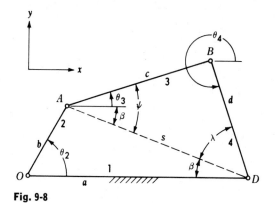

Fig. 9-8

and

$$\frac{\sin \beta}{b} = \frac{\sin \theta_2}{s}$$

Thus

$$\beta = \sin^{-1}\left(\frac{b}{s} \sin \theta_2\right) \tag{9-23}$$

Next, for triangle ABD,

$$d^2 = c^2 + s^2 - 2cs \cos \psi$$

Hence

$$\psi = \cos^{-1}\left(\frac{c^2 + s^2 - d^2}{2cs}\right) \tag{9-24}$$

Also, from triangle ABD,

$$\frac{\sin \lambda}{c} = \frac{\sin \psi}{d}$$

Thus

$$\lambda = \sin^{-1}\left(\frac{c}{d} \sin \psi\right) \tag{9-25}$$

The values of β, ψ, and λ can be computed using Eqs. (9-23) to (9-25). Then by observing the particular configuration of the linkage, the values of θ_3 and θ_4 will be apparent. For example, for the configuration in Fig. 9-8, $\theta_3 = \psi - \beta$ and $\theta_4 = 360 - (\lambda + \beta)$.

We will now consider velocities. In Fig. 9-9 a four-bar linkage is shown with the links represented by position vectors $\bar{a}$, $\bar{b}$, $\bar{c}$, and $\bar{d}$.   Then

$$\bar{b} + \bar{c} + \bar{d} = \bar{a} \tag{9-26}$$

Expressing these vectors in exponential form

$$\bar{a} = ae^{i\theta_1} = a$$

$$\bar{b} = be^{i\theta_2}$$

$$\bar{c} = ce^{i\theta_3} \tag{9-27}$$

$$\bar{d} = de^{i\theta_4}$$

Substituting Eqs. (9-27) into Eq. (9-26), we obtain

$$be^{i\theta_2} + ce^{i\theta_3} + de^{i\theta_4} = a$$

Next, if we differentiate this equation with respect to time and let

$$\omega_2 = \frac{d\theta_2}{dt} \qquad \omega_3 = \frac{d\theta_3}{dt} \qquad \omega_4 = \frac{d\theta_4}{dt}$$

then

$$ib\omega_2 e^{i\theta_2} + ic\omega_3 e^{i\theta_3} + i\, d\omega_4 e^{i\theta_4} = 0 \tag{9-28}$$

The real and imaginary parts of this equation are

$$\mathfrak{R}: \quad -b\omega_2 \sin\theta_2 - c\omega_3 \sin\theta_3 - d\omega_4 \sin\theta_4 = 0$$

$$\mathfrak{I}: \quad b\omega_2 \cos\theta_2 + c\omega_3 \cos\theta_3 + d\omega_4 \cos\theta_4 = 0 \tag{9-29}$$

Equations (9-29) can be solved for ω_3 and ω_4 as follows:

$$\omega_3 = -\frac{b \sin\delta}{c \sin\epsilon}\, \omega_2 \tag{9-30}$$

$$\omega_4 = \frac{b \sin\gamma}{d \sin\epsilon}\, \omega_2 \tag{9-31}$$

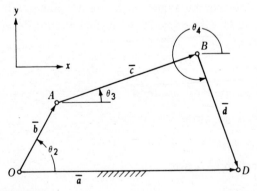

Fig. 9-9

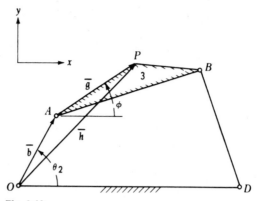

Fig. 9-10

where

$$\delta = \theta_2 - \theta_4$$
$$\epsilon = \theta_3 - \theta_4 \qquad (9\text{-}32)$$
$$\gamma = \theta_2 - \theta_3$$

Thus from Eqs. (9-30) and (9-31) the angular velocities of links 3 and 4 can be obtained from the known angular velocity of link 2.

Having found the angular velocity of link 3 we can compute the linear velocity of any point on the link as follows. In Fig. 9-10 let P be a point whose velocity is to be found. Position vector $\bar{h}$ is then

$$\bar{h} = \bar{b} + \bar{g}$$

or in exponential form

$$\bar{h} = be^{i\theta_2} + ge^{i\phi} \qquad (9\text{-}33)$$

where $b = OA$, $g = AP$, and ϕ is the angle which line AP makes with the x axis. Differentiating Eq. (9-33) with respect to time and denoting $d\phi/dt = \omega_3$, we obtain $\bar{v}_P$, the velocity of point P as follows:

$$\bar{v}_P = ib\omega_2 e^{i\theta_2} + ig\omega_3 e^{i\phi} \qquad (9\text{-}34)$$

The real and imaginary components are then

$$\mathcal{R}(\bar{v}_P) = -b\omega_2 \sin \theta_2 - g\omega_3 \sin \phi$$
$$\mathcal{I}(\bar{v}_P) = b\omega_2 \cos \theta_2 + g\omega_3 \cos \phi \qquad (9\text{-}35)$$

The magnitude and direction of $\bar{v}_P$ are

$$v_P = \sqrt{[\mathcal{R}(\bar{v}_P)]^2 + [\mathcal{I}(\bar{v}_P)]^2} \qquad (9\text{-}36)$$

and

$$\theta_P = \tan^{-1} \frac{\mathcal{I}(\bar{v}_P)}{\mathcal{R}(\bar{v}_P)} \qquad (9\text{-}37)$$

We will now consider accelerations. Differentiating Eq. (9-28) with respect to time and letting

$$\alpha_2 = \frac{d\omega_2}{dt} \qquad \alpha_3 = \frac{d\omega_3}{dt} \qquad \alpha_4 = \frac{d\omega_4}{dt}$$

we have

$$(\omega_2{}^2 - i\alpha_2)be^{i\theta_2} + (\omega_3{}^2 - i\alpha_3)ce^{i\theta_3} + (\omega_4{}^2 - i\alpha_4)\,de^{i\theta_4} = 0$$

Expanding this equation in terms of its real and imaginary parts, we obtain two equations similar to Eqs. (9-29), and they can be solved for α_3 and α_4. Doing this and substituting Eqs. (9-30) and (9-31) in the results, we obtain

$$\alpha_3 = \frac{\omega_3}{\omega_2}\alpha_2 - \frac{b\omega_2{}^2 \cos \delta + c\omega_3{}^2 \cos \epsilon + d\omega_4{}^2}{c \sin \epsilon} \qquad (9\text{-}38)$$

$$\alpha_4 = \frac{\omega_4}{\omega_2}\alpha_2 + \frac{b\omega_2{}^2 \cos \gamma + c\omega_3{}^2 + d\omega_4{}^2 \cos \epsilon}{d \sin \epsilon} \qquad (9\text{-}39)$$

Equations (9-38) and (9-39) give the angular accelerations of links 3 and 4, respectively.

The linear acceleration of a point P on link 3 in Fig. 9-10 is obtained by differentiation of Eq. (9-34) with respect to time. Denoting this acceleration as $\bar{a}_P$,

$$\bar{a}_P = (-\omega_2{}^2 + i\alpha_2)be^{i\theta_2} + (-\omega_3{}^2 + i\alpha_3)ge^{i\phi}$$

The real and imaginary parts are

$$\mathcal{R}(\bar{a}_P) = -b(\omega_2{}^2 \cos \theta_2 + \alpha_2 \sin \theta_2) - g(\omega_3{}^2 \cos \phi + \alpha_3 \sin \phi)$$
$$\mathcal{I}(\bar{a}_P) = -b(\omega_2{}^2 \sin \theta_2 - \alpha_2 \cos \theta_2) - g(\omega_3{}^2 \sin \phi - \alpha_3 \cos \phi) \qquad (9\text{-}40)$$

The magnitude and direction for $\bar{a}_P$ are then

$$a_P = \sqrt{[\mathcal{R}(\bar{a}_P)]^2 + [\mathcal{I}(\bar{a}_P)]^2}$$

and

$$\theta_P = \tan^{-1} \frac{(\mathcal{I}\bar{a}_P)}{\mathcal{R}(\bar{a}_P)}$$

EXAMPLE 9-1 In Fig. 9-11a the centerline of the cam is at 45° to line OD. The angular velocity of the cam is 5 rad/sec ccw and its angular acceleration

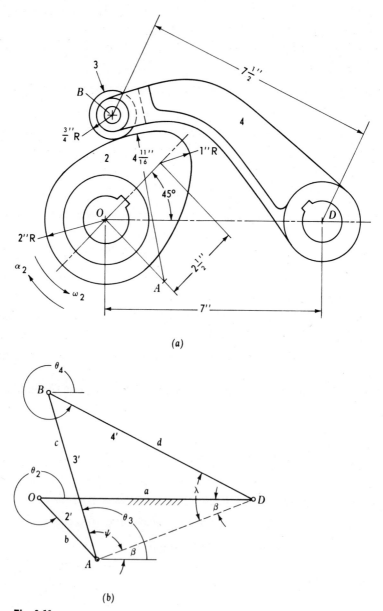

(a)

(b)

Fig. 9-11

is 2.5 rad/sec² cw.   Angular velocity and acceleration of the follower are to be found.

Solution   An equivalent four-bar linkage (as explained in Sec. 7-5) is shown in Fig. 9-11b.   Thus $\theta_2 = 315°$, $\omega_2 = 5$ rad/sec, and $\alpha_2 = -2.5$ rad/sec². Dimensions are $a = 7$ in., $b = 2.688$ in., $c = 5.438$ in., and $d = 7.500$ in.   Velocity ω_4 and acceleration α_4 are to be found.   Substituting into Eqs. (9-22) through (9-25), we have

$$s = \sqrt{7^2 + 2.69^2 - 2(7)(2.69)(0.707)}$$

$$= 5.45 \text{ in.}$$

$$\beta = \sin^{-1}\left[\frac{2.69}{5.45}(-0.707)\right] = 20.4°$$

$$\psi = \cos^{-1}\left[\frac{5.44^2 + 5.45^2 - 7.50^2}{2(5.44)5.45}\right] = 87.1°$$

$$\lambda = \sin^{-1}\left[\frac{5.44}{7.50}(0.999)\right] = 46.4°$$

From Fig. 9-11(b)

$$\theta_3 = \beta + \psi = 20.4 + 87.1 = 107.5°$$

$$\theta_4 = 360 - (\lambda - \beta) = 360 - (46.4 - 20.4) = 334.0°$$

From Eqs. (9-32)

$$\delta = 315 - 334.0 = -19.0°$$

$$\epsilon = 107.5 - 334.0 = -226.5°$$

$$\gamma = 315 - 107.5 = 207.5°$$

Substituting into Eqs. (9-30) and (9-31), we have

$$\omega_3 = -\frac{2.69(-0.326)5}{5.44(0.725)} = 1.11 \text{ rad/sec}$$

$$\omega_4 = \frac{2.69(-0.462)5}{7.50(0.725)} = -1.14 \text{ rad/sec}$$

Since the result has a minus sign, ω_4 is clockwise.   Substituting into Eq. (9-39) we obtain the angular acceleration of the follower as follows:

$$\alpha_4 = \frac{-1.14}{5}(-2.5) + \frac{2.69(25)(-0.886) + 5.44(1.23) + 7.50(1.30)(-0.686)}{7.50(0.727)}$$

$$= 0.570 - 10.92 = -10.35 \text{ rad/sec}^2$$

The minus sign indicates α_4 is clockwise.

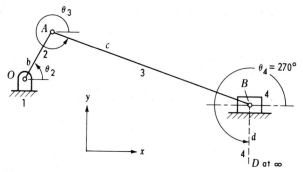

Fig. 9-12

Slider-crank mechanism

A special case of the four-bar linkage is the slider-crank mechanism shown in Fig. 9-12. That is, $\theta_4 = \text{constant} = 270°$, $d = \infty$, and thus $\omega_4 = \alpha_4 = 0$. Also $\delta = \theta_2 - \theta_4 = \theta_2 - 270°$, $\epsilon = \theta_3 - \theta_4 = \theta_3 - 270°$, and $\gamma = \theta_2 - \theta_3$. By substituting these values into Eqs. (9-30) and (9-38), we obtain the following expressions for the angular velocity and acceleration of the connecting rod:

$$\omega_3 = -\frac{b \cos \theta_2}{c \cos \theta_3}\omega_2 \tag{9-41}$$

$$\alpha_3 = \frac{\omega_3}{\omega_2}\alpha_2 + \frac{b\omega_2{}^2 \sin \theta_2 + c\omega_3{}^2 \sin \theta_3}{c \cos \theta_3} \tag{9-42}$$

These equations are valid whether or not the line of travel of point B passes through point O.

Crank-shaper mechanism

This linkage (Fig. 9-13) is an inversion of the slider-crank mechanism in which link 2 is made the fixed link. Note that the notation in Fig. 9-13 is

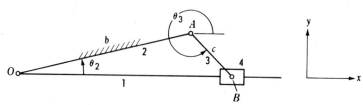

Fig. 9-13

the same as for Fig. 9-12. In Eq. (9-41) ω_3 and ω_2 can be written $\omega_{3\hat{1}}$ and $\omega_{2\hat{1}}$, which designate the angular velocities of links 3 and 2, respectively, relative to link 1. Equation (9-41) then becomes

$$\omega_{3\hat{1}} = -\frac{b \cos \theta_2}{c \cos \theta_3} \omega_{2\hat{1}}$$

or

$$\omega_3 - \omega_1 = -\frac{b \cos \theta_2}{c \cos \theta_3}(\omega_2 - \omega_1) \tag{9-43}$$

If we apply Eq. (9-43) to Fig. 9-13, then $\omega_2 = 0$ and we obtain

$$\omega_3 - \omega_1 = \frac{b \cos \theta_2}{c \cos \theta_3} \omega_1$$

or

$$\omega_1 = \frac{\omega_3}{1 + (b \cos \theta_2)/(c \cos \theta_3)} \tag{9-44}$$

Similarly, Eq. (9-42) may be written

$$\alpha_{3\hat{1}} = \frac{\omega_{3\hat{1}}}{\omega_{2\hat{1}}} \alpha_{2\hat{1}} + \frac{b\omega_{2\hat{1}}^2 \sin \theta_2 + c\omega_{3\hat{1}}^2 \sin \theta_3}{c \cos \theta_3}$$

or

$$\alpha_3 - \alpha_1 = \frac{\omega_3 - \omega_1}{\omega_2 - \omega_1}(\alpha_2 - \alpha_1) + \frac{b(\omega_2 - \omega_1)^2 \sin \theta_2 + c(\omega_3 - \omega_1)^2 \sin \theta_3}{c \cos \theta_3}$$

and since $\omega_2 = \alpha_2 = 0$,

$$\alpha_3 - \alpha_1 = -\alpha_1\left(\frac{\omega_3 - \omega_1}{-\omega_1}\right) + \frac{b\omega_1^2 \sin \theta_2 + c(\omega_3 - \omega_1)^2 \sin \theta_3}{c \cos \theta_3}$$

or

$$\alpha_1 = \frac{\alpha_3 - [b\omega_1^2 \sin \theta_2 + c(\omega_3 - \omega_1)^2 \sin \theta_3]/(c \cos \theta_3)}{1 + (\omega_3 - \omega_1)/\omega_1} \tag{9-45}$$

Equations (9-44) and (9-45) give the values of ω_1 and α_1 when ω_3 and α_3 are known, and vice versa.

EXAMPLE 9-2   In Fig. 9-14a the cam has an angular velocity of 3 rad/sec ccw and an angular acceleration of 1.5 rad/sec^2 cw. The angular velocity and acceleration of the follower are to be computed. Note: This is the same problem as Example 7-6, which was solved graphically.

Solution   An equivalent linkage is shown in Fig. 9-14b and follows the form illustrated in Fig. 7-18. Since our equivalent linkage is of the crank-shaper type, the notation has been made the same as in Fig. 9-13 in order

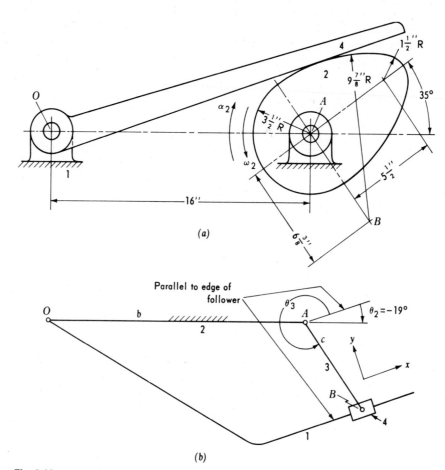

Fig. 9-14

that we can use the equations developed for the latter. Angle $\theta_2 = -19°$, $\theta_3 = 286°$, $\omega_3 = 3$ rad/sec, and $\alpha_3 = -1.5$ rad/sec². Substitution of the data into Eqs. (9-44) and (9-45) then gives

$$\omega_1 = \frac{3}{1 + [16(0.946)]/[6.375(0.276)]} = 0.313 \text{ rad/sec ccw}$$

$$\alpha_1 = \frac{-1.5 - [16(0.313)^2(-0.326) + 6.375(3 - 0.313)^2(-0.961)]/}{1 + (3 - 0.313)/0.313} \text{ [6.375(0.276)]}$$

$$= 2.50 \text{ rad/sec² ccw}$$

Thus in Fig. 9-14a $\omega_4 = 0.313$ rad/sec ccw and $\alpha_4 = 2.5$ rad/sec² ccw.

EXAMPLE 9-3   The cam in Fig. 9-15a has a constant angular velocity of 300 rpm ccw, and the velocity and acceleration of link 4 are to be found. C_2 is the center or curvature of body 2 at the point of contact.   An equivalent linkage, as explained in Sec. 7-5, is shown in Fig. 9-15b and is a Scotch-yoke mechanism.   Since $\omega_{2'}$ is constant, link 4' has simple harmonic motion, as explained in Sec. 2-7.   Thus

$$\omega_{2'} = \frac{2\pi(300)}{60} = 31.41 \text{ rad/sec}$$

Then for link 4' and thus for link 4,

$$V = -R\omega \sin \omega t = -R\omega \sin 30°$$
$$= -1\tfrac{11}{16}(31.41)0.5 = -10.81 \text{ ips}$$

and

$$A = -R\omega^2 \cos \omega t = -R\omega^2 \cos 30°$$
$$= -1\tfrac{11}{16}(31.41)^2\, 0.866 = -588 \text{ ips}^2$$

The minus signs in the results indicate the velocity and acceleration of the follower are upward.

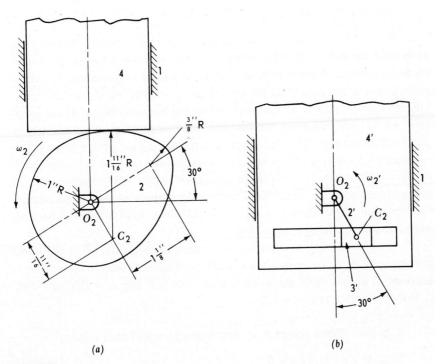

(a)   (b)

Fig. 9-15

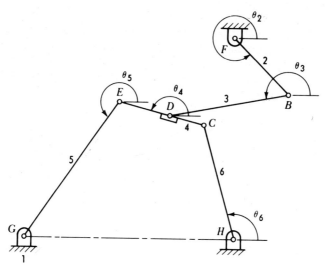

Fig. 9-16

Complex linkages

The linkage in Fig. 9-16 is an example of a complex mechanism. A determination of the velocities or accelerations in a complex linkage can be made by starting at some point of known velocity or acceleration, usually a point where these quantities are zero, and then traversing successive links. The relative-velocity or acceleration components expressed in exponential form are summed up along the way until another point of known velocity or acceleration is reached. The process is repeated until all the links in the mechanism have been traversed. As was done earlier, each vector equation is written in terms of its real and imaginary parts and thus provides two independent algebraic equations containing various unknown quantities. The foregoing procedure will provide a sufficient number of equations to permit a solution for the unknowns.

As an illustration of the procedure, consider Fig. 9-16. It is assumed that θ_2, θ_3, θ_4, θ_5, θ_6, ω_2, and α_2 are known. The angular velocities and accelerations of all the links are to be found. Let $r_2 = BF$, $r_3 = BD$, $r_4 = DE$, $r_5 = EG$, $r_6 = HC$, and $r_7 = CE$.

Velocities Since points F, G, and H are points of zero velocity,

$$V_{C/H} + V_{E/C} + V_{G/E} = 0$$

or

$$ir_6\omega_6e^{i\theta_6} + ir_7\omega_4e^{i\theta_4} + ir_5\omega_5e^{i\theta_5} = 0 \qquad (9\text{-}46)$$

Also

$$V_{B/F} \nrightarrow V_{D/B} \nrightarrow V_{E/D} \nrightarrow V_{G/E} = 0$$

or

$$ir_2\omega_2e^{i\theta_2} + ir_3\omega_3e^{i\theta_3} + ir_4\omega_4e^{i\theta_4} + ir_5\omega_5e^{i\theta_5} = 0 \qquad (9\text{-}47)$$

The sums of the real and imaginary parts of Eqs. (9-46) and (9-47) are then

$$-r_6\omega_6 \sin\theta_6 - r_7\omega_4 \sin\theta_4 - r_5\omega_5 \sin\theta_5 = 0$$

$$r_6\omega_6 \cos\theta_6 + r_7\omega_4 \cos\theta_4 + r_5\omega_5 \cos\theta_5 = 0$$

$$-r_2\omega_2 \sin\theta_2 - r_3\omega_3 \sin\theta_3 - r_4\omega_4 \sin\theta_4 - r_5\omega_5 \sin\theta_5 = 0$$

$$r_2\omega_2 \cos\theta_2 + r_3\omega_3 \cos\theta_3 + r_4\omega_4 \cos\theta_4 + r_5\omega_5 \cos\theta_5 = 0$$

After the numerical values of the known quantities are substituted into this set of equations, the latter can be solved for the values of ω_3, ω_4, ω_5, and ω_6.

Accelerations   Points F, G, and H have zero acceleration; thus

$$A^n_{C/H} \nrightarrow A^t_{C/H} \nrightarrow A^n_{E/C} \nrightarrow A^t_{E/C} \nrightarrow A^n_{G/E} \nrightarrow A^t_{G/E} = 0$$

or

$$-r_6\omega_6{}^2e^{i\theta_6} + r_6\alpha_6ie^{i\theta_6} - r_7\omega_4{}^2e^{i\theta_4} + r_7\alpha_4ie^{i\theta_4} - r_5\omega_5{}^2e^{i\theta_5}$$
$$+ r_5\alpha_5ie^{i\theta_5} = 0 \quad (9\text{-}48)$$

Also

$$A^n_{B/F} \nrightarrow A^t_{B/F} \nrightarrow A^n_{D/B} \nrightarrow A^t_{D/B} \nrightarrow A^n_{E/D} \nrightarrow A^t_{E/D} \nrightarrow A^n_{G/E} \nrightarrow A^t_{G/E} = 0$$

or

$$-r_2\omega_2{}^2e^{i\theta_2} + r_2\alpha_2ie^{i\theta_2} - r_3\omega_3{}^2e^{i\theta_3} + r_3\alpha_3ie^{i\theta_3} - r_4\omega_4{}^2e^{i\theta_4}$$
$$+ r_4\alpha_4ie^{i\theta_4} - r_5\omega_5{}^2e^{i\theta_5} + r_5\alpha_5ie^{i\theta_5} = 0 \quad (9\text{-}49)$$

Next, the sums of the real and imaginary parts of Eqs. (9-48) and (9-49) are

$$-r_6\omega_6{}^2 \cos\theta_6 - r_6\alpha_6 \sin\theta_6 - r_7\omega_4{}^2 \cos\theta_4 - r_7\alpha_4 \sin\theta_4$$
$$- r_5\omega_5{}^2 \cos\theta_5 - r_5\alpha_5 \sin\theta_5 = 0$$

$$-r_6\omega_6{}^2 \sin\theta_6 + r_6\alpha_6 \cos\theta_6 - r_7\omega_4{}^2 \sin\theta_4 + r_7\alpha_4 \cos\theta_4$$
$$- r_5\omega_5{}^2 \sin\theta_5 + r_5\alpha_5 \cos\theta_5 = 0$$

$$-r_2\omega_2{}^2 \cos\theta_2 - r_2\alpha_2 \sin\theta_2 - r_3\omega_3{}^2 \cos\theta_3 - r_3\alpha_3 \sin\theta_3 - r_4\omega_4{}^2 \cos\theta_4$$
$$- r_4\alpha_4 \sin\theta_4 - r_5\omega_5{}^2 \cos\theta_5 - r_5\alpha_5 \sin\theta_5 = 0$$

$$-r_2\omega_2{}^2 \sin\theta_2 + r_2\alpha_2 \cos\theta_2 - r_3\omega_3{}^2 \sin\theta_3 + r_3\alpha_3 \cos\theta_3 - r_4\omega_4{}^2 \sin\theta_4$$
$$+ r_4\alpha_4 \cos\theta_4 - r_5\omega_5{}^2 \sin\theta_5 + r_5\alpha_5 \cos\theta_5 = 0$$

The numerical values of all known quantities are then substituted into this set of four equations.   The latter can then be solved for the numerical values of α_3, α_4, α_5, and α_6.

PROBLEMS

9-1 Using the results of Sec. 9-2, compute the velocity in feet per second and acceleration in feet per second squared of the slider in Fig. P9-1 for $\theta = 0°$, $45°$, $90°$, $135°$, and $180°$.

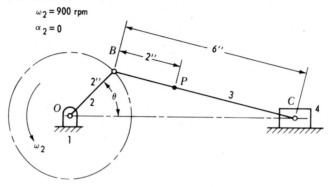

Fig. P9-1

9-2 Using the results of Sec. 9-2, find the magnitude of the velocity in feet per second and acceleration in feet per second squared of point P (Fig. P9-1) for $\theta = 90°$. In each case indicate the angle the resultant vector makes with the x axis.

9-3 A slider-crank mechanism has a crank length of 2 in., a connecting rod length of 6 in., and operates at 3,000 rpm.   Compute the maximum values of velocity in feet per second and acceleration in feet per second squared and determine at what crank angle these maximums occur.

9-4 A slider-crank mechanism similar to Fig. 9-1 has an $R = 2.5$ in. and an $L = 6$ in. The crank has a constant speed of 1,800 rpm.   Using the equations of Sec. 9-2, find the angular velocity in radians per second and the angular acceleration in radians per second squared of the connecting rod for a crank position $\theta = 30°$.

9-5 Use the equations of Sec. 9-4 to compute the acceleration of the piston in feet per second squared for the position shown in Fig. P9-5.

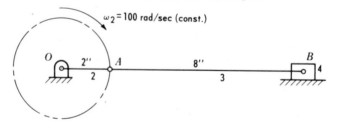

Fig. P9-5

9-6 (a) Draw an equivalent linkage of the slider-crank type (as explained in Sec. 7-5) for the cam mechanism in Fig. P9-6.   (b) Analyze the equivalent linkage, using the equations of Sec. 9-4.   Draw the x axis directed upward and the y axis directed to the left and label them.   Compute the velocity in inches per second and acceleration in inches per second squared of the follower.

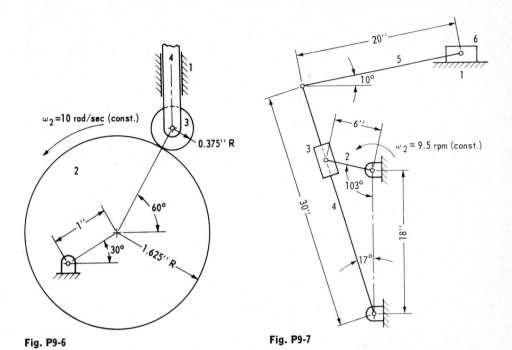

Fig. P9-6

Fig. P9-7

9-7 Use the equations of Sec. 9-4 to find the velocity in feet per minute and acceleration in feet per minute squared of slider 6 of the shaper mechanism in Fig. P9-7.

9-8 For the cam mechanism in Fig. P7-16 draw an equivalent linkage of the crank-
shaper type and compute the values of ω_4 in radians per second and α_4 in
radians per second squared using the equations of Sec. 9-4.

10

CAMS

10-1 INTRODUCTION

A *cam* is an irregular-shaped machine member which serves as a driving link and which imparts motion to a driven link called the *follower*, which either rolls or slides on the driver. Cams are very important mechanisms because they provide the simplest means of achieving almost any desired follower motion. Thus they are frequently occurring elements in many types of machines, especially in automatic machines such as machine tools, printing presses, internal-combustion engines, and mechanical calculators.

10-2 CAM TYPES

Some of the most common types of cams are the following:

1. Disk cams
2. Translation cams
3. Cylindrical cams

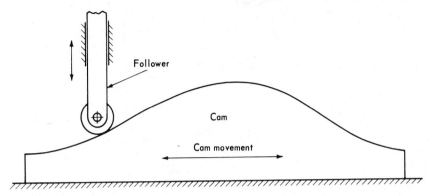

Fig. 10-1

A translation cam is shown in Fig. 10-1.   In Fig. 10-2, a disk cam is illustrated.
Disk cams usually rotate with constant velocity and throughout this chapter
we shall consider the angular velocity of the cam as constant.   Cylindrical
cams will be discussed later in the chapter.

10-3 DISPLACEMENT DIAGRAMS

A displacement diagram is a graph showing displacement of the follower
plotted as a function of time.   A displacement diagram is shown in Fig.
10-3a.   Degrees of cam rotation are plotted along the horizontal axis, and
the length of the diagram represents one revolution of the cam.   Since the
cam rpm is constant, equal angular divisions also represent equal time
increments.   Displacement of the follower is plotted along the vertical axis.
The displacement diagram determines the shape of the cam.   In the
analysis of an existing cam or in the design of a new one, the displacement
diagram is of primary interest.   Since this diagram is in reality a displacement-vs.-time graph, by successive differentiation we can obtain the
velocity-vs.-time and acceleration-vs.-time graphs. The derivative of
acceleration with respect to time is called *jerk* or *pulse*. The acceleration
of the follower is important in high-speed cams because it affects inertia
forces which result in vibration, noise, high stresses, and wear.   *Jerk* is a
measure of the time rate of change of inertia force and hence is an indica-
tion of the impact characteristics of the loading.   Experience has indicated
that infinite jerk causes vibrations in the follower system and affects the life
of the cam.

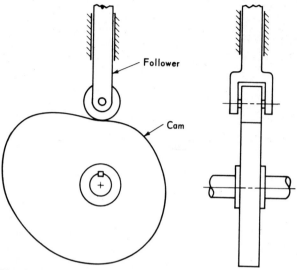

Fig. 10-2

Four common types of follower motion are the following:

1. Constant acceleration
2. Modified velocity
3. Simple harmonic
4. Cycloidal

10-4   CONSTANT ACCELERATION

The displacement of a body moving from rest with constant acceleration is

$$s = \tfrac{1}{2}At^2$$

where s = displacement
A = acceleration
t = time

The plot of the equation is a parabola, and hence the motion is often called *parabolic motion*. Since A is constant, the units of distance s traveled after time t will be proportional to t^2, as shown in Table 10-1 where the values represent equal time units. When constant-acceleration motion is used for the rise of a follower, it is used for half of the desired rise and then followed

Table 10-1

t	0	1	2	3	4	5	· · ·
s	0	1	4	9	16	25	· · ·

by a motion of constant deceleration for the remainder of the rise. The method of constructing the displacement diagram for this motion will be illustrated with an example.

EXAMPLE 10-1   The displacement diagram is shown in Fig. 10-3a.   A follower is to rise 1 in. with constant acceleration during 90° of cam rotation

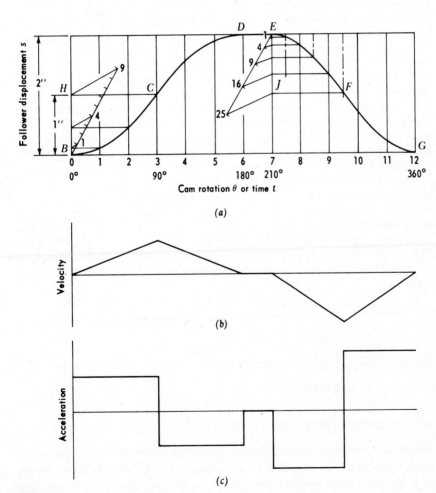

(a)

(b)

(c)

Fig. 10-3

and then is to rise an additional 1 in. with constant deceleration for the next 90°. From 180° to 210° the follower is to dwell and then from 210° to 360° the follower is to fall with constant acceleration followed by constant deceleration.

The 180° for the rise is divided into any number of equal divisions. An equal number is used so that half can be used for the acceleration B to C and the other half used for the deceleration C to D. Since three time units have been chosen for the rise from B to C, a total of nine equal divisions, each of any convenient length, are marked off along an inclined line drawn as shown at the left in the figure. Nine divisions are taken because from Table 10-1 we note that the total displacement after three equal units of time is 9. From point 9 on the inclined line, a line is drawn to the end of the 1-in. rise on the displacement axis. Points 4 and 1 are located on the displacement axis by drawing lines parallel to line 9-H as shown. The displacements 1, 4, and 9 are then projected horizontally to obtain points on the curve BC. Since the decelerated motion C to D is just the reverse of the motion from B to C, the ordinates at points 0, 1, 2, and 3 along the time axis can be laid off downward from the top of the diagram to obtain the C-to-D portion of the curve.

The fall of the follower from EG requires that an equal number of divisions be taken along the horizontal axis. Ten were chosen. For the five time intervals chosen for the accelerated motion E to F, we see from Table 10-1 that a total displacement of 25 equal units are indicated. Hence along an inclined line from E, 25 equal units of any convenient length are laid off. These are then transferred to the vertical line at E, as shown. Next, the vertical displacements are projected horizontally to obtain points on the E-to-F portion of the graph. The deceleration from F to G is the reverse of the acceleration from E to F.

The velocity and acceleration graphs are shown in Fig. 10-3b and c. The maximum acceleration occurs between E and G. If the cam were rotating at 120 rpm, then the elapse in time from E to F, which requires 75°, would be

$$^{75}\!\!/_{360} \times {}^{60}\!\!/_{120} = 0.104 \text{ sec}$$

$$A = \frac{2s}{t^2} = \frac{2 \times \frac{1}{12}}{(0.104)^2} = 15.4 \text{ fps}^2$$

The maximum velocity occurs at F and is

$$V = At = 15.4(0.104) = 1.60 \text{ fps}$$

10-5   MODIFIED CONSTANT VELOCITY

Motion with constant velocity means equal displacements for equal units of time. Therefore the displacement-vs.-time curve is a straight line. Figure 10-4a shows a displacement diagram for a follower which has a rise with

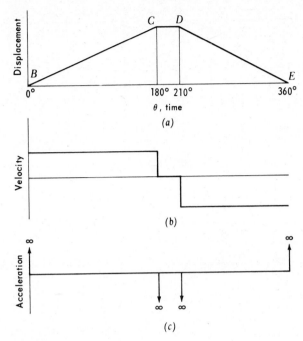

Fig. 10-4

constant velocity from B to C, dwells from C to D, and then falls with constant velocity from D to E. Theoretically this motion results in accelerations which are infinite at points B, C, D, and E, and this makes for impact loading. Thus the use of this type of motion should be avoided.   If such a diagram were used for an actual cam mechanism, the accelerations would not be infinite as indicated because the cam and follower would deform somewhat because of their elasticity, resulting in accelerations which are less.

Modified constant-velocity motion consists of introducing a period of constant acceleration before the constant velocity, and introducing a period of constant deceleration at the end of the constant velocity.   The method will be illustrated with an example.

EXAMPLE 10-2   The displacement diagram is shown in Fig. 10-5a. Suppose a follower is to rise 2 in. with modified constant velocity during 180° of cam rotation, is to dwell for 30°, and then is to fall a portion of the rise with constant acceleration for 60° followed by a fall for the remainder of the rise with constant deceleration during the last 90°.

The first step in laying out the modified constant-velocity motion is to arbitrarily decide what portion of the total time for this motion will be used

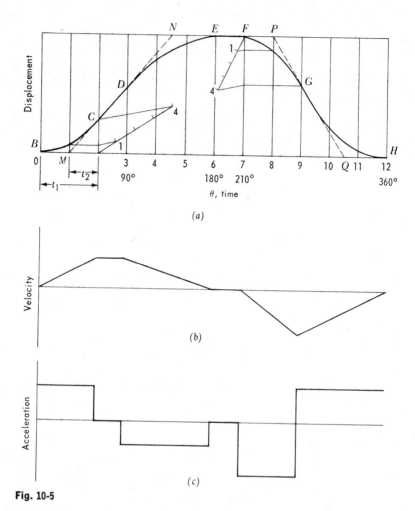

Fig. 10-5

for the constant velocity.   In Fig. 10-5a the 180° were divided into 60° for the acceleration, 30° for the constant velocity, and 90° for the deceleration.   The diagonal MN was drawn from the midpoint of the acceleration-time interval to the midpoint of the deceleration-time interval.   This makes the velocity at the end of curve BC equal to the velocity for the C-to-D line.   The proof that t_1 is twice t_2 is as follows.   Let s be the ordinate at point C and let V be the velocity at this point.   For constant acceleration B to C

$$s = \tfrac{1}{2}At_1^2$$

or

$$A = \frac{2s}{t_1^2}$$

and

$$V = At_1 = \frac{2s}{t_1^2}t_1 = \frac{2s}{t_1}$$

For constant velocity M to C

$$s = Vt_2$$

or

$$V = \frac{s}{t_2}$$

Equating the two expressions for V we obtain

$$\frac{2s}{t_1} = \frac{s}{t_2} \quad \text{or} \quad t_1 = 2t_2$$

When constructing the F-to-H portion of the curve, point G is located first by drawing line PQ. P is located at the midpoint of the time interval for the acceleration FG, and Q is located at the midpoint for the time interval for the deceleration GH.

10-6 SIMPLE HARMONIC MOTION

This motion was discussed in Sec. 2-7. The method of constructing the displacement diagram for simple harmonic motion will be illustrated with an example.

EXAMPLE 10-3 The displacement diagram is shown in Fig. 10-6a. A follower is to move outward 2 in. with simple harmonic motion during 180° of cam rotation, is to dwell for the next 60°, and is to return with simple harmonic motion during the last 120° of cam rotation. The construction is based on the idea that the projection on the diameter of a point moving along a circle with constant angular velocity represents simple harmonic motion. The 180° of cam rotation is divided into six equal divisions, and the semicircle at the left is divided into the corresponding number of sectors. Points on the semicircle are then projected horizontally as shown to obtain points on the curve BCD. The 120° of cam rotation used for the return of the follower are divided into four equal intervals, and the semicircle at the right is divided into a corresponding number of sectors.

The velocity and acceleration graphs are shown in Fig. 10-6b and c. The maximum velocity occurs at the 300° cam position and can be found as follows. Suppose the cam speed is 120 rpm.

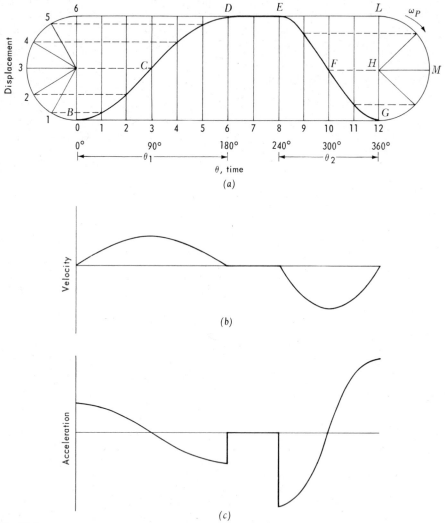

Fig. 10-6

Let ω_P represent the angular velocity of the point moving along the semicircle LMG. Then

$$\omega_P = \frac{\pi}{\text{time of fall}}$$

$$= \frac{\pi}{60/\text{rpm}(\theta_2/360)} = \frac{\pi}{(60/120)(120/360)} = 18.9 \text{ rad/sec}$$

The velocity is

$$V = R\omega_P \sin \omega_P t$$

Since we desire the velocity at the 300° position of the cam, the rotation of the point moving along the semicircle is angle $LHM = \omega_P t = 90°$. Hence

$$V_{max} = R\omega_P \sin 90°$$

$$= \tfrac{1}{12}(18.9) = 1.57 \text{ fps}$$

The equation for acceleration is

$$A = R\omega_P{}^2 \cos \omega_P t$$

and the maximum acceleration is at the 240° and 360° cam positions. At the 240° position the angle of rotation of the point on the semicircle is $\omega_P t = 0$. Hence

$$A_{max} = \tfrac{1}{12}(18.9)^2 \cos (0) = 29.8 \text{ fps}^2$$

10-7   CYCLOIDAL MOTION

The displacement diagram for cycloidal motion is obtained from a cycloid, which is the locus of a point on a circle as the circle rolls on a straight line. In Fig. 10-7a, the curve BDE is the displacement diagram for cycloidal motion having a total displacement h while the cam rotates an angle β. At the right, a circle, whose circumference is h, rolls on the straight line FE. A point on the circle describes the curve FHE, known as a cycloid. As the rolling circle rotates an angle ϕ, the cam rotates an angle θ. From the figure we note that the displacement s, which is the ordinate to point P on the graph, is

$$s = R\phi - R \sin \phi$$

$$= R(\phi - \sin \phi)$$

Since the circle makes one revolution for the total rise h,

$$\phi = 2\pi \frac{\theta}{\beta}$$

Also

$$R = \frac{h}{2\pi}$$

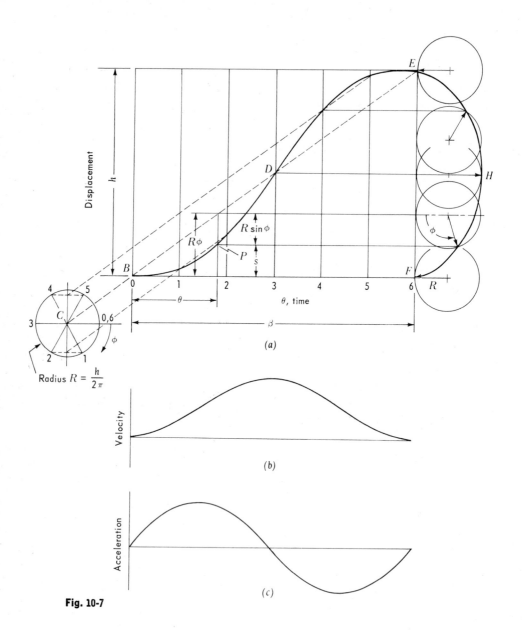

Displacement

h

$R\phi$

$R\sin\phi$

B

P

s

E

D

H

ϕ

F

R

θ

θ, time

β

C

Radius $R = \dfrac{h}{2\pi}$

ϕ

(a)

Velocity

(b)

Acceleration

(c)

Fig. 10-7

Substituting these last two equations into the equation for s, we obtain

$$s = \frac{h}{2\pi}\left(2\pi\frac{\theta}{\beta} - \sin 2\pi\frac{\theta}{\beta}\right)$$

$$= h\frac{\theta}{\beta} - \frac{h}{2\pi}\sin 2\pi\frac{\theta}{\beta} \tag{10-1}$$

A method for constructing the curve BDE consists of drawing the straight line BE. At any convenient distance to the left on this diagonal the center C of a circle is located. This circle is then divided into a number of equal sectors to correspond to the number of equal segments along the time axis of the diagram. Points on the circle are then projected, as shown, to a vertical line through C. Next, from the projection of each point on this vertical a line is drawn parallel to line BE in order to obtain the point of inter-section with the ordinate of corresponding number on the graph. Six inter-vals along the time axis were used here. By using more time intervals a more accurate graph can be obtained.

A proof that the construction just explained satisfies Eq. (10-1) is as follows. The first term in the equation represents the ordinate to the diag-onal BE. The second term represents the length which must then be sub-tracted in order to obtain the ordinate s.

By differentiating Eq. (10-1), and because the angular velocity ω of the cam is constant, the velocity and acceleration equations are found to be

$$V = \frac{h}{\beta}\omega\left(1 - \cos\frac{2\pi\theta}{\beta}\right) \tag{10-2}$$

$$A = \frac{2\pi h}{\beta^2}\omega^2\sin\frac{2\pi\theta}{\beta} \tag{10-3}$$

10-8 COMPARISON OF MOTION CURVES

The selection of the type of motion to be used for a cam follower depends on the speed of the cam, the noise and vibration permissible, and life expec-tancy. When operating speeds are low, selection of the motion is not critical.

For constant acceleration shown in Fig. 10-3, we note that the abrupt changes in the acceleration curve cause infinite jerk, which makes the motion unsuitable for high-speed applications. However, an advantage for constant-acceleration motion is that for a given rise in a given time, it gives the lowest value of acceleration.

In Fig. 10-4 the infinite accelerations which result from constant-velocity motion make it undesirable for high-speed cams.   Modified constant velocity (Fig. 10-5) was shown to be better than constant velocity because the infinite accelerations are eliminated; however, the abrupt changes in the acceleration curve make for infinite jerk.

Figure 10-6 shows that with simple harmonic motion the acceleration curve will be continuous only if the rise and fall periods are both 180°.   If these periods are unequal, or are preceded or followed by a dwell, then discontinuities occur in the acceleration curve and result in infinite jerk.

Cycloidal motion gives the highest peak acceleration for a given rise of any of the motions we have studied.   But from Fig. 10-7 we note that the acceleration curve for cycloidal motion connects with the acceleration curve for any other cycloidal motion or a dwell which may precede or follow it, and there will be no discontinuities to cause infinite jerk.   Thus, of the motions we have studied, cycloidal motion is best suited for high-speed cams.

Another motion curve which has proved useful for high-speed cams is the eighth-power polynomial curve.   This curve is discussed by Kloomok and Muffley.[1]

10-9 CONSTRUCTION OF CAM PROFILE

After the displacement diagram and type of follower have been selected to satisfy the requirements of the machine, the next step is to lay out the cam profile which will accomplish the motion.   The shape of the profile will depend on the size of the cam and the size, shape, and path of the follower. When laying out the cam profile, the principle of inversion is used.   With the cam held fixed, the frame and follower are rotated around the cam to bring the contact surface of the follower in its actual position relative to the cam for a large number of phases of the mechanism.   The cam profile is then drawn inside the envelope of these follower positions.   The methods illustrated in the following examples are not the only ways of constructing the cam profile.   Any method which will correctly position the follower relative to the cam can be used.

10-10 DISK CAM WITH RECIPROCATING KNIFE–EDGE FOLLOWER

In Fig. 10-8 this type of cam and follower is shown.   The cam is to rotate clockwise while the follower moves radially.   The displacement diagram for one revolution of the cam is shown.   The diagram has been divided into 12

[1] M. Kloomok and R. V. Muffley, Plate Cam Design—with Emphasis on Dynamic Effects, *Prod. Eng.*, February, 1955.

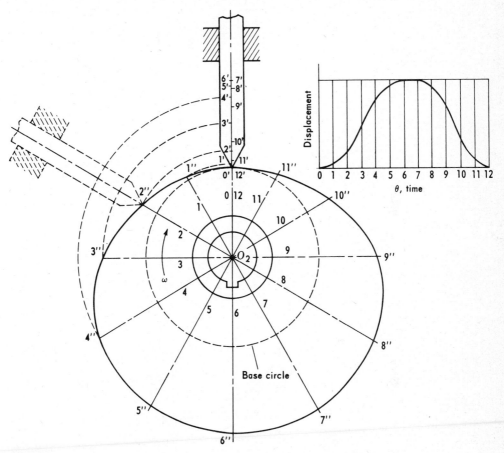

Fig. 10-8

equal intervals, and the cam has been divided into 12 corresponding equal angles.   The distance from the point of the follower in its lowest position to the center of rotation of the cam is the radius of the base circle.   Laid off along the follower are the ordinates from each position along the θ axis of the displacement diagram.   When the cam rotates two spaces clockwise, the edge of the follower is pushed upward from 0' to 2'.   To produce this same relative motion with the cam fixed, the frame is rotated two spaces counterclockwise, and the follower is moved outward a distance 0'2'.   Point 2'' is located by striking an arc from point 2' using O_2 as a center.   Points 1'', 3'', 4'', etc., around the cam are located in the same manner.   A smooth curve through points 0', 1'', 2'', 3'', etc., is then the desired cam profile.

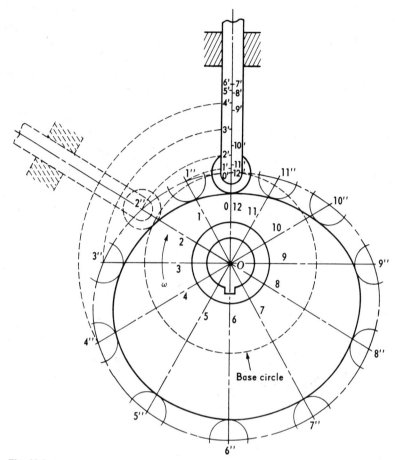

Fig. 10-9

A knife-edge follower is seldom used in practice because the small area of contact results in excessive wear.

10-11 DISK CAM WITH RECIPROCATING ROLLER FOLLOWER

This type of cam is shown in Fig. 10-9. The cam is to rotate clockwise, while the follower moves according to the scale on its centerline. The base circle passes through the axis of the roller when the follower is in its lowest position. Positions 1″, 2″, 3″, etc., of the roller axis are determined in the same manner as in the previous example. A smooth curve through these points is the pitch profile.   Arcs with radii equal to the roller radius are then struck from these points.   A smooth curve tangent to these arcs is the cam profile. It is necessary to determine the pitch profile first because the point of contact between the roller and cam does not lie on a radial line through the roller axis unless the follower is in a dwell position.   As the cam rotates, the contact point shifts from one side of this line to the other.

10-12 DISK CAM WITH OFFSET ROLLER FOLLOWER

In Fig. 10-10 a cam of this type is shown.   The distinguishing feature of this kind of cam is that the centerline of the follower does not pass through the center of the camshaft.   Sometimes the follower is offset so as to clear another part of the machine.   The main reason for this arrangement, however, is that by offsetting the follower, the side thrust on the follower is

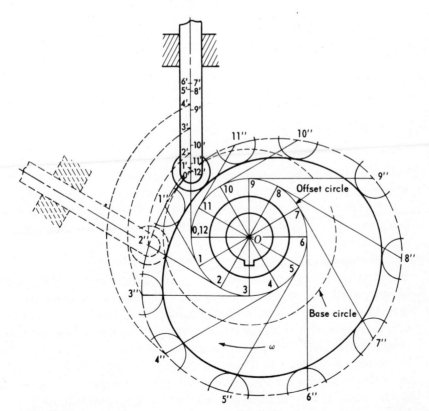

Fig. 10-10

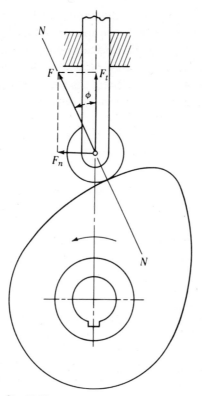

Fig. 10-11

reduced. Side thrust is discussed in the next section. When the offset is to the right, the cam should rotate counterclockwise. When the offset is to the left, the cam should rotate clockwise. These conditions will result in a smaller side thrust on the follower for a given rise in a given angle of cam rotation.

The cam in Fig. 10-10 is to rotate clockwise while the follower moves according to the scale on its centerline. The base-circle radius is the distance from the axis of rotation of the cam to the axis of the roller when it is in its lowest position. The perpendicular distance from the center of the cam to an extension of the centerline of the follower determines the radius of the offset circle. This distance is also called the *amount of the offset*. The offset circle is divided into 12 equal angles to correspond to the number of equal intervals along the time axis of the displacement diagram. A tangent is then drawn at each position on the offset circle to represent the position of the follower centerline as the cam is held stationary and the frame and follower are rotated counterclockwise about the cam. Then with O as a center, arcs are struck from 1', 2', 3', etc., to determine their points of inter- section 1'', 2'', 3'', etc., with the tangent lines. A smooth curve through points 1'', 2'', 3'', etc., is the pitch profile. Arcs with radii equal to the roller radius are then struck from these points. A smooth curve tangent to these arcs is the cam profile.

10-13 PRESSURE ANGLE

The angle which the common normal for the cam and follower makes with the path of the follower is called the *pressure angle* and is labeled ϕ in Fig. 10-11. In the figure, the force F which the cam exerts on the follower has components F_t and F_n, which are tangential and normal, respectively, to the path of follower motion. The normal component is an undesirable side thrust on the follower which tends to make it bind in its guide. It is evident that the side thrust can be reduced by reducing the pressure angle, and it is generally stated that for good performance the pressure angle should not exceed 30°. However, in special cases where the forces are small and the bearings are accurate, larger angles can be used.

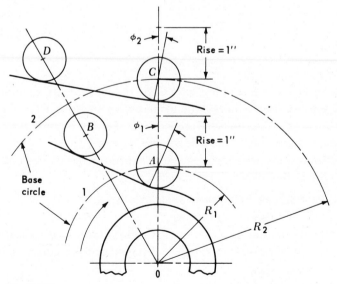

Fig. 10-12

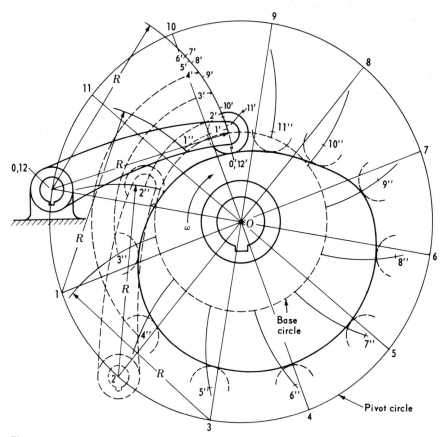

Fig. 10-13

An increase in the size of the base circle reduces the pressure angle. In Fig. 10-12 the rise from A to B is the same as the rise from CD. For the smaller base circle 1, the pressure angle is ϕ_1. For the larger base circle 2, the pressure angle is ϕ_2. It is evident that $\phi_2 < \phi_1$.

For a given displacement diagram the pressure angle can be reduced by one or more of the following methods:

1. Increase the diameter of the base circle.
2. Decrease the total rise of the follower.
3. Increase the amount of cam rotation for a given follower displacement.
4. Change the type of follower motion, i.e., constant velocity, constant acceleration, harmonic, etc.
5. Change the amount of follower offset.

10-14 DISK CAM WITH OSCILLATING ROLLER FOLLOWER

A mechanism of this type is shown in Fig. 10-13.   The cam is to rotate clockwise, and the follower is to oscillate according to the displacements indicated.   Angular displacements could be used for the follower, but it is
usually more convenient to work with arc lengths which represent the corresponding angular values.   Point O is the center of the camshaft, and the
distance from O to the axis of the roller when the follower is in its lowest
position is the radius of the base circle.   The pivot circle is drawn using
point O as a center and the distance from O to the pivot as the radius.   The
pivot circle is then divided into 12 equal sectors to correspond to the number
of equal intervals along the time axis of the displacement diagram.   Next,

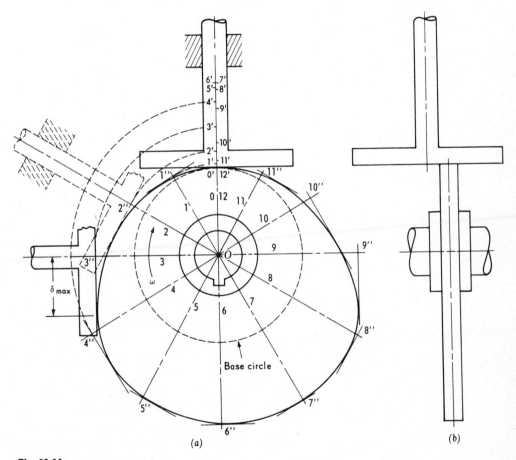

(a)   (b)

Fig. 10-14

the positions of the pivot as the follower is rotated around the cam are numbered 0, 1, 2, 3, etc.   Then using radius R, arcs are struck using points 1, 2, 3, etc., as centers.   Next, with the center of the compass at O, arcs are struck from points 1', 2', 3', etc., to locate points 1'', 2'', 3'', etc.   Then using the latter as centers and a radius equal to the roller radius, arcs are drawn to represent the various positions of the roller relative to the cam.   A smooth curve tangent to these arcs is the cam profile.

10-15   DISK CAM WITH RECIPROCATING FLAT–FACED FOLLOWER

This is shown in Fig. 10-14.   The cam is to rotate clockwise, and the follower is to move according to the displacements indicated along its centerline. Points 1'', 2'', 3'', etc., are located as in the previous examples.   At each of these points a perpendicular is drawn to the radial line on the cam.   The perpendiculars represent the face of the follower as it is rotated around the cam.   A smooth curve contacting each of these lines is the cam profile.

It can be seen from the figure that the point of contact shifts along the face of the follower.   By inspection the maximum deviation of the contact point from the follower centerline is found to occur at phase 3.   This is indicated as δ_{max}.   The radius of the circular face of the follower is actually made a little larger than what is required to accommodate δ_{max}.

In Fig. 10-14b an end view of the cam is shown.   The cam is often offset from the follower centerline in order that the follower stem will rotate. This distributes the contact over a larger area on the follower and reduces wear.

10-16   DISK CAM WITH OSCILLATING FLAT–FACED FOLLOWER

This type of cam mechanism is shown in Fig. 10-15.   The cam is to rotate clockwise and the follower is to oscillate according to the displacements indicated.   Point O is the center of the camshaft and point 0' is the lowest position for the face of the follower.   The radius of the base circle is the distance from O to 0'.   The pivot circle is drawn using the distance from O to the pivot as a radius.   Next, the positions of the pivot as the follower is rotated around the cam are numbered 1, 2, 3, etc.   Then using radius R, arcs are struck using points 1, 2, 3, etc., as centers.   Next, with the center of the compass at O, arcs are struck from points 1', 2', 3', etc., to locate points 1'', 2'', 3'', etc.   As shown at pivot position 0, the flat face of the follower,

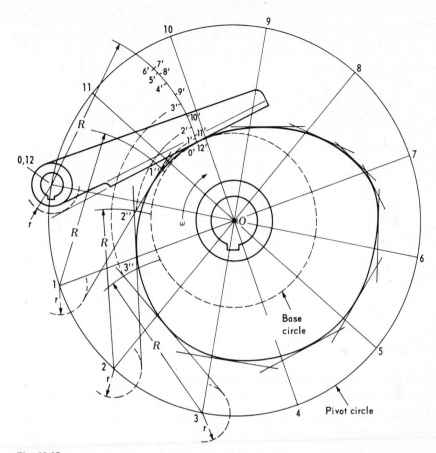

Fig. 10-15

when extended, is tangent to a circle of radius r.   As the follower is rotated around the cam, the face of the follower must be tangent to this circle of radius r and must pass through points 1″, 2″, 3″, etc., as shown.   A smooth curve contacting each position of the face of the follower is the cam profile.

10-17   DESIGN LIMITATIONS

In the design of a cam it is usual to assume the displacement diagram, the type of follower, and the size of the base circle.   Not always are these assumptions practical.

In Fig. 10-16a the assumptions resulted in a cam profile which does not contact all positions of the roller.   Thus near the top of the cam the

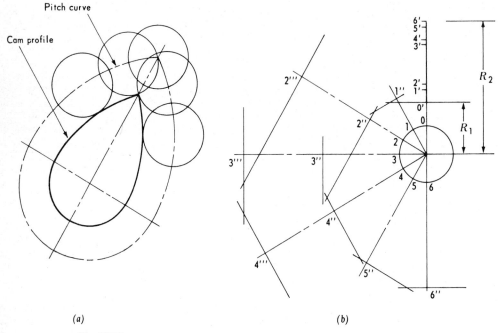

Pitch curve

Cam profile

R_2

R_1

(a)   (b)

Fig. 10-16

cam profile would not push the roller to the desired positions. This condition can be rectified by either increasing the size of the base circle or by reducing the radius of the roller. The contact stresses in the cam and roller are increased if the radii of curvature of either are reduced. Thus a roller of excessively small radius is to be avoided. Further, if the radius of curvature of the cam profile at some region is very small, the cam approaches a point, and this is not satisfactory except for very low speeds.

In Fig. 10-16b a cam with a reciprocating flat-faced follower is shown. When a base circle with radius R_1 is used, the follower positions are 1″, 2″, 3″, etc., and a smooth curve cannot be drawn to all these, since 3″ lies outside the intersection of 2″ and 4″. If the base-circle radius is increased to R_2, the new positions 2‴, 3‴, and 4‴ are satisfactory.

10-18 CIRCULAR–ARC DISK CAMS

Some cams are designed so that their profile consists of circular arcs joined by tangent lines as shown in Fig. 10-17a, or the entire cam profile may consist of tangent circular arcs as in Fig. 10-17b. In the design of cams of this type the follower displacements are not assumed and the cam profile then determined, but the follower displacements are determined from an assumed cam profile.   The advantages of this type of cam are the simplicity of the profile, the lower cost of manufacture, and ease with which dimensional accuracy can be checked.

The velocities and accelerations for the follower used with cams of this type can be found by the methods presented earlier in the chapters on velocity and acceleration analysis in mechanisms.

An undesirable characteristic of cams of this type is the abrupt change in the acceleration curve at points where arcs of different radii are joined. This gives rise to infinite jerk; in general, these cams should not be operated at high speed.

10-19   POSITIVE–RETURN CAM

With a disk cam and radial follower it is often required that a positive means be used to return the follower rather than to rely upon gravity or a spring. Figure 10-18 shows a cam of this type where the cam not only positively controls the follower during its outward motion but also on the return stroke. The displacements for the return stroke must be the same as for the outward motion but in the opposite direction.   The outward motion is according to the displacement scale, points $0'$ to $6'$.   The cam profile for this motion is determined in the usual manner.   Distance b is equal to the diameter of the base circle plus the total rise of the follower.   Points $7''$, $8''$, $9''$, etc., are located by making $1''7'' = 2''8'' = 3''9''$, etc., $= b$.   This con-

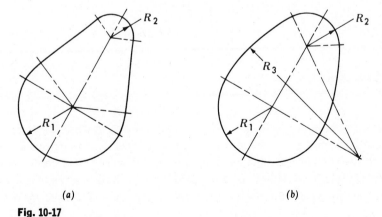

(a)   (b)

Fig. 10-17

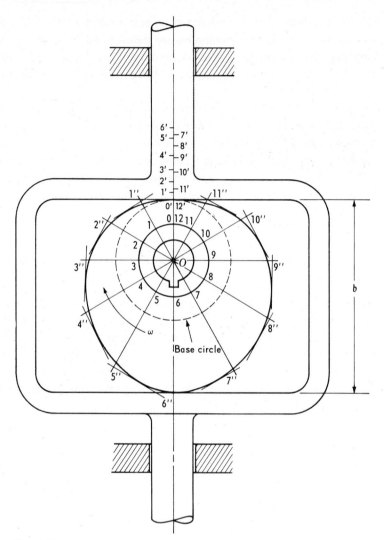

Fig. 10-18

struction results in a return motion for the follower which is the same as the outward motion. This cam is frequently called a *constant-breadth cam*.

If the disk in Fig. 10-18 were circular and if point O, the center of the camshaft, is located other than at the center of the circle, the result is known as an *eccentric cam*. Such a cam produces constrained simple harmonic motion.

The cam in Fig. 10-18 may also be designed using a follower with two rollers instead of two flat faces.   Then points 1″, 2″, 3″, etc., are positions of the roller axis.   Lengths 1″7″, 2″8″, 3″9″, etc., are made equal to the distance between the roller axes, which is a constant.   If it is necessary to have the return motion independent of the outward motion, then two cams keyed to the same shaft must be used.   One cam contacts roller 1 and gives the follower its outward motion.   The other cam contacts roller 2 and gives the follower its return motion.   These double disk cams may also be used with a follower having flat faces.

10-20 CYLINDRICAL CAM

Sometimes in a machine it is desired to have the axis of rotation of the cam parallel to the direction of follower motion.   Then in order to avoid complicating the system with gears, etc., as would be required if a disk cam were used, a cylindrical cam is employed.   This type of cam is shown in Fig. 10-19.   The cam is a cylinder which rotates completely about its axis, imparting motion to a follower, which is guided by a groove in the cylinder.   The follower may be either the type which reciprocates, such as the upper one in the figure, or it may be the type which oscillates, shown in the lower part of the figure.

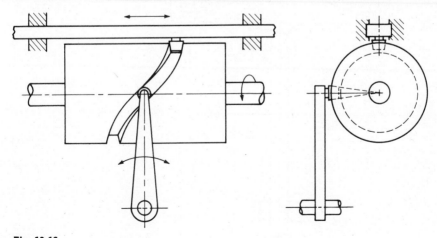

Fig. 10-19

Cylindrical cams find many applications, particularly on machine tools. Another common application is the level winding mechanism on a fishing reel.

10-21 INVERSE CAM

Another cam mechanism is the inverse cam, Fig. 10-20.   In this device the functions of the parts are reversed; the body with the groove is the driven member, and the roller is the driver.   The driving crank may either oscillate or make complete revolutions.   The groove for the roller may be shaped to give any desired motion for 180° of rotation of the driver.   This mechanism is used on sewing machines and in other applications where the load is

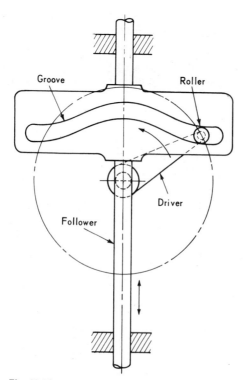

Fig. 10-20

light.   The procedure for laying out the groove that will impart a desired motion to the follower is as usual.   We assume a number of positions of the driver at various time intervals.   The corresponding positions of the follower are determined for the kind of motion desired.   Points are then projected in such a way as to determine the centerline of the groove.

PROBLEMS

Problems 10-1 to 10-4 consist of drawing the displacement diagrams for the motions specified.   In each case the length of the diagram is to be $4\frac{1}{2}$ in. and unless otherwise stated 30° intervals ($\frac{3}{8}$ in. = 30°) are to be used along the θ axis.   The height of the diagram is to equal the maximum follower displacement drawn full size.

10-1   (a) Rise $\frac{3}{4}$ in. with constant acceleration in 90° followed by a rise of $\frac{3}{4}$ in. with constant deceleration in 90°.   (b) Dwell 30°.   (c) Fall $\frac{3}{4}$ in. with constant acceleration in 60° followed by a fall of $\frac{3}{4}$ in. with constant deceleration in 60°.   (d) Dwell 30°.

10-2   (a) Rise $1\frac{1}{2}$ in. with modified constant velocity in 180°.   Use constant acceleration for the first 60° (use 15° intervals along θ axis), then constant velocity for 60°, followed by constant deceleration for 60° (use 15° intervals along θ axis).   (b) Dwell 30°.   (c) Fall $1\frac{1}{2}$ in. with modified constant velocity in 150°.   Use constant acceleration for the first 30° (use 15° intervals along θ axis), then constant velocity for 90°, followed by constant deceleration for 30° (use 15° intervals along θ axis).

10-3   (a) Rise $1\frac{1}{2}$ in. with simple harmonic motion in 180°.   (b) Dwell 60°.   (c) Fall $1\frac{1}{2}$ in. with simple harmonic motion in 120° (use 15° intervals along θ axis).

10-4   (a) Rise $1\frac{1}{2}$ in. with cycloidal motion in 180°.   (b) Fall $1\frac{1}{2}$ in. with cycloidal motion in 120° (use 15° intervals along θ axis).   (c) Dwell 60°.

In the following problems determine the maximum velocity in feet per second and maximum acceleration in feet per second squared of the follower during the rise and fall:

10-5   Prob. 10-1.   Speed of cam is 300 rpm.
10-6   Prob. 10-2.   Speed of cam is 450 rpm.
10-7   Prob. 10-3.   Speed of cam is 200 rpm.
10-8   Prob. 10-4.   Speed of cam is 240 rpm.

The following problems are designed to fit on an $8\frac{1}{2} \times 11$ in. sheet with the $8\frac{1}{2}$-in. dimension in the horizontal direction.   The center of the cam should be located at the center of the sheet.

10-9   Lay out the cam profile for the cam in Fig. P10-9.   Use the displacement diagram from Prob. 10-1.   Measure the maximum pressure angle, label it ϕ_{max}, and record its value.

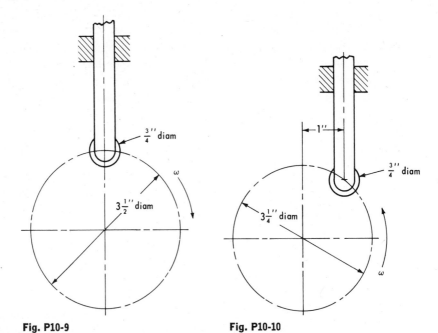

Fig. P10-9   Fig. P10-10

10-10 Lay out the cam profile for the cam in Fig. P10-10. Use the displacement diagram from Prob. 10-2. Measure the maximum pressure angle, label it $\phi_{\max}$, and record its value.

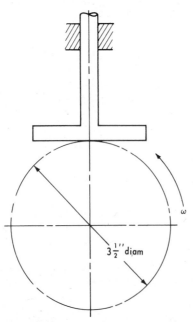

Fig. P10-11

10-11 Lay out the cam profile for the cam in Fig. P10-11. Use the displacement diagram from Prob. 10-3. Measure the maximum distance the contact point shifts to left or right of the follower centerline, label it δ_{max}, and record its value.

10-12 Lay out the cam profile for the cam in Fig. P10-12. Use the displacement diagram from Prob. 10-2. Measure the maximum distances the contact point shifts to the left and right of the point at which the follower contacts the base circle. Label these lengths δ_{max} (left) and δ_{max} (right) and record their values.

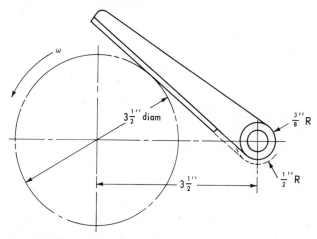

Fig. P10-12

11

ROLLING CONTACT

11-1 INTRODUCTION

It is of importance to study members having rolling contact because such members can be used to transmit power, to produce desired motions, and because of their use in ball and roller bearings.   Further, a study of rolling contact is basic to the development of the theory of gears.

11-2 CONDITIONS FOR ROLLING

In Sec. 2-15 it was shown that rolling contact exists in a direct-contact mechanism only if the linear velocities of the bodies at their point of contact are identical.   This requires that the point of contact lie on the line of centers. In Sec. 4-6 it was shown that the instant center for the bodies lies at their point of contact.

11-3 ROLLING CYLINDERS

Rolling external cylinders are shown in Fig. 11-1 and an external and internal cylinder in Fig. 11-2.   In these figures either body 2 or 3 can be the driver.

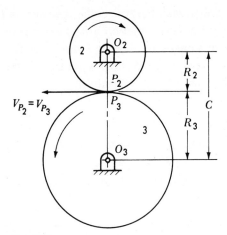

Fig. 11-1

Positive drive exists if motion of the driving link compels the driven to move. In Sec. 2-16 it was shown that for positive drive the common normal through the point of contact must not pass through either or both of the centers of rotation. In Figs. 11-1 and 11-2 positive drive does not exist because the common normal through the point of contact passes through the centers of rotation O_2 and O_3. In these figures motion is transmitted from one body to the other only if there is sufficient friction at the contacting surfaces. Hence these mechanisms are called *friction drives*. Friction drives are desirable in some machines and undesirable in others. In an automobile engine a friction drive cannot be used to drive the camshaft. A friction drive is desirable in some applications because if an over-

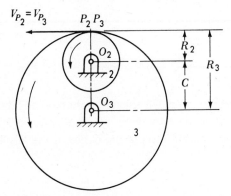

Fig. 11-2

load comes on the shaft of the driven member, the bodies can slip and thus prevent damage to the parts. In general, friction drives are not used to transmit large forces because this necessitates high bearing loads.

In Sec. 2-15 it was shown that for bodies having rolling contact, the angular-velocity ratio is inversely as the contact radii

$$\frac{\omega_2}{\omega_3} = \frac{R_3}{R_2} \tag{11-1}$$

If the center distance C and the angular-velocity ratio are known, the radii of the cylinders can be found as follows. If the cylinders are to rotate in opposite directions, external cylinders must be used (Fig. 11-1) and

$$C = R_2 + R_3$$

or

$$R_3 = C - R_2$$

Substituting this in Eq. (11-1) gives

$$\frac{\omega_2}{\omega_3} = \frac{C - R_2}{R_2} = \frac{C}{R_2} - 1$$

Then

$$\frac{C}{R_2} = \frac{\omega_2}{\omega_3} + 1$$

and

$$R_2 = \frac{C}{(\omega_2/\omega_3) + 1} \tag{11-2}$$

If the cylinders are to rotate in the same direction, an external and internal cylinder must be used (Fig. 11-2). Following the same procedure as for external cylinders, we obtain

$$R_2 = \frac{C}{(\omega_2/\omega_3) - 1}$$

11-4 ROLLING CONES

Cones can be used to transmit motion between shafts whose axes intersect. If the shafts are to rotate in opposite directions, external cones (Fig. 11-3) must be used. Let BP and CP be the radii of the cones at the large end. Then for rolling contact

$$\frac{\omega_2}{\omega_3} = \frac{CP}{BP}$$

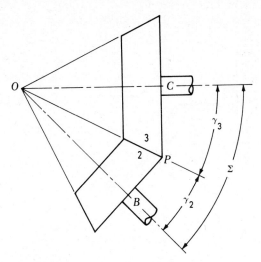

Fig. 11-3

For rolling contact at all points along line PO the ratio of the radii at these points must be the same as at the large ends of the cones. Hence the cones must have a common apex O.

The usual design problem is to determine the cone angles γ_2 and γ_3 when the angle between shafts Σ and the velocity ratio are specified.

The cone angles can be expressed as

$$\sin \gamma_2 = \frac{BP}{OP} \quad \text{and} \quad \sin \gamma_3 = \frac{CP}{OP}$$

Then

$$\frac{\sin \gamma_2}{\sin \gamma_3} = \frac{BP}{CP} = \frac{\omega_3}{\omega_2} = \frac{\sin (\Sigma - \gamma_3)}{\sin \gamma_3}$$

This can be written

$$\frac{\sin \Sigma \cos \gamma_3 - \cos \Sigma \sin \gamma_3}{\sin \gamma_3} = \frac{\omega_3}{\omega_2}$$

Dividing numerator and denominator by $\cos \gamma_3$ and solving for $\tan \gamma_3$ gives

$$\tan \gamma_3 = \frac{\sin \Sigma}{\omega_3/\omega_2 + \cos \Sigma} \tag{11-3}$$

If the shafts are to rotate in the same direction, an external and internal cone (Fig. 11-4) must be used. For this case the relationship is

$$\tan \gamma_3 = \frac{\sin \Sigma}{\omega_3/\omega_2 - \cos \Sigma} \tag{11-4}$$

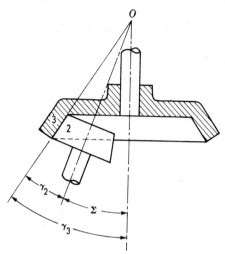

Fig. 11-4

The cone angles can also be determined graphically. In Fig. 11-5 external cones whose shafts form an angle Σ are to be constructed such that

$$\frac{\omega_2}{\omega_3} = \frac{5}{3}$$

Since

$$\frac{\omega_2}{\omega_3} = \frac{R_3}{R_2}$$

then

$$\frac{R_3}{R_2} = \frac{5}{3}$$

At two convenient points A and B erect perpendiculars to the respective shafts. Lay off five units on the perpendicular to shaft 3 and three units on the perpendicular to shaft 2 and draw lines X-X and Y-Y parallel to lines $O2$ and $O3$. These will intersect at P, thus giving the required cone angles.

In Fig. 11-6 the construction for an external and internal cone is shown for a ratio

$$\frac{\omega_2}{\omega_3} = \frac{7}{3}$$

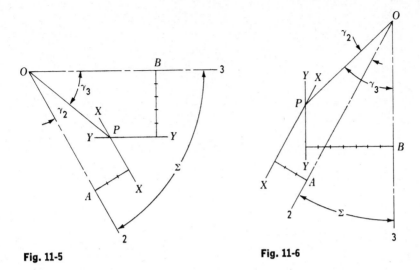

Fig. 11-5   **Fig. 11-6**

11-5 ROLLING HYPERBOLOIDS

In Fig. 11-7 the two contacting surfaces are hyperboloids. A hyperboloid can be generated by rotating a straight line about an axis which it does not intersect and to which it is not parallel.   In the figure, straight line A-A is rotated about axis B-B holding R_2 and ψ_2 constant to generate the hyperboloid 2.   By rotating line A-A about axis C-C holding R_3 and ψ_3 constant, hyperboloid 3 is generated.   Hence line A-A lies on the surfaces of both hyperboloids and is their common surface element.   The minimum circles of radii R_2 and R_3 are called the *gorge circles*.   When two hyperboloids as shown are rotated about their axes, they will have rolling contact in a direction normal to their common surface element, but they will slide on one another along the element.   Portions of rolling hyperboloids are used for the pitch surfaces of hypoid gears.

11-6 ROLLING ELLIPSES

In Fig. 11-8a, e and f are identical ellipses.   Points O_2, B, O_4, and C are the foci.   The major axes EP and PH are made equal to the distance between centers O_2 and O_4.   A property of the ellipse is that the sum of the distances

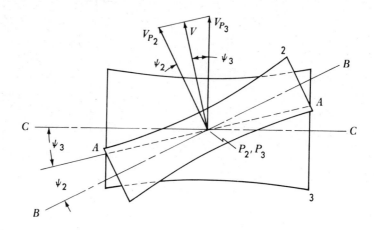

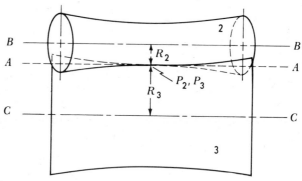

Fig. 11-7

from the foci to a point on the profile is equal to the major axis. Hence $O_2P_2 + BP_2 = O_4P_4 + CP_4 = $ major axis $= O_2O_4$.

If the ellipses are positioned so that arc $PP_2 = $ arc PP_4, then since the ellipses are identical, $O_2P_2 = CP_4$ and $BP_2 = O_4P_4$. Thus

$$O_2P_2 + O_4P_4 = O_2O_4$$

and the point P will lie on the line of centers O_2O_4 for all positions of the ellipses. Hence the ellipses will roll on one another.

In Fig. 11-8b the ellipses are shown with 2 rotated an angle θ_2. Then 4 will have rotated an angle θ_4. If one of the ellipses rotates at a constant angular velocity, the other will have a variable angular velocity.

If the ellipses are replaced by the four-bar linkage O_2BCO_4 as shown in Fig. 11-8*b*, then cranks O_2B and O_4C will have the same relative motion as the ellipses 2 and 4, respectively. This is because length $BP_2 + P_4C$ is a constant, and link BC always intersects the line of centers O_2O_4 where the ellipses are in contact. Thus in Fig. 11-8*b*, point P_2P_4 is the common instant center 24 for either the rolling ellipses or the four-bar linkage.

Rolling ellipses give positive drive for only 180° of rotation. Thus in Fig. 11-8*a*, if 4 is the driver and rotates counterclockwise, then 2 will be compelled to rotate only for the first 180° of rotation of 4. To assure positive drive for all positions, gear teeth can be cut on the ellipses. They are then called *elliptical gears*. Elliptical gears have been used for quick-return mechanisms.

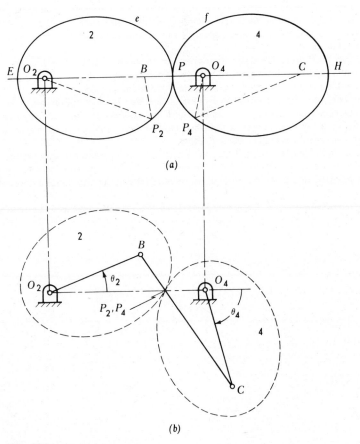

(a)

(b)

Fig. 11-8

11-7   GENERAL CASE OF ROLLING CURVES

Rolling curves can be designed having profiles which will give a desired angular displacement relationship between the driver and follower.   The procedure is discussed in Chap. 14, which deals with function generators.

PROBLEMS

11-1   Power is to be transmitted from one shaft to another which is parallel.   The shafts are 12 in. apart and are to rotate in opposite directions with an angular-velocity ratio of 1.5.   Determine the diameters of the rolling cylinders.

11-2   Same as Prob. 11-1 except the shafts are to rotate in the same direction.

11-3   Power is to be transmitted by means of two rolling cones whose shafts intersect at 45°.   The shafts are to rotate in opposite directions with an angular-velocity ratio of 1.5, and the maximum diameter of the larger cone is to be 8 in.   Determine graphically the cone angles γ_2 and γ_3 and the maximum diameter of the smaller cone.

11-4   Same as Prob. 11-3 except the cones are to rotate in the same direction.

11-5   Same as Prob. 11-3 except obtain the results analytically.

11-6   Same as Prob. 11-4 except obtain the results analytically.

11-7   Power is to be transmitted by two equal ellipses which roll on one another.   The major axis is equal to 4 in., and the minor axis equals $2\frac{1}{2}$ in.   (a) Determine the distance from the center of the ellipse to the foci.   (b) Determine the maximum and minimum angular-velocity ratios.

11-8   Two equal ellipses are to be used to transmit power between two parallel shafts which are 10 in. apart.   If the maximum and minimum angular-velocity ratios are to be 7 and 0.143, determine the values of the major and minor axes required.

12

GEARS

12-1 INTRODUCTION

In the preceding chapter mechanisms consisting of bodies in rolling contact were discussed. The power which can be transmitted by rolling bodies is limited by the friction which can be developed at their surfaces. If the load is excessive, slippage occurs and in order to provide positive drive, teeth are placed on the contacting members. The resulting members are then called *gears*.

Gears are used to transmit motion from a rotating shaft to another which rotates, or from a rotating shaft to a body which translates and which can be considered as rotating about an axis at infinity.

In this chapter we shall consider only those gears which provide a constant angular-velocity ratio. With exception to worm gears, the gears discussed will be equivalent to rolling bodies. For example, spur gears (Fig. 12-1) are used to transmit power between parallel shafts. A pair of spur gears is again shown in Fig. 12-2. The gears in Fig. 12-2*a* give the same motion to the shafts as the pair of equivalent rolling cylinders in Fig. 12-2*b*.

Fig. 12-1.   *(Boston Gear Works.)*

12-2   FUNDAMENTAL LAW OF GEARING

In Fig. 12-3 a direct-contact mechanism is shown, and line N-N, the common normal to the contacting surfaces, intersects the line of centers O_2O_3 at point P.   In Sec. 2-12 it was shown that the angular-velocity ratio for bodies 2 and 3 is inversely as the segments into which the normal cuts the line of centers.   That is

$$\frac{\omega_2}{\omega_3} = \frac{O_3P}{O_2P}$$

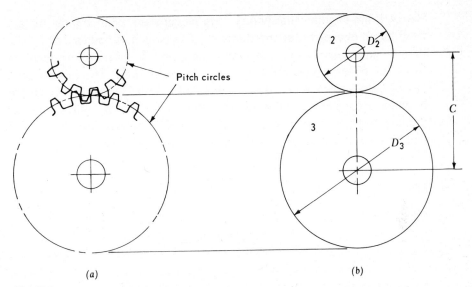

Fig. 12-2

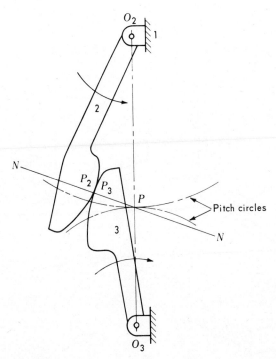

Fig. 12-3

Hence, if the angular-velocity ratio is to be constant, point P must remain fixed for all phases of the mechanism. The motion of bodies 2 and 3 will then be equivalent to two imaginary rolling circles which are in contact at point P. These circles are known as the *pitch circles*. The fundamental law of gearing for circular gears states that in order for a pair of gears to transmit a constant angular-velocity ratio, the shape of their contacting profiles must be such that the common normal passes through a fixed point on the line of centers. This is point P, the pitch point.

12-3 NOMENCLATURE

Spur gears have pitch surfaces which are cylindrical in form; they are used to transmit power between parallel shafts. Their teeth are straight and are parallel to the axes. Spur gears are the simplest type of gears and hence we shall study them first. Many of the definitions and nomenclature for spur gears (Fig. 12-4) are basic to all types of gears.

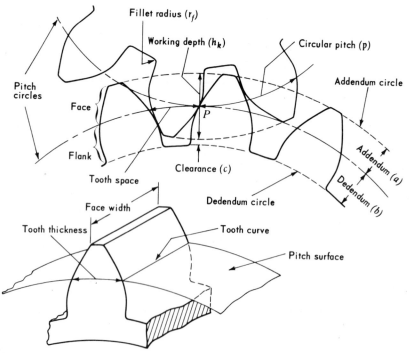

Fig. 12-4

The *pitch diameter D* is the diameter of the pitch circle.   The pitch circle is a theoretical circle upon which all computations are made.

The *pitch surface* is a cylinder whose diameter is the pitch diameter.

The *circular pitch p* is the distance from a point on one tooth to the corresponding point on the next tooth measured along the pitch circle.

The *diametral pitch P* is the ratio of the number of teeth on a gear to its pitch diameter in inches.   Let N be the number of teeth and D the diameter of the pitch circle.   Then

$$P = \frac{N}{D} \tag{12-1}$$

Since circular pitch is

$$p = \frac{\pi D}{N} \tag{12-2}$$

then from these last two equations

$$pP = \pi \tag{12-3}$$

This equation can be used to determine either pitch if the other is known. Though either pitch is a measure of tooth size, diametral pitch is the one which is usually used.

The *addendum a* is the radial distance from the pitch circle to the addendum circle or outside circle.

The *dedendum b* is the radial distance from the pitch circle to the dedendum circle or root circle.

The *working depth* h_k is the depth of engagement of a pair of gears; that is, it is the sum of their addendums.

The *whole depth* h_t is the full depth of a gear tooth and is the sum of its addendum and dedendum.

The *clearance c* is the amount by which the dedendum of a gear exceeds the addendum of its mating gear.

The *fillet* is the concave curve where the bottom of the tooth joins the dedendum circle.

The *fillet radius* r_f is the radius of the fillet curve.   In generated teeth, this is an approximate radius of curvature.

The *tooth thickness* is the thickness of tooth measured along the pitch circle.

The *width of tooth space* is the width of the space between teeth measured along the pitch circle.

The *center distance C* is the distance from center to center for two mating gears (see Fig. 12-2).   Thus

$$C = \frac{D_2 + D_3}{2}$$

(12-4)

Backlash is the amount by which the width of tooth space on a gear exceeds the tooth thickness on the mating gear, measured along the pitch circles.   Theoretically, backlash should be zero, but in practice some back-lash must be allowed to prevent jamming of the teeth due to tooth errors and thermal expansion.   Unless otherwise stated, throughout this text zero backlash will be assumed.   For a set of mating gears, backlash can be provided by either mounting the gears at a center distance greater than the theoretical or by feeding the cutter deeper than standard.

The *pinion* is the smaller of two gears in mesh.

The *gear* is the larger of two gears in mesh.

The angular-velocity ratio for a pair of spur gears is inversely as the diameters of their pitch circles or inversely as their numbers of teeth (see Fig. 12-2).   Thus

$$\frac{\omega_2}{\omega_3} = \frac{D_3}{D_2} = \frac{N_3}{N_2}$$

(12-5)

The *gear ratio m_G* is the ratio of the larger to the smaller number of teeth on a pair of gears.

12-4   INVOLUTE GEAR TEETH

Any two mating tooth profiles which satisfy the fundamental law of gearing as explained in Sec. 12-2 are called *conjugate profiles*.   Almost any reasonable curve could be selected for the profile on a given gear, and then the con-jugate profile for the mating gear could be determined by applying the con-dition that the common normal must pass through a fixed point on the line of centers.   Though many tooth shapes are possible, the cycloid and invo-lute have been standardized.   The cycloid was used first and still is used in clocks and watches.   The involute has several advantages, the most important of which are its ease of manufacture and the fact that the center distance for a pair of involute gears can be varied without changing the velocity ratio.   An exhaustive treatment of involute gearing will not be given here, but the fundamentals required for the elementary problems usually encountered in the design of machinery will be presented.   Further, sufficient background will be provided to enable the student to study more advanced works on gearing.

How involute gear teeth satisfy the law of gearing can be understood by considering Fig. 12-5. The figure shows a string having its end points attached to two rotating cylinders. Let N represent a fixed point or knot on the string. Then if the string is kept taut as the disks rotate, point N will trace curve AB on the upper disk and curve CD on the lower disk. These curves are known as *involute curves;* the circles are called the *base circles* and the string is known as the *generating line.* Let point E be the point of tangency for the string and disk 2. Since the disk is rolling on the string, point E is their common instant center and line EN is the radius of rotation for point N as it describes path AB. Because the motion of point N is always perpendicular to its radius of rotation, the string will be normal to curve AB for all positions as the knot moves along the curve. This is an important property of the involute, that the generating line is perpendicular to the involute curve for all positions. It can be seen that the same remarks apply to the string and involute curve CD. Furthermore, line EF maintains a fixed position. Thus if the involute curves AB and CD are used for mating tooth profiles, their common normal EF will pass through a fixed point P on the line of centers O_2O_3, and the law of gearing will be satisfied.

12-5 CONSTRUCTION OF AN INVOLUTE

We often want to draw an involute curve so that it will pass through a given point. This is shown in Fig. 12-6. Let it be assumed that we have the base circle and that we wish to construct an involute curve passing through point P. Through P a tangent is drawn to the circle. This locates point Q. The line PQ is then divided into some number of equal lengths, and these same lengths are laid off on the circle to the left and right of Q. Next, tangents are drawn to the points on the circle, and divisions of the same length are laid off along them to determine points on the involute.

12-6 INVOLUTE NOMENCLATURE

A pair of involute gears in mesh is illustrated in Fig. 12-7. The upper gear (gear 2) is the pinion and is the driver. The pinion rotates clockwise and the lower gear (gear 3) rotates counterclockwise. Point P is the pitch point

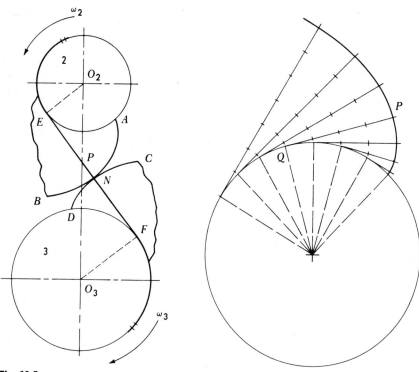

Fig. 12-5   Fig. 12-6

and line EF represents the string wrapped around the base circles. The point of contact for any two teeth is always along this line. If we were to neglect friction between the contacting tooth surfaces, then the force which the driving gear exerts on the driven would be along this line, which is called the *line of action*. The angle which the line of action makes with a perpendicular to the line of centers O_2O_3 is called the *pressure angle ϕ*. Though in practice this angle is called the pressure angle, the true pressure angle deviates from it somewhat because of the friction force. The pressure angle is constant for involute gears, and from the figure we note that the angle made by the line of centers, with a line from the center of each gear to the point where the line of action is tangent to the base circles, is equal to the pressure angle.

If R_2 and R_3, which are the radii of the pitch circles, are known for a pair of gears, and if the pressure angle is known, then the radii of the base circles can be found by the following construction. Line O_2O_3 can be drawn and point P located.   A perpendicular can be drawn to line O_2O_3, passing through P.   From this perpendicular the pressure angle ϕ can be laid off, and the line of action can be drawn.   Then by dropping a perpendicular from the center of each gear to the line of action, r_2 and r_3, which are the radii of the base circles, are determined.   From the figure it can be seen that the radius of the base circle for either gear can be found by the following equation

$$r = R \cos \phi \tag{12-6}$$

where r and R are the radii of the base circle and pitch circle, respectively.

The *base pitch*, denoted by p_b, is defined as the distance from a point on one tooth to the corresponding point on the next tooth measured along

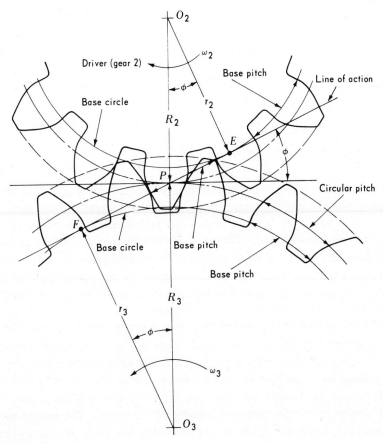

Fig. 12-7

the base circle. It is also the normal distance between the corresponding sides of two adjacent teeth as shown in the center of the figure. For this reason it is also sometimes called the *normal pitch*. The base pitch can be found by dividing the circumference of the base circle by the number of teeth. Since the circular pitch is equal to the circumference of the pitch circle divided by the number of teeth, then from Eq. (12-2) we can obtain the following relationship between base pitch and circular pitch.

$$p_b = p \cos \phi \qquad (12\text{-}7)$$

The base circle and the base pitch are properties of a single gear and are fixed once the gear is made. However, the diameters of the pitch circles and the circular pitch are determined by the center distance at which a pair of gears in mesh is mounted. From the Fig. 12-7 it can be seen that the pressure angle is also determined by the center distance. The base pitch is an important property of involute gears and must be the same for two gears if they are to mesh correctly.

The tooth profile from the base circle to the dedendum circle is usually drawn as a radial line with a fillet used at the dedendum circle in order to relieve stress concentration at that point.

12-7 INVOLUTE–GEAR–TOOTH ACTION

Portions of a pair of involute gears in mesh are shown in Fig. 12-8, where a pair of mating teeth is shown in three phase positions. The teeth first come into contact at point A, where the addendum circle of the driven gear cuts the line of action. Contact follows the line of action through point P, and contact ceases at point B, where the addendum circle of the driving gear cuts the line of action. Line AB is called the path of the point of contact and its length is the length of the path of contact. Point C is the intersection of the tooth profile on gear 2 with its pitch circle at the beginning of contact, and point G is the same point on the profile when the tooth is at the end of contact. Points D and H are the corresponding points for gear 3. Arcs CPG and DPH are the arcs on the pitch circles through which the mating tooth profiles move as they pass from the initial to the final point of contact. These arcs are known as the *arcs of action*. Since the pitch circles roll on one another, these arcs are equal. The angles which subtend these arcs are θ_2 and θ_3 and are called the *angles of action*. These angles are not equal unless the gears have equal pitch diameters. The

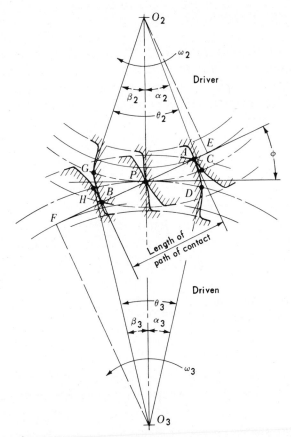

Fig. 12-8

angles of action are divided into two parts called the *angles of approach* and *angles of recess* and are shown in Fig. 12-8 as α and β with the subscripts pertaining to each gear. The *angle of approach* is defined as the angle through which a gear rotates from the instant a pair of teeth come into contact until the teeth are in contact at the pitch point. The *angle of recess* is the angle through which a gear rotates from the instant the teeth are in contact at the pitch point until contact is broken. In general the angle of approach is not equal to the angle of recess. Experience has indicated that gear-tooth action is smoother in recess than in approach. The values of the angles of approach and recess can be determined graphically from a layout of the gears or they can be computed by the method explained in Sec. 12-9.

12-8 INVOLUTE RACK AND PINION

An involute rack and pinion are shown in Fig. 12-10. A *rack* is a portion of a gear having an infinite pitch diameter; thus its pitch circle is a straight line called the *pitch line*. The line of action is tangent to the base circle at infinity; hence the involute profile of the rack is a straight line and is perpendicular to the line of action.

12-9 CONTACT RATIO

The *contact ratio* is defined as the average number of pairs of teeth which are in contact. This can be found by noting how many times the base pitch as shown along the line of action in Fig. 12-7 fits into the length of the path of contact AB shown in Fig. 12-8. Thus the contact ratio (m_c) can be expressed as follows:

$$m_c = \frac{\text{length of path of contact}}{\text{base pitch}}$$

$$= \frac{AB}{p_b} \tag{12-8}$$

The length AB can be determined graphically from a layout as shown in Fig. 12-9, or it can be computed from the following relationships which apply to the figure:

$$\beta_3 = \sin^{-1}\left(\frac{PO_3 \sin \alpha}{AO_3}\right) \tag{12-9}$$

$$\theta_3 = 180° - (\alpha + \beta_3) \tag{12-10}$$

$$AP = \frac{AO_3 \sin \theta_3}{\sin \alpha} \tag{12-11}$$

$$\beta_2 = \sin^{-1}\left(\frac{PO_2 \sin \alpha}{BO_2}\right) \tag{12-12}$$

$$\theta_2 = 180° - (\alpha + \beta_2) \tag{12-13}$$

$$PB = \frac{BO_2 \sin \theta_2}{\sin \alpha} \tag{12-14}$$

The contact ratio usually is not a whole number. If the ratio is 1.6, it does not mean that there are 1.6 pairs of teeth in contact. It means that there are alternately one pair and two pairs of teeth in contact, and on a

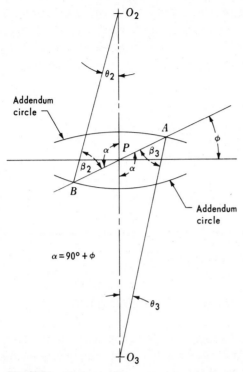

$$\alpha = 90° + \phi$$

Fig. 12-9

time basis the average is 1.6. The theoretical minimum value for the contact ratio is 1.00; that is, there must always be at least one pair of teeth in contact for continuous action. In practice 1.4 has been recommended as a minimum. The larger the contact ratio, the quieter the gears will operate. From Fig. 12-9 it can be seen that the length of the path of contact can be increased, and hence the contact ratio increased, by increasing the addendums of the gears. However, there are limitations to the amount the addendums can be increased, as will be discussed in the next section.

The angles of approach and recess may be computed as follows. Consider Fig. 12-8. Length AP is equal to the length of arc along the base circle of gear 2, which extends between the tooth profiles of this gear, which pass through points A and P. The angle subtended by this arc is equal to the angle of approach α_2. Hence

$$\alpha_2 = \frac{AP}{r_2} \tag{12-15}$$

where r_2 is the base-circle radius of gear 2. Length PB is equal to the length of arc along the base circle of gear 2, which extends between the tooth profiles of this gear, which pass through points P and B. The angle subtended by this arc is equal to the angle of recess β_2. Hence

$$\beta_2 = \frac{PB}{r_2} \qquad (12\text{-}16)$$

Similarly, the angles of approach and recess for gear 3 are

$$\alpha_3 = \frac{AP}{r_3} \qquad (12\text{-}17)$$

and

$$\beta_3 = \frac{PB}{r_3} \qquad (12\text{-}18)$$

where r_3 is the base-circle radius of gear 3.

12-10 INVOLUTE INTERFERENCE AND UNDERCUTTING

In Fig. 12-10 the addendum of a rack has been chosen so that the first point of contact will occur at point E, which is the point of tangency for the line of action and the base circle of the pinion. The involute profile on the pinion cannot extend inside the base circle. From the base circle inward the pinion profile is drawn as a radial line. Then the maximum length of the line of approach is EP. The maximum addendum which should be used on the rack is a. In order to investigate what happens if the addendum of the rack were larger, it is shown as a'. Then if the pitch circle of the pinion and the pitch line of the rack were rolled to the right, the rack-and pinion-tooth positions would be as shown by the dotted profiles, and it is found that the rack tooth overlaps or interferes with the pinion tooth. If interference occurs when the pinion teeth are cut with a hob as explained in Sec. 12-16, then the hob undercuts the teeth as shown in Fig. 12-10. This weakens the teeth and removes a part of the involute profile, which shortens the path of contact.

There can be no interference due to a large addendum on the pinion since the point of tangency of the line of action with the base circle for the rack lies to the left at infinity. The maximum length for the path of recess is PB' and results when the addendum of the pinion is increased to the extent that the pinion tooth becomes pointed.

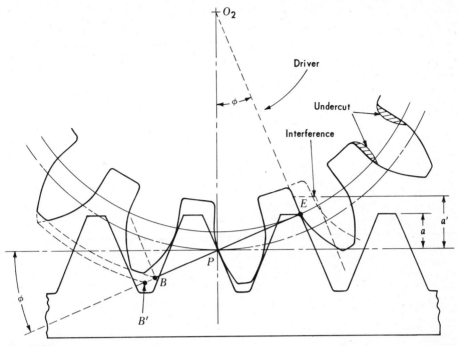

Fig. 12-10

If a rack will mesh with a pinion without interference, then any finite external gear having the same addendum as the rack will mesh with the pinion without interference. This can be seen from Fig. 12-10. The addendum circle of any finite gear will intersect the line of action to the left of E.

12-11   CHECKING FOR INTERFERENCE

We can check any pair of involute gears to determine whether interference occurs by considering Fig. 12-11. Points E and F are the points of tangency for the line of action and the base circles and are known as the *interference points*. If either point A or B lies outside of points E and F, respectively, then there is interference.

In Sec. 12-10 we considered the conditions for interference of an involute rack and pinion. From a consideration of Fig. 12-10 we note that for a given rack the possibility of interference will be greater the smaller the

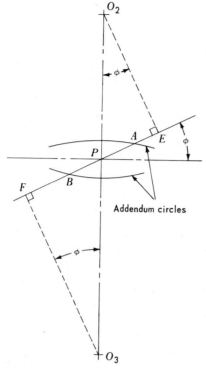

Fig. 12-11

pinion. That is, the initial point of contact, which is the point where the addendum line of the rack intersects the line of action, is more likely to fall to the right of the interference point E the smaller we make the pinion.

Since the pinion and rack must have the same pitch, the problem of determining the smallest pinion which can be used without interference is the same as that of determining the minimum number of teeth which can be used on the pinion. The limiting condition or smallest pinion which can be used is shown in Fig. 12-12, where the pinion shown has its interference point at the initial point of contact A. The addendum a of the rack and the pressure angle ϕ are known. Then the minimum number of teeth which can be used on the pinion can be calculated as follows. From the figure

$$\sin \phi = \frac{AP}{R}$$

where R is the radius of the pitch circle of the pinion.   Also

$$\sin \phi = \frac{a}{AP}$$

Multiplying these two equations, we obtain

$$\sin^2 \phi = \frac{a}{R}$$

But the addendum dimension a, as shown in Table 12-2, is a constant k divided by the diametral pitch P.   Hence

$$\sin^2 \phi = \frac{k}{PR}$$

However,

$$P = \frac{N}{D} = \frac{N}{2R}$$

where N is the number of teeth.   Thus

$$\sin^2 \phi = \frac{2k}{N}$$

and

$$N = \frac{2k}{\sin^2 \phi} \tag{12-19}$$

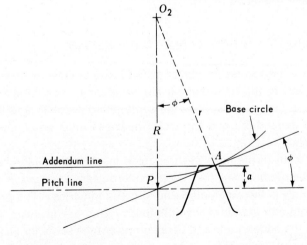

Fig. 12-12

Table 12-1

	14½° FULL DEPTH	20° FULL DEPTH	25° FULL DEPTH	20° STUB
N	32	18	12	14

From this equation the smallest number of teeth for a pinion that will mesh with a rack without interference can be computed for any given standard tooth system. These are shown in Table 12-1 for four common systems. When the value of N obtained by the equation is not a whole number, then since a gear cannot have a fraction of a tooth, the next-largest whole number must be used.

It is sometimes desired to find the smallest pinion that will mesh with a given gear without interference. In Fig. 12-13, O_3P is the pitch-circle radius, O_3A is the addendum circle radius, and ϕ the pressure angle for the given gear. The base-circle radius of the smallest pinion that will mesh without interference is shown as line $O_2'A$, which has been drawn perpendicular to line PA. Any pinion having a larger base-circle radius O_2E will not have interference. From the figure we can see that the pitch radius PO_2' can be computed as follows.

$$PO_2' = \frac{AP}{\sin \phi} \tag{12-20}$$

where length AP is obtained by Eq. (12-11).

12-12 STANDARD INTERCHANGEABLE TOOTH FORMS

A set of gears is interchangeable when any two gears selected from the set will mesh and satisfy the fundamental law of gearing. For interchangeability all gears of the set must have the same pitch, pressure angle, and addendum and dedendum, and the tooth thickness must equal one-half the circular pitch.

Standard tooth forms have been adopted in order that interchangeable gears will be readily available. Nonstandard gears are used in some applications mainly in the automotive and aircraft industries. The proportions of four common standard involute-tooth forms are given in Table 12-2. Note that the addendums and dedendums are the same for all gears in each system and are expressed as a function of the diametral pitch.

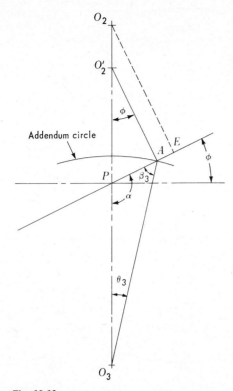

Fig. 12-13

The cutting tools used in cutting the teeth are specified in terms of the diametral pitch. Commonly used diametral pitches are as follows:

1, 1¼, 1½, 1¾, 2, 2¼, 2½, 2¾, 3, 3½, 4, 5, 6, 7, 8, 9, 10, 12, 14, 16, 18, 20, 22, 24, 26, 28, 30, 48, 64, 72, 80, 96, 120

Table 12-2

	14½° FULL DEPTH	20° FULL DEPTH	25° FULL DEPTH	20° STUB
Addendum	$1.000/P$	$1.000/P$	$1.000/P$	$0.800/P$
Dedendum	$1.157/P$	$1.250/P$	$1.250/P$	$1.000/P$
Clearance	$0.157/P$	$0.250/P$	$0.250/P$	$0.200/P$
Fillet radius (basic rack)	$0.209/P$	$0.300/P$	$0.300/P$	$0.300/P$

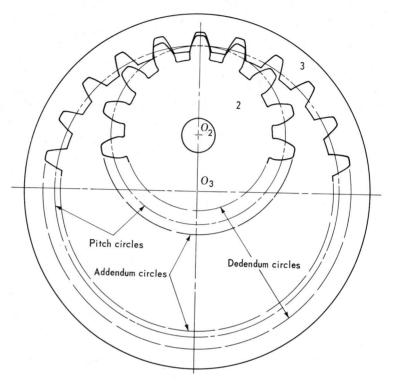

Fig. 12-14

The American Gear Manufacturers' Association (AGMA) recognizes the 20° and 25° full-depth systems as standards for new applications. The 14½° full-depth and 20° stub-tooth systems are obsolete, but are still used for replacement purposes.

The 20° full-depth system has several advantages when compared with the 25° full-depth system. The lower pressure angle gives a higher contact ratio which results in quieter operation and reduced wear. Further, because of the lower pressure angle, the normal force on the teeth for a given torque is less. This results in lower bearing loads.

The 25° full-depth system has the following advantages when compared with the 20° full-depth system. For a given circular pitch, the larger pressure angle results in teeth which are broader at the base and hence are stronger in bending. Another advantage is that fewer teeth may be used on the pinion without the pinion teeth being undercut.

12-13 INTERNAL INVOLUTE GEARS

An *internal* or *annular gear* is a gear having teeth cut on the inside of the rim. Figure 12-14 shows an involute internal gear and pinion.   The pinion is usually the driver.   The basic theory is the same as for external gears; the tooth action is illustrated in Fig. 12-15.   Point E is the point of tangency of the line of action and the base circle of the pinion and is the interference point for the pinion.   Note that the addendum of the gear must not extend below this point; otherwise the face of the gear tooth will interfere with the flank of the pinion tooth which is noninvolute from its base circle inward. Because the addendum of the gear must be shorter than those given in

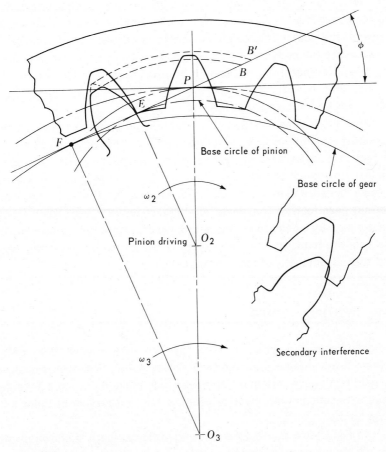

Fig. 12-15

Table 12-2, standard tooth proportions are not used on annular gears. As for external gears the initial point of contact occurs where the addendum circle of the driven gear cuts the line of action. In the figure the addendum of the gear has been made so that this occurs at point E. There is no interference point for the gear. This is because the entire tooth profile on the gear is an involute curve, and hence regardless of the size of the pinion addendum the face of the pinion tooth will always contact an involute profile on the gear. As for external gears, the final point of contact is where the addendum of the driver cuts the line of action and is at point B. The length of the line of action can be increased by increasing the addendum on the pinion. The limiting condition is reached when the pinion tooth becomes pointed, and then the final point of contact occurs at B'.

When too large a pinion is used with an internal gear of given size, the teeth will "foul" each other as a pinion tooth enters or leaves a tooth space on the gear. This condition, which is peculiar to internal gears, is shown in the figure and is called *secondary interference*. In order to prevent this, the difference in the number of teeth on the gear and pinion should be at least 12 for the 14½° full-depth system, 10 for the 20° full-depth system, and 8 for the 20° stub-tooth form.

The primary advantage of an internal gear-and-pinion drive is that it is more compact than an external gear drive. Other advantages are lower contact stresses because of a convex surface contacting a concave surface, relatively lower tooth-sliding velocities, and a greater length of contact being possible because there is no limit on the involute profile on the flank of the internal gear tooth. The beam strength of annular gear teeth is much greater than for a corresponding external gear. However, unless the pinion and gear are made of different materials, the pinion is always the weaker, and this advantage is of little consequence.

12-14 CYCLOIDAL GEARS

Cycloidal gears, though used in the past, are seldom used today because of their disadvantages in relation to involute gears. However, since cycloidal gear teeth have certain advantages which make them more suitable for some applications than involute gears, a brief discussion of them will be given here.

In Fig. 12-16 a pair of gears having cycloidal teeth is shown in mesh. The pitch circles and generating circles are in contact at the pitch point P.

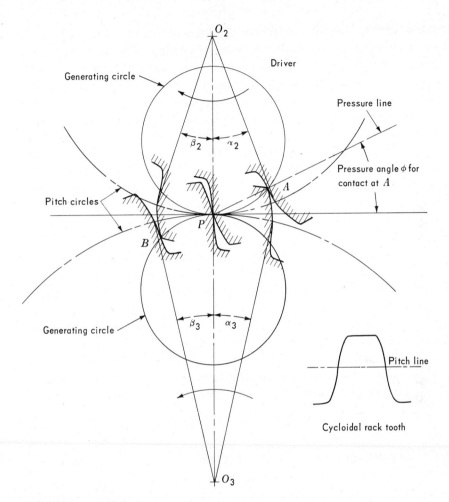

Fig. 12-16

It is to be remembered that the portion of the tooth profile which lies outside of the pitch circle is called the *face*, and the portion inside the pitch
circle is the *flank*. The cycloidal profile is formed by a point on a circle as
it rolls on the pitch circle. The rolling circle is called the *generating circle*,
and if the generating circle rolls on the outside of the pitch circle, it forms
the face of the tooth. When the generating circle rolls on the inside of the
pitch circle, it forms the flank of the tooth. The path of contact for the
meshing teeth is along the generating circles and is the arc AP followed by
arc PB. A normal to the contacting tooth profiles is the pressure line, and

it passes through the fixed point P for all phases of contact. Thus cycloidal-tooth profiles satisfy the fundamental law of gearing. The pressure angle is the angle which the pressure line makes with a perpendicular to the line of centers. The pressure angle decreases as the teeth move from their initial point of contact A, reaching a value of zero when they are in contact at P, and then increases as they move from P to their final point of contact B.

A cycloidal rack does not have straight-tooth profiles like an involute rack. In Fig. 12-16, if the lower gear were made infinite, its pitch circle would become a straight line called the *pitch line*. A cycloidal rack tooth is shown in the figure. When the upper generating circle rolls on the pitch line, it forms the face of the rack tooth, which is a convex curve. When the lower generating circle rolls on the pitch line, it forms the flank of the rack tooth, which is concave.

The disadvantages of cycloidal teeth are that they are difficult to man-ufacture and will transmit a constant angular-velocity ratio only if they are mounted at their theoretically correct center distance. Interference is not encountered with cycloidal gears, and this is their primary advantage. They are used extensively in clocks, watches, and certain instruments because pinions having low numbers of teeth (as low as 6 or 7) can be used to achieve a large reduction ratio without the problem of interference and the weakening effect of undercutting. Further, cycloidal teeth have less sliding action than involute teeth, and this results in less wear.

12-15   ADVANTAGES OF INVOLUTE TEETH

It was explained in Sec. 12-8 that an involute-rack profile is straight. Cutting tools and grinding wheels used in the manufacture of gears and cutters are straight, and this makes involute gears more economical to manufacture than other types.

Another advantage of involute gearing is that the center distance can be changed and yet the gears will transmit a constant angular-velocity ratio; i.e., they will satisfy the fundamental law of gearing. In Fig. 12-17 a pair of involute gears with centers O_2 and O_3 is shown in mesh, and P is the pitch point. A pair of teeth is shown in contact at point C, and the angular-veloc-ity ratio is equal to O_3P/O_2P. If the center of gear 3 is moved from O_3 to O_3', contact is at C'. The normal to the tooth profiles is tangent to the base cir-cles and intersects the line of centers at P'. Triangles O_2PE and O_3PF are

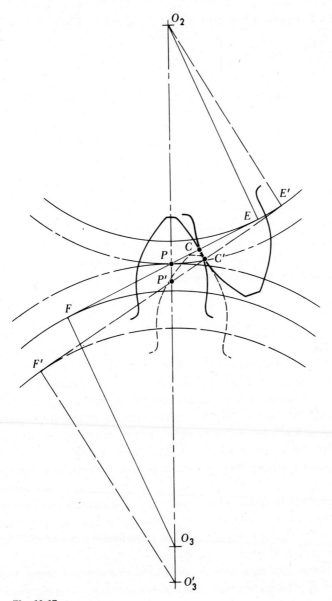

Fig. 12-17

similar.   Also triangles $O_2P'E'$ and $O_3'P'F'$ are similar.   Thus

$$\frac{O_3P}{O_2P} = \frac{FO_3}{EO_2} \quad \text{and} \quad \frac{O_3'P'}{O_2P'} = \frac{F'O_3'}{E'O_2}$$

But

$$F'O_3' = FO_3 \quad \text{and} \quad E'O_2 = EO_2$$

Hence

$$\frac{O_3P}{O_2P} = \frac{O_3'P'}{O_2P'}$$

The velocity ratio is not changed. An increase in the center distance increases the pressure angle and backlash and reduces the length of the path of contact. The base circle is a fixed property of a gear. The pitch circle is nonexistent until the gear is meshed with another gear. When a pitch circle is specified for a gear, the pressure angle must also be specified. This then determines the size of the base circle.

12-16   METHODS OF GEAR MANUFACTURE

Gear teeth can be cut on a milling machine using a form cutter having a profile which is the shape of the tooth space being cut. Accurate teeth can be produced in this manner only if a given cutter is used to cut a particular number of teeth on the gear blank. In practice a given cutter is used to cut gears having tooth numbers within a certain range. Gears produced by this method are extensively used, but because their tooth profiles are not the correct shape, they are suitable only for relatively low-speed operation.

Accurate involute gears are produced by the generating process. The generating process can be easily understood by imagining a gear blank made of clay with its pitch line rolling on the pitch line of a steel rack. Accurate teeth which would have conjugate action would be produced on the clay blank. If the steel rack were relieved to form a shaping cutter, then it could be used to cut accurate teeth in a steel gear blank. The cutter would be made to reciprocate parallel to the axis of the gear blank, while the blank is made to rotate so that its pitch circle will roll on the pitch line of the rack. A cutter of this type is known as a *rack-shaped cutter*. Clearance on the gear is produced by providing an extension on the addendum of the cutter. The primary advantage of the generating method of cutting gear

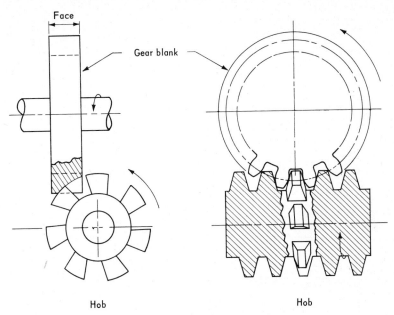

Face

Gear blank

Hob   Hob

Fig. 12-18

teeth is that a given cutter will cut teeth of a certain pitch and pressure angle regardless of the number of teeth to be cut.

The most common methods of generating gear teeth are *hobbing* and *shaping*. The hobbing process is the most rapid method and is illustrated in Fig. 12-18. The hob resembles a worm or screw with gashes cut axially in the threads to form cutting edges. The shape of the thread in the plane of the blank is that of a rack so that the hob acts like a rack-shaped cutter. As the hob and gear blank are made to rotate at the proper angular-velocity ratio, the hob is slowly fed parallel to the axis of the blank. By the time the hob has crossed the face width of the blank, the teeth will be fully cut, and the hob and blank will have made many revolutions.

A Fellows gear shaper uses a cutter (Fig. 12-19) which resembles a pinion except that the teeth are relieved to form cutting edges. At first neither the cutter nor blank are made to rotate; the cutter reciprocates across the face width of the blank and is slowly fed radially into the blank until its pitch circle comes tangent to the pitch line of the blank. The radial motion is then stopped. As the cutter continues to reciprocate, the cutter and blank are driven so that their pitch circles roll slowly on one another. By the time the blank makes a complete revolution, all the teeth will be cut.

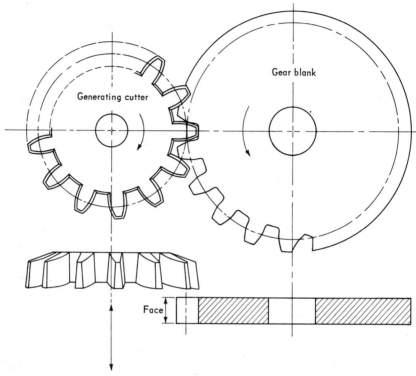

Gear blank

Generating cutter

Face

Fig. 12-19

12-17   UNEQUAL ADDENDUM GEARS

When a large gear is to mesh with a small gear, interference often occurs, and the pinion will be undercut. In order to avoid undercutting, a non-standard system of gears has been developed known as *unequal addendum gears*. The advantages of this system can be easily seen if we consider a 12-toothed 20° full-depth pinion and rack as shown in Fig. 12-20a. If the pinion teeth are generated with a rack cutter or hob, they will be undercut as shown and weakened. The path of contact extends from L to M. The rack tooth is too long by an amount y. In Fig. 12-20b the rack has been withdrawn by an amount y, and the pinion addendum increased by the same amount. The resulting pinion tooth is shown. The advantages gained are a stronger pinion tooth and an increase in the length of the path of contact to EN. Hence the contact ratio is increased. The base

circle, pitch circle, and pressure angle remain the same.   Unequal addendum gears are not interchangeable.

The method of using standard cutters to produce unequal addendum
gears is to decrease the radius of the gear blank so that interference will
be eliminated.   The gear teeth are then cut to the standard full depth.
The radius of the pinion blank is increased by the same amount that the
gear blank was decreased, and the teeth are cut to the standard full depth.

12-18   PARALLEL HELICAL GEARS

A *stepped gear*, Fig. 12-21, consists of two or more spur gears fastened together
as shown.   Each is advanced relative to the adjacent one by an amount
equal to the circular pitch divided by the number of gears.   When a pair of
conventional spur gears is in operation, the load comes upon the tip of the
tooth along the full-face width.   With stepped gears the load is applied
first to a portion of the face width, later to the next portion or step, etc.   As

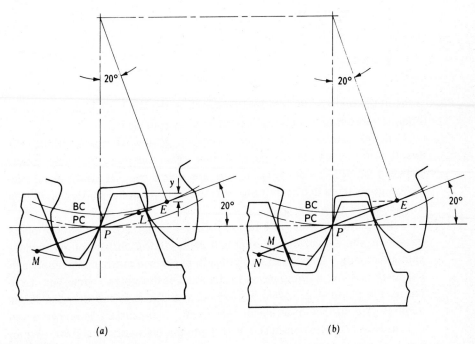

(a)   (b)

Fig. 12-20

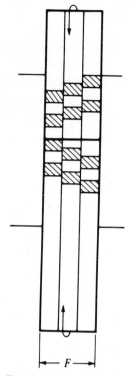

Fig. 12-21

a result the teeth come into contact with less impact.   Thus stepped gears operate more quietly and are smoother than spur gears.

If the number of steps on a stepped gear were made infinite, the result would be a helical gear. Helical gears used to transmit power between parallel shafts (Fig. 12-22a) are called *parallel helical gears*, and helical gears used to transmit power between nonparallel shafts (Fig. 12-22b) are called *crossed helical gears.*

If a plane is rolled on a base cylinder (Fig. 12-23a), a line in the plane parallel to the axis of the cylinder will generate the surface of an involute spur gear tooth.   Thus when a pair of spur gears is in mesh, contact between the teeth is along a line parallel to the axes of the gears.   However, if the line in the plane (Fig. 12-23b) is inclined to the axis of the cylinder, it will generate the surface of a helical gear tooth.   This surface is known as an *involute helicoid* and is composed of straight-line elements as shown.   When a pair of parallel helical gears is in mesh there is line contact between the teeth along one of these elements.   When spur gear teeth come into contact, there is contact across the entire face width, whereas with helical gears contact begins at one end of the tooth and progresses across the tooth. This is shown in Fig. 12-24, where the straight lines a-a, b-b, c-c, etc., are the successive lines of contact.   These lines are the elements of the involute helicoid shown in Fig. 12-23b.   This gradual contact across the teeth results in less impact loading, and thus helical gears operate more quietly than spur gears, have longer life, and are stronger.

Figure 12-25 shows the pitch helix, which is the helix formed by the intersection of the helical tooth and the pitch cylinder whose diameter is D.

(a)

(b)

Fig. 12-22.   *(Boston Gear Works.)*

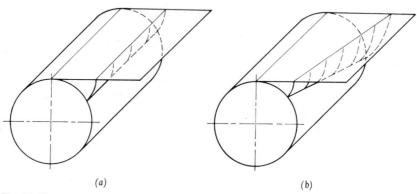

(a)   (b)

Fig. 12-23

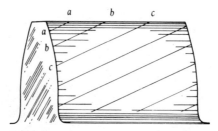

Fig. 12-24

F is the face width of the gear. Let us consider a horizontal plane passing through the axis of the cylinder. Then at the point where the helix pierces this plane, let a tangent to the helix be drawn as shown. The angle which the tangent makes with the horizontal plane is called the *helix angle* ψ.

The direction in which the teeth of a helical gear slope is known as the *hand* of the gear. To determine the hand of a helical gear, place it on its side as shown in Fig. 12-26. If the teeth slope upward to the right, it is called a *gear of right hand,* and if the teeth slope upward to the left it is called a *gear of left hand.* Helical gears connecting parallel shafts are of opposite hand.

In Fig. 12-27 the helical-tooth elements lying on the pitch cylinder are shown. ψ is the helix angle. p and P are the circular and diametral pitch in the plane of rotation, respectively, and are defined in the same manner as for spur gears. Thus

$$p = \frac{\pi D}{N} \qquad P = \frac{N}{D}$$

and

$$pP = \pi$$

(12-21)

In parallel helical gears the face width as shown in Fig. 12-27 is made large enough so that for a given helix angle the tooth advance is greater than the circular pitch. This provides overlapping action. That is, the leading end of one tooth will come into contact before the trailing end of the adjacent tooth goes out of contact. It can be seen from the figure that if the tooth advance were just equal to the circular pitch, then the face width would be equal to $p/\tan\psi$. In order to provide overlapping action, the AGMA† recommends that the limiting face width be increased by at least 15 percent, which results in the following equation:

$$F > \frac{1.15p}{\tan\psi} \qquad\qquad (12\text{-}22)$$

†American Gear Manufacturers Association.

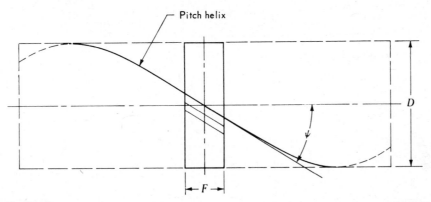

Fig. 12-25

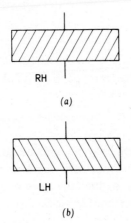

Fig. 12-26

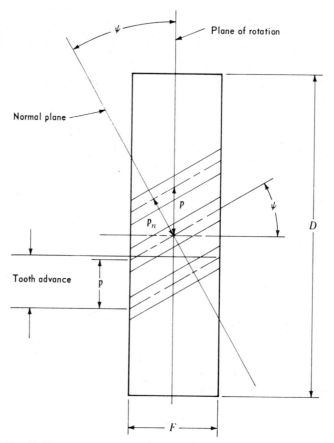

Fig. 12-27

The circular pitch in the normal plane is p_n and is the arc length along the pitch cylinder from a point on one tooth to the corresponding point on the next tooth in a plane normal to the helix. From the figure,

$$p_n = p \cos \psi \tag{12-23}$$

P_n is the diametral pitch in the normal plane and may be considered as the number of teeth per inch of diameter of an imaginary circle whose circumference is Np_n. Since p_n is smaller than p by the factor $\cos \psi$, and since P_n is larger than P by a factor $1/\cos \psi$, then from Eq. (12-21)

$$p_n P_n = \pi \tag{12-24}$$

Substituting Eqs.(12-24) and (12-21) into Eq. (12-23), we obtain

$$\frac{\pi}{P_n} = \frac{\pi}{P} \cos \psi$$

or

$$P = P_n \cos \psi \tag{12-25}$$

The same hob used to cut a spur gear is used to cut a helical gear. In order to cut the helical gear, the axis of the hob must be tilted an amount equal to the helix angle from the position used if it were cutting a spur gear.   Spur gear teeth cut with a hob have a standard value of P.   A helical gear cut with this same hob will have a P_n of the same value.   When a helical gear is to be cut with a hob, the pitch and pressure angle are specified in the normal plane.   Helical gears can also be cut with a gear shaper. The gear shaper used to cut a spur gear cannot be used to cut a helical gear.   The gear shaper used to cut a helical gear has helical teeth and resembles a helical pinion.   When the teeth are cut with a gear shaper, the pitch and pressure angle are specified in the plane of rotation.

The same equations which are used for spur gears give the angular-velocity ratio and center distance for a pair of parallel helical gears (Fig. 12-28).   Thus

$$\frac{\omega_2}{\omega_3} = \frac{D_3}{D_2} = \frac{N_3}{N_2} \tag{12-26}$$

and

$$C = \frac{D_1 + D_2}{2} \tag{12-27}$$

An undesirable feature of helical gears is that they produce end thrust as shown in Fig. 12-28.   This is usually not serious.   If ball or tapered roller bearings are used to carry the radial load, they will usually carry the thrust also.

Thrust is eliminated when herringbone gears are used.   In Fig. 12-29, a herringbone pinion is used on the lower front shaft.   A herringbone gear is equivalent to two helical gears of opposite hand side by side.

12-19 CROSSED HELICAL GEARS

Helical gears used to transmit power between nonparallel, nonintersecting shafts are called *crossed helical gears* and may be of the same or opposite hand as shown in Fig. 12-30.   The dashed line represents a tooth on the back

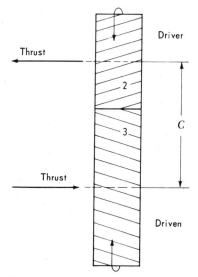

Fig. 12-28

Fig. 12-29. *(Link-Belt Company.)*

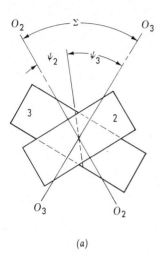

(a)

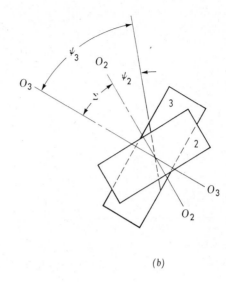

(b)

Fig. 12-30

of gear 2 and a tooth on the front of gear 3. If the gears are of the same hand as in Fig. 12-30a, then Σ, the angle between shafts, is

$$\Sigma = \psi_2 + \psi_3 \tag{12-28}$$

and for gears of opposite hand as in Fig. 12-30b,

$$\Sigma = \psi_3 - \psi_2 \tag{12-29}$$

where ψ_2 and ψ_3 are the helix angles.

A pair of parallel helical gears are of opposite hand, their pitch cylinders and teeth have line contact, and there is no sliding along the tooth elements. The individual gears of a crossed helical pair are the same as parallel helical gears. When helical gears are meshed on nonparallel shafts, their tooth action is different. Their pitch cylinders and teeth have point contact, and there is sliding along the elements of the teeth. In Fig. 12-31, the teeth are in contact at P_2, P_3. The dashed line through P_2, P_3 represents a tooth on the back of gear 2 and a tooth on the front of gear 3 in contact with it. V_{P_2} and V_{P_3} are the velocities, and they must have a common component V along the normal. The difference between V_{P_2} and V_{P_3} is the velocity with which the teeth slide on one another. Because of the point contact the power which can be transmitted is less than for parallel helical gears.

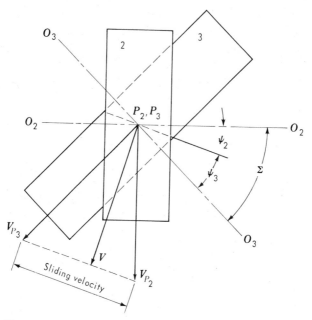

Fig. 12-31

The angular-velocity ratio for helical gears on nonparallel shafts is

$$\frac{\omega_2}{\omega_3} = \frac{N_3}{N_2} = \frac{D_3 P_3}{D_2 P_2} = \frac{D_3 P_{n_3} \cos \psi_3}{D_2 P_{n_2} \cos \psi_2}$$

where N_2 and N_3 = numbers of teeth

$\quad\quad P_2$ and P_3 = diametral pitches in plane of rotation

$\quad\quad P_{n_2}$ and P_{n_3} = diametral pitches in normal plane

Any pair of helical gears in mesh must have the same circular and diametral pitch in the normal plane. Hence the last equation becomes

$$\frac{\omega_2}{\omega_3} = \frac{N_3}{N_2} = \frac{D_3 \cos \psi_3}{D_2 \cos \psi_2} \tag{12-30}$$

It is to be noted from Eq. (12-30) that the angular-velocity ratio for helical gears on nonparallel shafts will vary inversely with their pitch diameters only if their helix angles are equal.

The center distance for crossed helical gears is

$$C = \frac{D_2 + D_3}{2} \tag{12-31}$$

12-20   WORM GEARS

Worm gears, Fig. 12-32, are similar to crossed helical gears.   The *pinion* or *worm* has a small number of teeth, usually one to four, and since they completely wrap around the pitch cylinder they are called *threads*.   Its mating gear is called a *worm gear*, which is not a true helical gear.   A worm and worm gear are used to provide a high angular-velocity reduction between nonintersecting shafts which usually are at right angles.   The worm gear is not a helical gear because its face is made concave to fit the curvature of the worm in order to provide line contact instead of point contact.   Because of the line contact, worm gearing can transmit high tooth loads.   However, a disadvantage of worm gearing is the high sliding velocities across the teeth, the same as with crossed helical gears.

Three worms are shown in Fig. 12-33; *a* has one thread and is called a *single-thread worm; b* has two threads and is called a *double-thread worm; c* is a *triple-thread worm.*   The *axial pitch* of a worm, or just *pitch* as it is called, is *p,* and is defined as the distance from a point on one thread profile to the same point on an adjacent profile measured along the pitch cylinder. The *lead l* is the amount the helix advances axially for one turn about the

Fig. 12-32.   *(Cleveland Worm & Gear Division, Eaton Manufacturing Co.)*

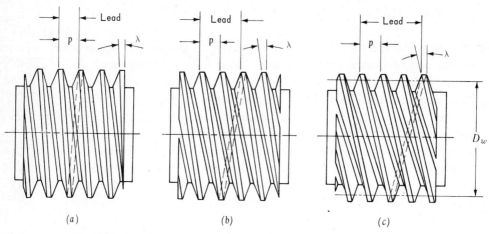

(a)   (b)   (c)

Fig. 12-33

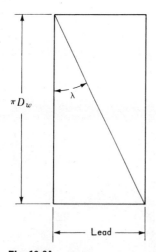

Fig. 12-34

pitch cylinder and is equal to the number of threads times the pitch. The slope of the threads is called the *lead angle* λ. Figure 12-34 shows a development of the pitch helix where D_w is the pitch diameter of the worm. From the figure,

$$\tan \lambda = \frac{l}{\pi D_w} = \frac{N_w p}{\pi D_w} \qquad (12\text{-}32)$$

Additional nomenclature for worm gearing appears in Fig. 12-35. The pitch diameters for the worm and worm gear are D_w and D_g, and C is the center distance. Then

$$C = \frac{D_w + D_g}{2} \tag{12-33}$$

The shafts of the worm and gear are usually at 90°. Then the circular pitch p of the worm gear must be the same as the axial pitch of the worm and can be expressed in the same manner as for a spur gear. Hence

$$p = \frac{\pi D_g}{N_g} \tag{12-34}$$

where N_g is the number of teeth on the worm gear.

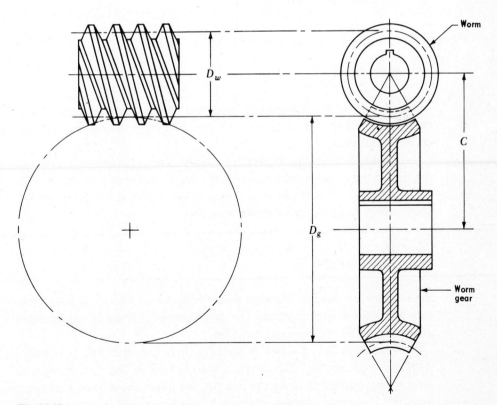

Fig. 12-35

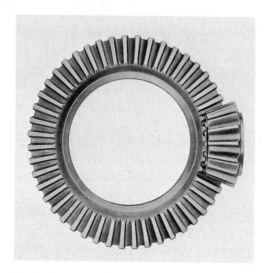

Fig. 12-36.   *(Gleason Works.)*

The velocity ratio for a worm drive is

$$\frac{\omega_w}{\omega_g} = \frac{N_g}{N_w} \tag{12-35}$$

where ω_w and ω_g are the angular velocities of the worm and worm gear. N_w is the number of threads on the worm, and N_g is the number of teeth on the worm gear.

The teeth on the worm gear are cut with a hob, but instead of moving the hob in the direction of the axis of the blank, the hob is moved radially into the blank.   This produces a concave face on the worm gear (see Fig. 12-35), which gives line contact with the worm.

12-21   BEVEL GEARS

Bevel gears are used to transmit power between shafts whose axes intersect and whose pitch surfaces are rolling cones.   Straight-toothed bevel gears (Fig. 12-36) are the most common.

A cross section of a pair of bevel gears is shown in Fig. 12-37 where AOP and BOP are the pitch cones.   As explained in Sec. 11-4, in order to have rolling contact all along the line OP, the cones must have a common apex O.   The angle between shafts is Σ, and though 90° is most common, almost any angle can be used. D_2 and D_3, the pitch diameters, are the diameters of the pitch cones at the large end. The diametral pitch P and circular pitch p for a bevel gear are defined in the same manner as for a spur gear and indicate the tooth size at the large end of the pitch cone. Thus

$$P = \frac{N}{D} \qquad p = \frac{\pi D}{N} \qquad pP = \pi$$

where N is the number of teeth.

The angular-velocity ratio is the inverse of the ratio of pitch diameters or numbers of teeth. Thus

$$\frac{\omega_2}{\omega_3} = \frac{D_3}{D_2} = \frac{N_3}{N_2} \qquad\qquad (12\text{-}36)$$

In Chap. 11 it is explained how the pitch-cone angles γ_2 and γ_3 are determined when Σ and the angular-velocity ratio are known.

A cross section of a bevel gear showing the terminology appears in Fig. 12-38. The tooth elements are straight lines which converge at the cone apex, and the face width is usually limited to one-third of the cone distance.

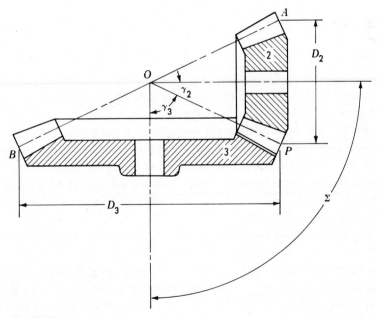

Fig. 12-37

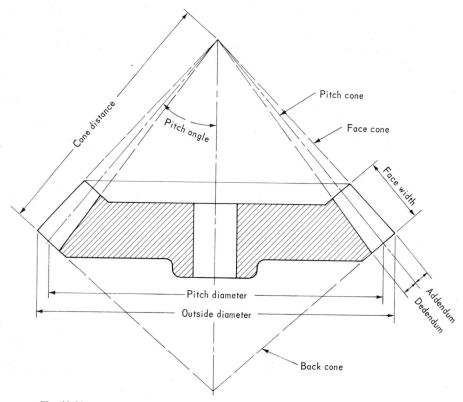

Fig. 12-38

The smaller portion of the tooth, which is omitted, would carry very little of the load and is difficult to manufacture.

In Fig. 12-39 an involute tooth developed from a base cone is shown. Imagine a piece of paper wrapped around a cone and let OA be a slit in the paper. Then if the paper is unwrapped from the cone and is kept taut, line OA will move to position OB. Point B will always be a fixed distance from O, and therefore lies on the surface of a sphere. Curve AB is known as a *spherical involute*.

In studying bevel gear tooth action, it is difficult to work with the true tooth profile since it lies on the surface of a sphere. Thus an approximation in a plane is made. In Fig. 12-40 a pair of bevel gears in mesh is shown along with their back cones. The elements of the back cone are perpendicular to the pitch-cone elements. The surface of the back cone approximates the spherical surface, and in order to study the bevel gear tooth action, the cones are laid out flat, as shown at the right in the figure.   Each cone forms a portion of a spur gear whose pitch-circle radius is equal to the back-cone radius of the bevel gear.   This is known as Tredgold's approximation.   The tooth action for the bevel gears is the same as that for the larger spur gears.   The number of teeth N_b on the spur gear is known as the *formative number of teeth*, and in terms of the number of teeth N on the bevel gear, is

$$N_b = N\frac{R_b}{R} = \frac{N}{\cos \gamma} \tag{12-37}$$

The formative number of teeth is usually not an integer.

Bevel gears are usually made with unequal addendums in order to avoid interference and to gain other advantages, which were discussed earlier.   Because of this, and because their pitch cones must have a common apex, bevel gears are designed as a pair and are not interchangeable. Since the elements of the teeth of bevel gears are not parallel to the gear axis, there is end thrust, which must be carried by the bearings.

Spiral bevel gears (Fig. 12-41) have curved teeth.   The tooth elements are theoretically spirals, but in practice they are made circular arcs because of the ease of manufacture.   Because of the curved teeth, spiral bevel gears

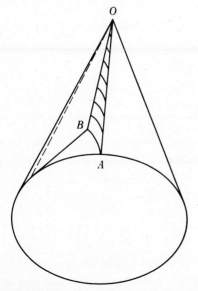

Fig. 12-39

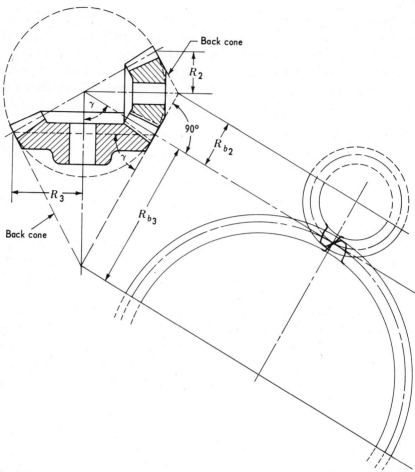

Fig. 12-40

have the same advantages over straight-toothed bevel gears as helical gears have over spur gears. That is, because of the progressive contact, they operate more quietly and have greater tooth strength.

12-22 HYPOID GEARS

Hypoid gears (Fig. 12-42) are used to transmit power between nonparallel, nonintersecting shafts. They are usually made for shafts which are at 90°, are made in pairs, and are not interchangeable.

The pitch surfaces are portions of rolling hyperboloids as described in Sec. 11-5 which have sliding along their line of contact. The gear teeth are parallel to the line of contact; hence there is sliding along the tooth elements, which is a disadvantage. However, for a given gear ratio a hypoid pinion is larger than an equivalent spiral bevel pinion, and this results in stronger teeth. Further, hypoid gears operate more quietly than spiral

Fig. 12-41. *(Gleason Works.)*

Fig. 12-42. *(Gleason Works.)*

bevel gears. Hypoid gears have been widely used for automotive rear-axle drives because the offset pinion permits lowering the drive shaft and use of a lower body.

PROBLEMS

12-1 A pair of meshing spur gears has 22 and 38 teeth, a diametral pitch of 8, and a pinion running at 1,800 rpm. Determine the following: (a) pitch diameters, (b) center distance, (c) circular pitch, (d) pitch-line velocity in feet per minute, and (e) rpm of the gear.

12-2 A pair of meshing spur gears has a diametral pitch of 10, a center distance of 2.6 in., and a velocity ratio of 1.6. Determine the number of teeth on each gear.

12-3 A spur gear having $14\frac{1}{2}°$ full-depth involute teeth has an outside diameter of 7.500 in. and a diametral pitch of 4. Determine the number of teeth.

12-4 A spur gear has forty-eight 20°-stub involute teeth and an outside diameter measuring 8.266 in. Determine the diametral pitch and the circular pitch.

12-5 A pair of spur gears has 15 and 22 teeth, a diametral pitch of 2, 20° pressure angle, 0.50-in. addendum, and 0.625-in. dedendum. The pinion drives in a clockwise direction. Make a full-size drawing similar to Fig. 12-8. Use a fillet radius equal to the clearance. Compute (a) the circular pitch, (b) base pitch, (c) pitch-circle radii, (d) base-circle radii, (e) length of path of contact, (f) contact ratio, and (g) angles of approach and recess for pinion and gear. Label the pitch point and the first and last points of contact. Use the figure to check your work graphically.

12-6 A pair of spur gears has 16 and 18 teeth, a diametral pitch of 2, addendum of $\frac{1}{2}$ in., and pressure angle of $14\frac{1}{2}°$. Show that the gears have interference. Determine graphically the amount by which the addendums must be reduced if the interference is to be eliminated. Measure the length of the path of contact for this new condition and compute the contact ratio.

12-7 Determine the number of teeth on the smallest pinion that will mesh with a rack having a $22\frac{1}{2}°$ pressure angle and an addendum equal to $1/P$.

12-8 Same as Prob. 12-7 except the rack has a 25° pressure angle.

12-9 A spur gear having 32 teeth and a pressure angle of 20° has a diametral pitch of 4 and an addendum of 0.200 in. Determine the number of teeth on the smallest pinion that will mesh with the gear without interference.

2-10 A pair of involute spur gears has 15 and 18 teeth and a diametral pitch of 2. Determine (a) the circular pitch and base pitch, and the contact ratio by finding graphically the length of the path of contact if the teeth are standard $14\frac{1}{2}°$ full-depth, and (b) same as (a) except the teeth are standard 20° stub.

12-11 An internal spur gear having 200 teeth meshes with a pinion having 40 teeth and a diametral pitch of 10. Determine (a) the velocity ratio if the pinion is the driver and (b) the center distance.

12-12 A pair of spur gears has 16 and 24 teeth, a diametral pitch of 4, and a 20° pressure angle. (a) Determine the center distance. (b) If the center distance is increased 0.125 in., find the resulting pressure angle.

12-13 A pair of spur gears is to have 10 and 35 teeth. They are to be cut with a 20° full-depth cutter of 2.5 diametral pitch. (a) Determine graphically the amount by which the addendum of the gear must be decreased to eliminate interference.

The addendum of the pinion is to be increased the same amount. (b) Find graphically the length of the path of contact and determine the contact ratio.

12-14   A pair of helical gears having 30 and 48 teeth and a 23° helix angle transmits power between parallel shafts. The diametral pitch in the normal plane is 8 and the pressure angle in this plane is 20°. Determine (a) the diametral pitch in the plane of rotation, (b) pitch diameters, (c) center distance, (d) circular pitch in the normal plane, (e) circular pitch in the plane of rotation, and (f) face width if the minimum recommended by the AGMA is used.

12-15   A pair of crossed helical gears connects shafts making an angle of 45°. The right-hand pinion has 36 teeth and a helix angle of 20°. The right-hand gear has 48 teeth and its diametral pitch in the normal plane is 10. Determine (a) the helix angle of the gear, (b) circular pitch in the normal plane, (c) diametral pitch of the pinion in its plane of rotation, (d) diametral pitch of the gear in its plane of rotation, and (e) center distance, and (f) make a drawing half-scale, similar to Fig. 12-31, and compute V_{P_2} in feet per minute for a pinion speed of 400 rpm. Find graphically the value of the sliding velocity.

12-16   A worm and worm gear have axes at 90° and give a speed reduction of 15 to 1. The triple-thread worm has a lead angle of 20° and an axial pitch of 0.4 in. Determine the following for the worm gear: (a) number of teeth, (b) pitch diameter, and (c) helix angle. (d) Determine the pitch diameter of the worm. (e) Compute the center distance.

12-17   A pair of straight-toothed bevel gears connecting shafts at 90° has 18 and 36 teeth and a diametral pitch of 6. The gears have 20° involute teeth with un-equal addendums. Determine (a) the angular-velocity ratio and (b) pitch diameter of the pinion and gear. Determine the following for the gear: (c) pitch angle, (d) cone distance, (e) back-cone radius, and (f) formative number of teeth.

13

GEAR TRAINS,
TRANSLATION SCREWS,
MECHANICAL ADVANTAGE

13-1 ORDINARY GEAR TRAINS

A *gear train* is composed of two or more gears in mesh for the purpose of transmitting motion from one shaft to another. Ordinary gear trains are those in which the axes of none of the gears move relative to the frame and are of two types, simple gear trains and compound gear trains.

A *simple gear train* is one in which there is only one gear on each shaft such as in Fig. 13-1, where the gears are represented by their pitch circles. Gear A drives B, B drives C, C drives D, and D drives E. Let the numbers of teeth on the gears be N_A, N_B, etc. The ratio of the angular velocities of any pair of gears in mesh is the inverse of their numbers of teeth.

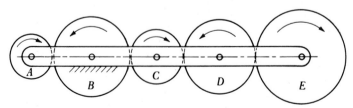

Fig. 13-1

Thus

$$\frac{\omega_A}{\omega_B} = \frac{N_B}{N_A} \qquad \frac{\omega_B}{\omega_C} = \frac{N_C}{N_B} \qquad \frac{\omega_C}{\omega_D} = \frac{N_D}{N_C} \qquad \frac{\omega_D}{\omega_E} = \frac{N_E}{N_D} \qquad (13\text{-}1)$$

The *velocity ratio* (VR) of a gear train is the ratio of the angular velocity of the first gear in the train to the angular velocity of the last gear.   For the train in Fig. 13-1 the velocity ratio is

$$VR = \frac{\omega_A}{\omega_E} = \frac{\omega_A}{\omega_B} \times \frac{\omega_B}{\omega_C} \times \frac{\omega_C}{\omega_D} \times \frac{\omega_D}{\omega_E}$$

and by substitution of Eqs. (13-1)

$$VR = \frac{N_B}{N_A} \times \frac{N_C}{N_B} \times \frac{N_D}{N_C} \times \frac{N_E}{N_D} \qquad (13\text{-}2)$$

The sign of the velocity ratio is considered positive if the first and last gears rotate in the same direction and negative if they rotate in opposite directions.   The easiest way to keep track of the directions of rotation is to place arrows on the gears.   Later we will use signs to indicate directions of rotation.   A plus sign will denote counterclockwise rotation, and a minus sign clockwise rotation.

From Eq. (13-2) we see that the velocity ratio for a simple gear train depends only on the number of teeth on the last and first gears in the train because the tooth numbers on the intermediate gears cancel out.   The latter are called *idler gears*.   It is also obvious that the velocity ratio depends only on the numbers of teeth on the last and first gears because the pitch circles roll on one another and all the gears have the same pitch-line velocity.   Idler gears are used for two purposes: to connect gears where a large center distance is required, and to control the directional relationship between gears such as A and E in Fig. 13-1.   Note that each idler added in the figure changes the direction of rotation of the last gear in the train.

A pair of gears are compounded if they have a common axis and are integral, e.g., gears B and C or D and E in Fig. 13-2a.   A *compound gear train* is a gear train containing compound gears.   In the compound gear train shown in Fig. 13-2a, the numbers indicate the number of teeth on each gear.   If the speed of A is 1,600 rpm ccw, the speeds of the various gears will be

$$\omega_A = +1{,}600 \text{ rpm}$$

$$\omega_B = \omega_C = {}^{30}\!\!/_{50} \times 1{,}600 = -960 \text{ rpm}$$

$$\omega_D = \omega_E = {}^{20}\!\!/_{40} \times 960 = 480 \text{ rpm}$$

$$\omega_F = {}^{18}\!\!/_{36} \times 480 = -240 \text{ rpm}$$

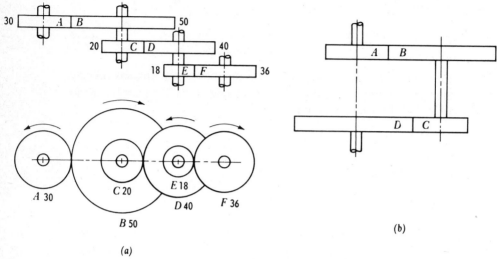

(a)

(b)

Fig. 13-2

The velocity ratio is

$$VR = \frac{\omega_A}{\omega_F} = -\frac{1,600}{240} = -2\%_3$$

For a compound gear train the velocity ratio can be written

$$VR = \frac{\text{product of teeth on driven gears}}{\text{product of teeth on driving gears}} \qquad (13\text{-}3)$$

$$= {}^{5}\%_{30} \times {}^{4}\%_{20} \times {}^{3}\%_{18} = -2\%_3$$

The compound train in Fig. 13-2b is known as a *reverted gear train* because the first and last gears are coaxial. Reverted gear trains are used in automobile transmissions, lathe back gears, industrial speed reducers, and in clocks (where the minute and hour hand shafts are coaxial).

The advantage of a compound train over a simple gear train is that a much larger speed reduction from the first to the last shaft can be obtained with small gears. If a simple train were used to give a large speed reduction, the last gear would have to be large. Usually for a speed reduction in excess of 7 to 1 a simple train is not used; instead a compound train or worm gearing is employed.

13-2 AUTOMOBILE TRANSMISSION

A conventional three-speed automobile transmission is shown diagrammatically in Fig. 13-3. Gear A is driven by the engine. Gears D, E, F, and G rotate as a unit. Gear H is an idler that meshes with G. These gears are always in motion when the shaft to the engine is in motion. Gears B and C can slide axially on the splined shaft to the rear wheels. The transmission is shown in neutral, and thus there is no connection between the shaft to the engine and the shaft to the rear wheels.

In first or low speed, gear C is shifted to the left to mesh with gear F. The velocity ratio of $ADFC$ is

$$VR = {}^{31}\!/_{14} \times {}^{27}\!/_{18} = 3.32$$

For second or intermediate speed, gear B is shifted to the right to mesh with gear E. The velocity ratio of $ADEB$ is

$$VR = {}^{31}\!/_{14} \times {}^{20}\!/_{25} = 1.77$$

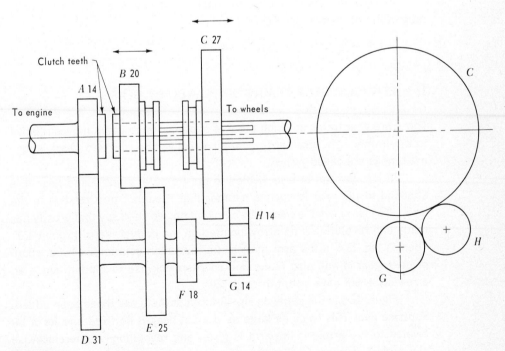

Fig. 13-3

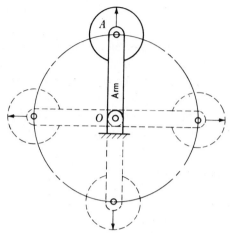

Fig. 13-4

For third or high gear, gear B is shifted to the left to engage its clutch teeth with those of gear A. This gives a direct drive with a velocity ratio of 1.

For reverse, C is shifted to the right to mesh with idler H. The velocity ratio of the drive through $ADGHC$ is

$$VR = {}^{31}\!/_{14} \times {}^{14}\!/_{14} \times {}^{27}\!/_{14} = -4.27$$

13-3 EPICYCLIC GEAR TRAINS OR PLANETARY GEARS

These are gear trains in which the axis of one or more gears moves relative to the frame. The gear at the center is called the *sun*, and gears whose axes move are called *planets*.

In Fig. 13-4 the arm is pivoted to the frame at point O and gear A is attached to the arm in such a manner that it cannot turn relative to the arm. To clarify what is meant by a revolution, we shall say that a body has made one revolution if an arrow painted on the body sweeps through 360°. Then in Fig. 13-4, if the arm makes one revolution counterclockwise about pivot O, gear A will also make one counterclockwise revolution, since an arrow on A will have swept through 360°.

Figure 13-5 is the same as Fig. 13-4 except that gear B has been added. Suppose gear B is twice as large as A. Let gear B be fixed and let A be pivoted to the arm. If the arm is given one revolution counterclockwise about pivot O, gear A will roll on fixed gear B, and A will make a total of

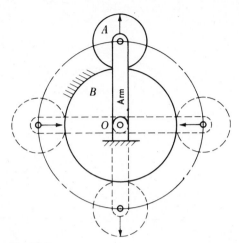

Fig. 13-5

three revolutions counterclockwise.   This can be explained as follows.   If gear A had not been able to rotate relative to the arm and if A and B had no teeth but were smooth cylinders so that A could slide on B, then A would have made one revolution counterclockwise as in Fig. 13-4.   However, because of the rolling of gear A on gear B in Fig. 13-5, the circumference of A is unrolled twice on the circumference of B.   This gives A two additional revolutions counterclockwise.   Thus A makes a total of three revolutions counterclockwise.

The same result can be found by the method of superposition, which states that the resultant revolutions or turns of any gear may be found by taking the number of turns it makes with the arm plus the number of turns it makes relative to the arm.   The method will be illustrated by applying it to Fig. 13-5.   It is convenient to make a table similar to Table 13-1.   The

Table 13-1

MEMBER	ARM	A	B
Train locked, arm given one positive turn	+1	+1	+1
Arm fixed, B given one negative turn	0	$+(N_B/N_A) = +2$	−1
Resultant turns	+1	+3	0

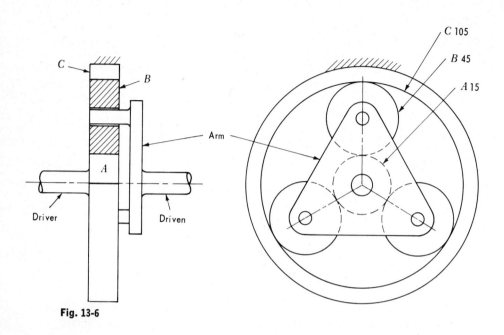

Fig. 13-6

various links or members are listed across the top of the table as shown. First imagine the train to be locked (i.e., all the members welded to the arm) and let the arm be given one turn counterclockwise. Then each member, including gear B, will make one positive turn. Hence we enter a $+1$ for each member in the first row of the table. Remember that counterclockwise is considered positive and clockwise is considered negative. Next, we consider the train to be no longer locked, but with the arm held fixed we give B one negative turn (i.e., one turn clockwise) to bring it back where it should be, since B is a fixed gear in the actual device. Thus we enter a zero in the second row under the heading "Arm" and a -1 under B. We then complete row 2 by asking ourselves how many turns will gear A make when the arm is fixed and B has been given the one negative turn. Gear A will make $N_B/N_A = 2$ ccw turns, i.e., two positive turns where N_B and N_A indicate the numbers of teeth. The resultant turns are shown in the table. Thus for one positive turn of the arm, gear A makes three positive turns.

EXAMPLE 13-1 In the planetary gear train shown in Fig. 13-6, gear A is keyed to the driving shaft; C is an internal gear which is fixed, and the arm is integral with the driven shaft. As the driver A rotates, it causes gears B

Table 13-2

MEMBER	ARM	A	B	C
Train locked, arm given one positive turn	+1	+1	+1	+1
Arm fixed, C given one negative turn	0	$+ (^{105}\!/_{45} \times ^{45}\!/_{15})$	$-^{105}\!/_{45}$	-1
Resultant turns	+1	+8	$-1\frac{1}{3}$	0

to roll on gear C, resulting in rotation of the arm.   The tabulation is shown in Table 13-2.   The results in the table indicate that for eight positive turns of A (the driver), the arm (the driven member) makes one positive turn. Since the resultant turns for gear A and the arm have the same sign, the driving and driven shafts turn in the same direction.   The speed reduction is 8 to 1.   Only one gear B is kinematically necessary, but the others are included to maintain balance and to distribute the load over more gears, permitting smaller gears to be used.

EXAMPLE 13-2   The planetary gear train shown in Fig. 13-7 is the same as in Fig. 13-6, except that C is keyed to the driver, and A instead of C is the fixed gear.   The tabulation appears in Table 13-3.   The results indicate that

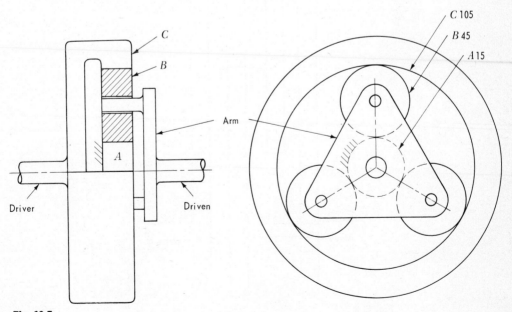

Fig. 13-7

Table 13-3

MEMBER	ARM	A	B	C
Train locked, arm given one positive turn	+1	+1	+1	+1
Arm fixed, A given one negative turn	0	−1	$+15\frac{4}{4}5$	$+ (15\frac{4}{4}5 \times 45\frac{4}{1}05)$
Resultant turns	+1	0	$+1\frac{1}{3}$	$+1\frac{1}{7}$

C (the driver) makes $1\frac{1}{7}$ positive turns for 1 positive turn of the arm (the driven). Thus the speed reduction is $1\frac{1}{7}$ to 1.

EXAMPLE 13-3   In the planetary gear train shown in Fig. 13-8, gear A is the driver and gears B and D are compounded; i.e., they are integral. C and E are internal gears and C is fixed. When considering directions of rotation, let us consider a right-end view. The tabulation appears in Table

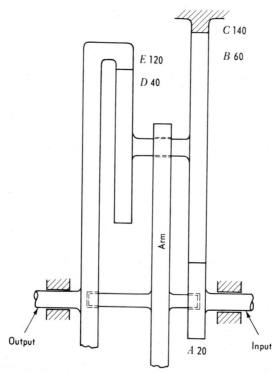

Fig. 13-8

Table 13-4

MEMBER	ARM	A	B	C	D	E
Train locked, arm given one positive turn	+1	+1	+1	+1	+1	+1
Arm fixed, C given one negative turn	0	$+ (^{140}\!/_{60} \times ^{69}\!/_{20})$	$-^{140}\!/_{60}$	-1	$-^{140}\!/_{60}$	$- (^{140}\!/_{60} \times ^{49}\!/_{120})$
Resultant turns	+1	+8	$-^{5}\!/_{3}$	0	$-^{5}\!/_{3}$	$+^{2}\!/_{9}$

13-4. The results indicate that for eight positive turns of A (the driver), E (the driven) makes $^{2}\!/_{9}$ of a positive turn. Hence the speed reduction is 8 to $^{2}\!/_{9}$, or 36 to 1.

13-4 EPICYCLICS WITH TWO INPUTS

A gear train of this type is shown in Fig. 13-9. Let n_1, n_2, and n_o represent the turns of input 1, input 2, and the output, respectively. By superposition, the number of turns of the output equals the output turns due to input 1 plus the output turns due to input 2. This can be expressed in equation form as follows:

$$n_o = \underbrace{n_1 \left(\frac{n_o}{n_1}\right)_{\substack{\text{Input 2} \\ \text{held fixed}}}}_{\text{I}} + \underbrace{n_2 \left(\frac{n_o}{n_2}\right)_{\substack{\text{Input 1} \\ \text{held fixed}}}}_{\text{II}} \qquad (13\text{-}4)$$

The application of Eq. (13-4) will be illustrated by the following example.

EXAMPLE 13-4   In Fig. 13-9 suppose input 1 turns at 120 rpm ccw, input 2 turns at 360 rpm cw, and the speed and direction of rotation of the output shaft are to be determined.

In order to evaluate item I in Eq. (13-4), we construct Table 13-5.   With input 2 held fixed, B and C are fixed and the rest of the system behaves as a planetary gear train in which the arm is the driver and C is the fixed gear. From the results in the table

$$\left(\frac{n_o}{n_1}\right)_{\substack{\text{Input 2} \\ \text{held fixed}}} = \frac{n_F}{n_{\text{arm}}} = \frac{+^{5}\!/_{3}}{+1} = +^{5}\!/_{3}$$

Next, when evaluating item II in Eq. (13-4), we do not construct a table, because with input 1 held fixed, the rest of the system behaves as an ordinary gear train.   Hence

$$\left(\frac{n_o}{n_2}\right)_{\substack{\text{Input 1} \\ \text{held fixed}}} = \frac{n_F}{n_A} = ^{20}\!/_{32} \times ^{48}\!/_{24} \times ^{36}\!/_{108} = +^{5}\!/_{12}$$

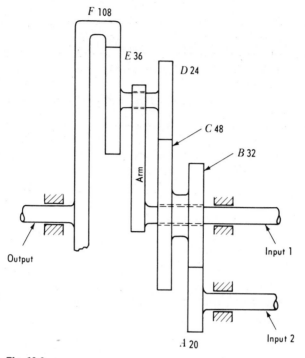

Fig. 13-9

Note that the sign is positive because gears F and A turn in the same directions. Substitution of the values found for I and II into Eq. (13-4) then gives

$$n_o = +\tfrac{5}{3}n_1 + \tfrac{5}{12}n_2$$
$$= +\tfrac{5}{3}(+120) + \tfrac{5}{12}(-360)$$
$$= +200 - 150 = +50$$

Thus the speed of the output shaft is 50 rpm ccw.

Table 13-5

MEMBER	ARM	C	D, E	F
Train locked, arm given one positive turn	+1	+1	+1	+1
Arm fixed, C given one negative turn	0	−1	$+\tfrac{48}{24}$	$+(\tfrac{48}{24} \times \tfrac{36}{108})$
Resultant turns	+1	0	+3	$+\tfrac{5}{3}$

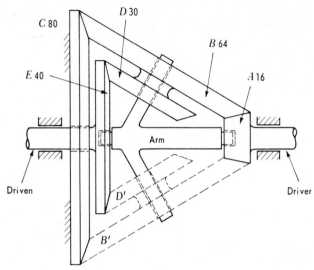

Fig. 13-10

13-5 EPICYCLIC BEVEL GEAR TRAINS

Bevel gears can be used to make a more compact epicyclic system and they permit a very high-speed reduction with few gears.

    EXAMPLE 13-5 In Fig. 13-10, A is the driver and E is the driven. C is a fixed gear and B and D are compound gears which turn freely on the arm. Gears B' and D' are shown dotted because kinematically they are not necessary. It is desired to find the relationship between the turns of A and E. When constructing a table for a planetary system containing bevel gears, the method is similar to that for spur gears. The only exception is that columns in the table for bevel gears whose axes are not parallel

Table 13-6

MEMBER	ARM	A	B	C	D	E
Train locked, arm given one positive turn	$+1$	$+1$	$\cdots$	$+1$	$\cdots$	$+1$
Arm fixed, C given one negative turn	0	$+{}^{80}\!/_{16}$	$\cdots$	-1	$\cdots$	$-({}^{80}\!/_{64} \times {}^{30}\!/_{40})$
Resultant turns	$+1$	$+6$	$\cdots$	0	$\cdots$	$+1\!/_{16}$

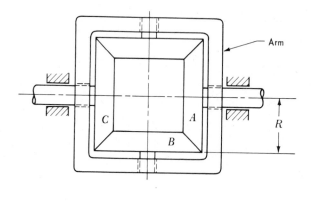

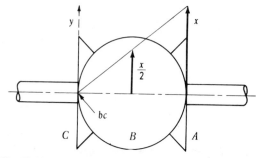

Fig. 13-11

to those of the driver and driven are left blank. This is done because clockwise and counterclockwise have no meaning for such gears. The tabulation is shown in Table 13-6. From the results in the table we note that for six positive turns of A (the driver), E (the driven) makes $\frac{1}{16}$ of a positive turn. Thus the speed reduction is 6 to $\frac{1}{16}$, or 96 to 1.

13-6   BEVEL GEAR DIFFERENTIAL

The bevel gear differential (Fig. 13-11) is a mechanism used for adding and subtracting two variables. Gears A and C are the same size. The operation of this mechanism can be most easily understood by considering instantaneous velocities. Let x be the velocity of the point of contact of gears A and B, and assume that gear C is stationary. Then since bc is the instant center for gears B and C, the velocity of the center of gear B, which may also be considered as a point on an extension of the arm, is $x/2$. Then

$$\omega_A = \frac{x}{R} \quad \text{and} \quad \omega_{\text{arm}} = \frac{x}{2R}$$

Similarly, if gear A is held fixed and a velocity y is assumed, the velocity of the center of gear B will be $y/2$ and

$$\omega_C = \frac{y}{R} \quad \text{and} \quad \omega_{\text{arm}} = \frac{y}{2R}$$

Next, if both gears A and C are rotating, note that velocity $y/2$ will add or subtract from $x/2$, depending on the directions of rotation of A and C. Then

$$\frac{x/2 + y/2}{R} = \omega_{\text{arm}}$$

or

$$\frac{\omega_A}{2} + \frac{\omega_C}{2} = \omega_{\text{arm}}$$

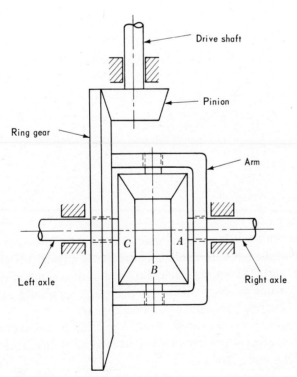

Fig. 13-12

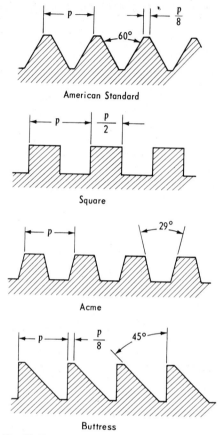

Fig. 13-13

and

$$\omega_A + \omega_C = 2\omega_{\mathrm{arm}} \tag{13-5}$$

This is the general equation for a bevel gear differential. The device is called a *differential* because it differentiates between the angular velocities of A and C. From Fig. 13-11 we note that if ω_A and ω_C are equal in magnitude and have the same direction, then $\omega_A = \omega_C = \omega_{\mathrm{arm}}$. If ω_A and ω_B are equal in magnitude but have opposite directions, then $\omega_{\mathrm{arm}} = 0$.

   Figure 13-12 shows a bevel gear differential as used in automobiles. For straightforward motion of the car there is no relative motion between gears A, B, C, and the arm. Thus

$$\omega_A = \omega_C = \omega_{\mathrm{arm}}$$

When the car makes a turn, the outside rear wheel must increase in speed and the inner wheel must decrease in speed by the same amount in order that the tires will not slip. From Eq. (13-5) we note that for a given value of ω_{arm}, if ω_A is increased, then ω_C will decrease by the same amount. Thus the differential automatically allows one rear wheel to slow down as the other speeds up. If the wheel connected to gear C is on dry pavement and the wheel connected to gear A is on ice, the wheel on the pavement will remain at rest while the other turns at twice the speed of the arm. The torques in shafts A and C must always be equal. Since there is practically no torque in shaft A, there is practically no torque in shaft C. Thus the car does not move.

13-7 TRANSLATION SCREWS

Bolts, cap screws, and *studs* are threaded members used as fasteners, i.e., for the purpose of holding parts together. Screws are also used to produce motion. The latter are called *translation screws.* The American Standard thread shown in Fig. 13-13 is used for screw fasteners. The other thread forms are used for translation screws.

The *pitch* and *lead* for a screw have the same meaning as for a worm, which was discussed earlier. Unless otherwise specified, it is to be assumed that a screw is of right hand and has a single thread. The press in Fig. 13-14 has a single-thread screw with a pitch of ¼ in. For one turn

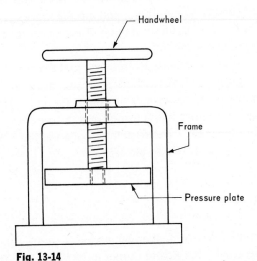

Fig. 13-14

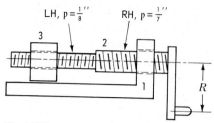

Fig. 13-15

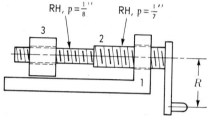

Fig. 13-16

clockwise of the handwheel when viewed from above, the pressure plate will move down ¼ in. relative to the frame.

A *compound screw* consists of two screws arranged so that the resultant motion is the sum of the individual motions. In Fig. 13-15 the large screw has 7 threads per inch and the small screw has 8 threads per inch. For one turn of the crank clockwise when viewed from the right end, the screw moves 1⁄7 in. to the left relative to the frame, and block 3 moves ⅛ in. to the left relative to the screw. The resultant motion of the block relative to the frame is 1⁄7 + ⅛ = 0.2679 in. to the left.

A *differential screw* consists of two screws arranged so that the resultant motion is the difference of the individual motions. Let the crank in Fig. 13-16 be given one turn clockwise when viewed from the right end. Then the screw moves 1⁄7 in. to the left relative to the frame, and the block moves ⅛ in. to the right relative to the screw. The resultant motion of the block relative to the frame is 1⁄7 − ⅛ = 0.0179 in. to the left.

13-8   MECHANICAL ADVANTAGE

Let Fig. 13-17 illustrate schematically any mechanism and let s_i and s_o represent the distances which the input force F_i and output force F_o move. The efficiency E is defined as

$$E = \frac{W_o}{W_i}$$

where W_o and W_i are the work output and work input.   Then

$$E = \frac{F_o s_o}{F_i s_i}$$

or

$$\frac{F_o}{F_i} = E \frac{s_i}{s_o} \tag{13-6}$$

If we divide numerator and denominator by time t,

$$\frac{F_o}{F_i} = E \frac{V_i}{V_o} \tag{13-7}$$

where V_i and V_o are the velocities of the input and output forces.
The mechanical advantage (MA) of a mechanism is defined as follows:

$$MA = \frac{F_o}{F_i} \tag{13-8}$$

Then from Eqs. (13-6) to (13-8)

$$MA = E \frac{s_i}{s_o} \tag{13-9}$$

and

$$MA = E \frac{V_i}{V_o} \tag{13-10}$$

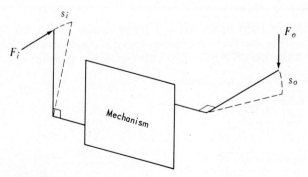

Fig. 13-17

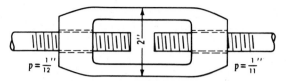

Fig. 13-18

If there were no friction, the efficiency would be 1.00. Since the influence of friction cannot be determined until the link and joint proportions are known, friction is usually neglected in the first stages of design of a mechanism. Unless otherwise stated, we shall assume an efficiency of 1.00 in problems. In mechanisms where the ratio s_i/s_o is constant, Eq. (13-9) is convenient for determining the mechanical advantage. For many mechanisms the ratio s_i/s_o is a variable. Then the mechanical advantage for any desired phase of the mechanism can be determined by making a velocity analysis for that phase and by substituting the values of V_i and V_o into Eq. (13-10).

EXAMPLE 13-6 The mechanical advantage is to be determined for the turnbuckle in Fig. 13-18, first if the screws are of opposite hand and second if they are of the same hand. The input force acts perpendicular to the paper and tangent to the handle at a 1-in. radius. If the handle is given one turn counterclockwise when viewed from the right end, this force moves through a distance $s_i = 2\pi R = 2\pi(1) = 6.283$ in. If the left screw is of left hand, it moves to the right $\frac{1}{12}$ in. relative to the handle. If the right screw is of right hand, it moves to the left $\frac{1}{11}$ in. relative to the handle. The output force acts along the axis of the screws and moves through a distance equal to the change in the gap between the ends of the screws. Hence $s_o = \frac{1}{12} + \frac{1}{11} = \frac{23}{132}$ in. Then from Eq. (13-9)

$$\text{MA} = E\frac{s_i}{s_o} = (1)\ \frac{6.283}{\frac{23}{132}} = 36.1$$

If both screws are of right hand, the left screw moves to the left $\frac{1}{12}$ in. relative to the handle, and the right screw moves $\frac{1}{11}$ in. to the left relative to the handle. Then the change in gap between screw ends is

$$s_o = \frac{1}{11} - \frac{1}{12} = \frac{1}{132}\text{ in.}\quad \text{Then}$$

$$\text{MA} = (1)\ \frac{6.283}{\frac{1}{132}} = 830$$

PROBLEMS

13-1 Determine the rpm and direction of rotation of gear G in the gear train shown in Fig. P13-1.

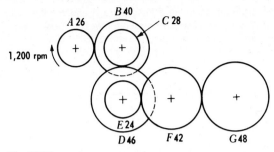

Fig. P13-1

13-2 Determine the rpm of gear F in Fig. P13-2. What is its direction of rotation when viewed from the right end?

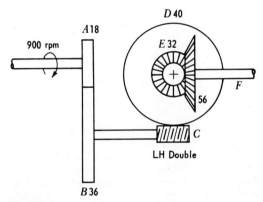

Fig. P13-2

13-3 Determine the number of teeth on sprocket D in Fig. P13-3 if the cable speed is to be 2 fps approximately. What is the direction of rotation of A when viewed from the right end?

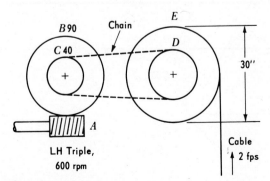

Fig. P13-3

13-4   In the reverted gear train shown in Fig. P13-4, gears A and B have a diametral pitch of 10 and gears C and D have a diametral pitch of 12.   Determine suitable tooth numbers for these gears if the velocity ratio is to be 11.4.   The number of teeth on each gear is to be a minimum, but not less than 28.

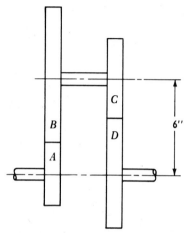

Fig. P13-4

13-5   The overdrive mechanism of an automobile is illustrated in Fig. P13-5.   When the overdrive is "in," the arm receives power from the engine and turns at engine speed.   Gear A is fixed.   The internal gear C is directly connected to and turns with the drive shaft which connects the transmission and the differential.   The rear-axle ratio is 3.5, and the outside diameter of the tires is 27 in.   At a car speed of 60 mph, determine the speed of the engine (a) when the overdrive is not "in" (direct-drive engine to differential), and (b) when the overdrive is "in."

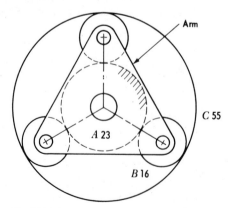

Fig. P13-5

13-6 Determine the speed reduction between the input and output shafts in Fig. P13-6. If the input shaft rotates clockwise when viewed from the right end, what is the direction of rotation of the output shaft?

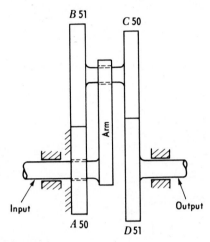

Fig. P13-6

13-7 Determine the speed of rotation of the output shaft in Fig. P13-7 and its direction of rotation when viewed from the right end.

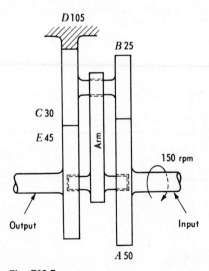

Fig. P13-7

13-8 Determine the speed and direction of rotation of the output shaft F in Fig. P13-8.

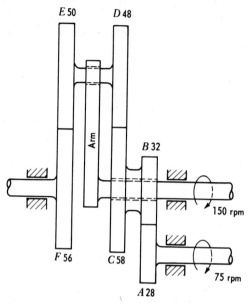

Fig. P13-8

13-9 In the gear train in Fig. P13-9, gear A is fixed, the arm is the driver, and gear D is the driven member. (*a*) Determine the speed reduction for the gear train. If the arm rotates counterclockwise when viewed from the right end, what is the direction of rotation of gear D? (*b*) Same as (*a*) except the numbers of teeth on gears A and D are interchanged.

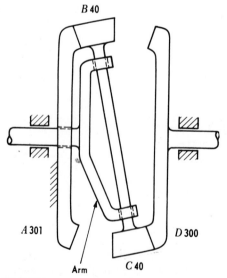

Fig. P13-9

13-10 In Fig. P13-10, the shaft with the arm is the output shaft.   For one counterclockwise turn of the lower shaft, which contains gears A and H, determine
the number of turns and the direction of rotation for the output shaft.

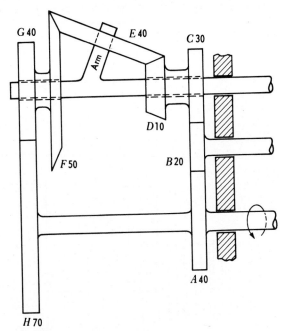

Fig. P13-10

13-11 Determine the mechanical advantage for the pulley system (block and tackle)
shown in Fig. P13-11.   Hint: If the load W is moved upward 1 ft, how many
feet will the applied force F move?

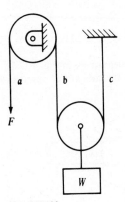

Fig. P13-11

13-12 Determine the mechanical advantage for the differential hoist (Fig. P13-12). The two pulleys D_1 and D_2 turn as a unit.   An endless chain or rope passes onto the larger pulley at a, off at b, onto the smaller pulley at c, and off at d. Hint: For one turn of the upper pulleys find the distance the applied force F moves and the amount the loop $defa$ shortens; then find amount W moves up.

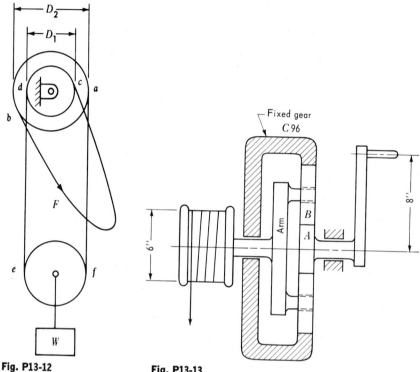

Fig. P13-12   Fig. P13-13

13-13 The mechanical advantage of the hoist shown in Fig. P13-13 is to be 24.   Determine the number of teeth on gears A and B.

13-14 Determine the mechanical advantage for the turnbuckle in Fig. P13-14 (a) if the threads are of the same hand and (b) if the threads are of opposite hand.

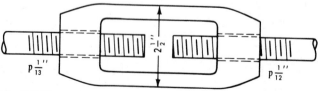

Fig. P13-14

14

SYNTHESIS
OF MECHANISMS

14-1 INTRODUCTION

Synthesis is the design of a mechanism to produce a desired output motion for a given input motion. Up until now we have analyzed mechanisms; that is, given a mechanism of definite proportions, we have analyzed its motion, finding displacements, velocities, and accelerations. Synthesis is the opposite of analysis; that is, given the input and output motions, we determine the mechanism required. We have done some of this earlier when we designed a cam to give a follower given displacements and when we determined the numbers of teeth on the members in a gear train in order to produce a desired velocity ratio.

Various types of mechanisms such as bar linkages, cams, or rolling surfaces, including gears, can be used to obtain a desired output from a given input. Thus it is important to the designer to be familiar with the methods of synthesis of these different types. Further, it is impossible to solve the problem without first assuming the type of mechanism which is to be used.

In the early development of synthesis, graphical methods using trial

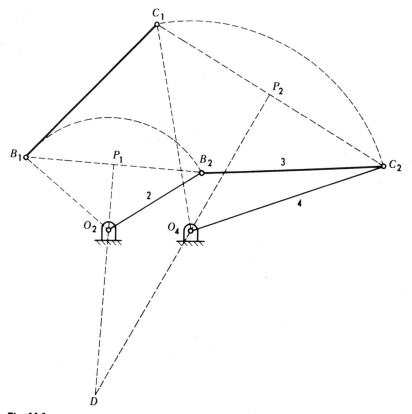

Fig. 14-1

and error and intuition predominated.   Such methods still have merit and are used in problems where analytical procedures have not or cannot be developed.   Some graphical methods giving direct results have been available for many years, but the increased need for solutions to more difficult problems of synthesis has led to the development of analytical methods. The mathematics of some problems in synthesis is formidable even for the four-bar linkage, and it becomes more so as the number of links is increased.   However, the wide use of digital computers has been an aid in this respect.

One of the major reasons for the increased interest in kinematic synthesis has been the need for mechanisms used in analog computers.   In this chapter several methods of synthesis will be discussed and will serve as an introduction.

14-2 DESIGN FOR COUPLER POSITION

In Fig. 14-1, B_1C_1 represents one position of link 3 and B_2C_2 a second position. A four-bar linkage can be designed to produce this displacement by arbitrarily locating points O_2 and O_4 along the perpendicular bisectors of B_1B_2 and C_1C_2, respectively.

   If it is desired that the velocity of link 3 be zero for each of the two positions shown in Fig. 14-1, this can be accomplished by modifying the linkage. The method is explained by Hinkle[1] and is as follows. Since O_2 and O_4 can lie anywhere along the perpendicular bisectors, we can choose both of them at D, the point where the bisectors intersect. The linkage is then shown in Fig. 14-2. Links 2, 3, and 4 will then have no relative motion,

[1] R. T. Hinkle, "Kinematics of Machines," 2d ed., pp. 254 and 255, Prentice-Hall, Inc., Englewood Cliffs, N.J., 1960.

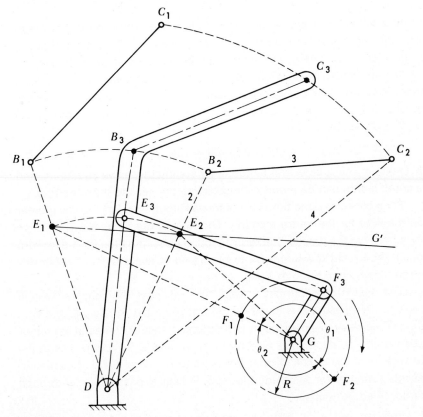

Fig. 14-2

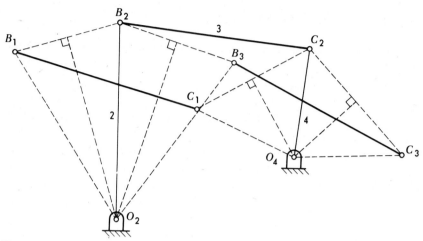

Fig. 14-3

and thus will act as a single link. DB_3C_3 has been drawn in the figure to represent this link in an intermediate, third position. This is not a specified position but is shown merely to add clarity. Two additional links, EF and FG, are added to provide zero velocity at positions B_1C_1 and B_2C_2. Length DE_2 and the position of G can be arbitrarily chosen. Lengths E_1G and E_2G are fixed and can be measured from the drawing. Let $E_1F_1 = E_2F_2 = EF$. Then

$$EF + R = E_1G \qquad \text{and} \qquad EF - R = E_2G$$

The length of the coupler can be found by adding these two equations, and the crank length can be found by subtracting the second from the first.

For a constant ω the time for the forward motion of link DBC is greater than the time for the return motion. The ratio of the times is θ_1/θ_2. If we desired to have the time for the return motion equal the time for the forward motion, then point G would have to be placed at some point G' arbitrarily chosen along a line through E_1E_2. In this case $\theta_1 = \theta_2$.

The design of a four-bar linkage that will pass the coupler through three specified positions is illustrated in Fig. 14-3. Let B_1C_1, B_2C_2, and B_3C_3 be the three desired positions of the coupler. Center O_2 lies at the intersection of the perpendicular bisectors to B_1B_2 and B_2B_3, and O_4 lies at the intersection of the perpendicular bisectors to C_1C_2 and C_2C_3. Rosenauer[1] presents a method for designing a four-bar linkage to pass a coupler through four specified positions.

[1] N. Rosenauer and A. H. Willis, "Kinematics of Machines," Associated General Publications, Sydney, Australia, 1953.

14-3 COUPLER–POINT CURVES

It is often desired to have a mechanism guide a point along a specified path. The paths generated by points on the coupler link of a four-bar linkage have been termed *coupler curves* and the generating point is called a *coupler point*.

Figure 14-4 shows a four-bar linkage used in a motion-picture projector to give the film intermittent motion.   Point E on the coupler traces the coupler curve shown.   As the driving crank rotates, the catcher moves into a film slot, pulls the film down one frame, moves out of the slot, and then moves up preparatory to engaging the film again.

The straight-line mechanisms discussed in Sec. 3-9 and shown in Figs. 3-13 to 3-17 are also examples of the use of coupler curves.   Other applications of coupler curves are in such automatic machines as those for wrapping, packaging, printing, weaving, farming, vending, and conveying.

A mechanism for rapidly drawing coupler curves is illustrated in Fig. 14-5.[1] The lengths of the four links AD, AB, BC, and CD can be changed by adjustments at B, C, and D. An adjusting screw at F permits length EF to be changed, as well as the angular position of line EF relative to line BC. As the handwheel which is integral with crank AB is rotated, a pencil at E traces a coupler curve. The proportions of the mechanism can be

[1] Joseph Boehm, Four-bar Linkages, *Mach. Des.*, August, 1952.

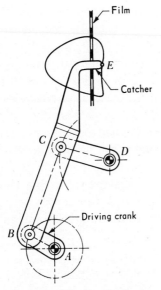

Fig. 14-4. *(Machine Design.)*

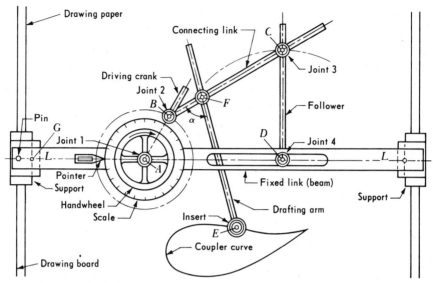

Fig. 14-5. *(Machine Design.)*

changed until a satisfactory curve is obtained.   Figure 14-6 shows a variety of curves which can be generated.

A very useful aid to the designer in selecting the proper linkage to obtain a desired coupler curve is the Hrones-Nelson catalog.[1]   This is a large book with pages 11×17 in. containing over 7,000 coupler curves plotted with an adjustable linkage designed for the purpose.   The mechanism is a four-bar linkage with a rotating input crank and oscillating output crank. The nomenclature is as shown in Fig. 14-7.   The input crank is the shortest link; its length is unchanged and is unity.   Each of links A and B has a length varying from 1.5 to 4, and link C from 1.5 to 6.5, all at intervals of 0.5. This gives a large variety of motion possibilities.   A plate is attached to the coupler and contains a rectangular grid with 50 point locations as shown in Fig. 14-8.   For each combination of link lengths the trace for each of the points on the plate is drawn.   The 50 curves are presented on five pages with 10 curves on each page.   By scanning through the catalog the designer can quickly find the linkage to solve his problem.   Hrones and Nelson have also included in the catalog a chapter with some discussion and sample design problems.

[1] J. A. Hrones and G. L. Nelson, "Analysis of the Four-bar Linkage," The Technology Press of The Massachusetts Institute of Technology and John Wiley & Sons, Inc., New York, 1951.

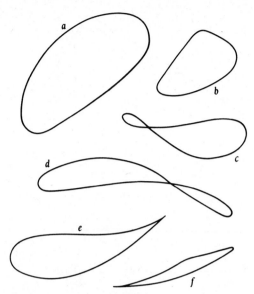

Fig. 14-6. *(Machine Design.)*

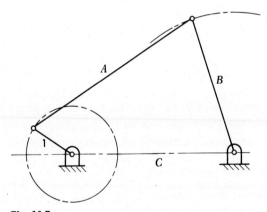

Fig. 14-7

14-4 FOUR–BAR FUNCTION GENERATOR BY OVERLAY METHOD

The overlay method is treated extensively by Nickson,[1] and is a graphical means for determining approximately the proportions between the lengths of the links of a four-bar linkage in order that the angular displacements of

[1] P. T. Nickson, A Simplified Approach to Linkage Designs, *Mach. Des.*, December, 1953.

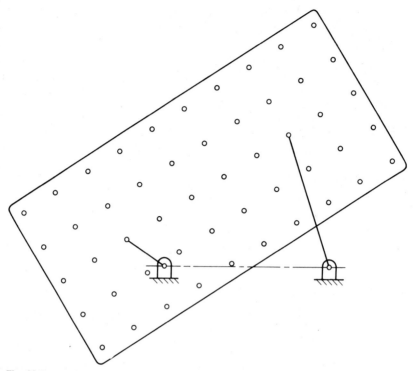

Fig. 14-8

driver and follower will satisfy a specific schedule or mathematical relationship. Suppose a four-bar mechanism is to be designed so that the
displacements of the two cranks will satisfy the following relationship:

	ROTATION FROM START, deg	
POSITION	DRIVING CRANK	DRIVEN CRANK
0	0	0
1	25	16
2	50	34
3	75	56
4	100	84

The procedure consists of laying out on paper Fig. 14-9a, the positions
of the driving crank, using any convenient length. Then with an assumed

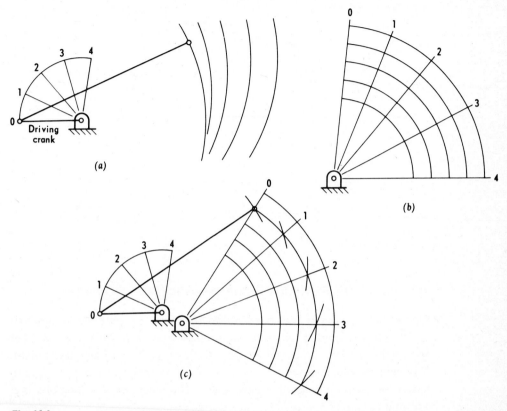

Fig. 14-9

length for the coupler as a radius, and with the center of the compass at each crankpin position of the driver, an arc is swung. On a sheet of transparent paper a layout, Fig. 14-9*b*, is made of the positions of the driven crank, showing a number of possible lengths for it. Next, the transparent layout is placed over the first drawing as shown in Fig. 14-9*c* and shifted about with an attempt to make the arcs of the first intersect in proper sequence one of the possible series of crankpin positions of the driven crank. Different coupler lengths may have to be tried, redrawing the first layout, before a satisfactory fit is obtained. In some instances a solution may not be possible. Ranges for rotation of the cranks which have been successful in practice are from 60 to 100°.

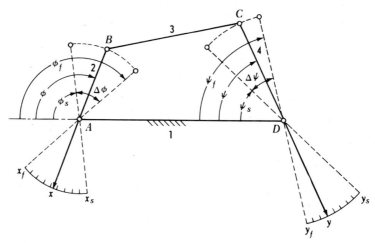

Fig. 14-10

14-5   FUNCTION GENERATOR CRANK–ANGLE RELATIONSHIPS

Figure 14-10 shows a four-bar linkage arranged to generate a function $y = f(x)$ over a limited range; for example, $y = x^2$ from $x = 1$ to $x = 5$. Let the range in x be $x_f - x_s$ (subscript s signifies start and f signifies finish). The corresponding range in ϕ is $\phi_f - \phi_s$. The range in y is $y_f - y_s$, and the corresponding range in ψ is $\psi_f - \psi_s$. The mechanical analog of x is ϕ and the analog of y is ψ.

If we merely wished to design a device for indicating corresponding values of x and y, we could use a pointer and a double scale as shown in Fig. 14-11. There the x scale is uniform, but the y scale is not. A more important requirement is to design a mechanism having uniform scales for both x and y. There are advantages in having a uniform scale. Such a scale is easier to graduate. Further, in control instruments used in industrial processes, the effects produced by the instruments are usually proportional to their movements.

Figure 14-12 shows a linear relationship between x and ϕ. From the figure,

$$\phi = \phi_s + \frac{\phi_f - \phi_s}{x_f - x_s}(x - x_s) \tag{14-1}$$

and

$$x = x_s + \frac{x_f - x_s}{\phi_f - \phi_s}(\phi - \phi_s) \tag{14-2}$$

A linear relationship between y and ψ can be shown by the same type of figure, and similarly,

$$\psi = \psi_s + \frac{\psi_f - \psi_s}{y_f - y_s}(y - y_s) \tag{14-3}$$

and

$$y = y_s + \frac{y_f - y_s}{\psi_f - \psi_s}(\psi - \psi_s) \tag{14-4}$$

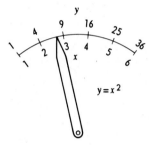

Fig. 14-11

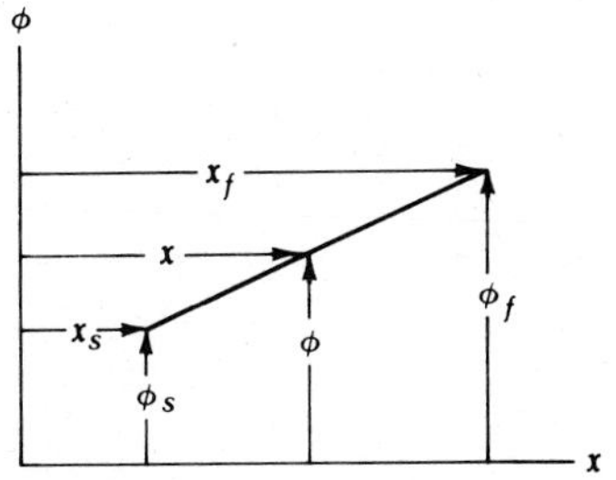

Fig. 14-12

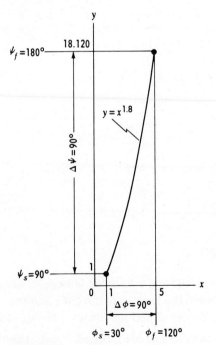

Fig. 14-13

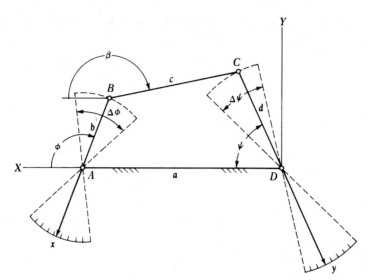

Fig. 14-14

EXAMPLE 14-1   Determine the relationship between ϕ and ψ for a function generator to generate $y = x^{1.8}$ when x varies from 1 to 5.   Let $\phi_s = 30°$, $\Delta\phi = 90°$, $\psi_s = 90°$, and $\Delta\psi = 90°$.   Solution: The graph of the function is shown in Fig. 14-13.   Equation (14-2) gives

$$x = 1 + \frac{5 - 1}{90°}(\phi - 30°) = 1 + \frac{4}{90°}(\phi - 30°) \tag{14-5}$$

and Eq. (14-4) gives

$$y = 1 + \frac{18.1195 - 1}{90°}(\psi - 90°) = 1 + \frac{17.1195}{90°}(\psi - 90°) \tag{14-6}$$

Substitution of expressions (14-5) and (14-6) into $y = x^{1.8}$ yields

$$\psi = 84.7503 + 5.2576\left[1 + \frac{4}{90°}(\phi - 30°)\right]^{1.8} \tag{14-7}$$

14-6   FREUDENSTEIN'S METHOD[1]

This is an analytical method for designing a four-bar linkage to generate a given function.   A relation between the lengths of the links and the crank angles in a four-bar linkage can be developed as follows.   In Fig. 14-14 let

[1] F. Freudenstein, Approximate Synthesis of Four-bar Linkages, *Trans. ASME*, pp. 853–861, August, 1955.

the links be considered as vectors. The sum of the X components must be zero.

$$b \cos \phi + c \cos \beta - d \cos \psi + a = 0 \qquad (14\text{-}8)$$

Also the sum of the Y components must be zero.

$$b \sin \phi + c \sin \beta - d \sin \psi = 0 \qquad (14\text{-}9)$$

Rearranging and squaring both sides of Eqs. (14-8) and (14-9) yields

$$c^2 \cos^2 \beta = (d \cos \psi - a - b \cos \phi)^2 \qquad (14\text{-}10)$$

$$c^2 \sin^2 \beta = (d \sin \psi - b \sin \phi)^2 \qquad (14\text{-}11)$$

Expanding the right sides of this pair of equations and then adding gives

$$c^2 = a^2 + b^2 + d^2 - 2ad \cos \psi - 2bd \cos \phi \cos \psi$$
$$- 2bd \sin \phi \sin \psi + 2ab \cos \phi \qquad (14\text{-}12)$$

This may be written

$$R_1 \cos \phi - R_2 \cos \psi + R_3 = \cos (\phi - \psi) \qquad (14\text{-}13)$$

where

$$R_1 = \frac{a}{d} \qquad R_2 = \frac{a}{b}$$
$$\qquad\qquad\qquad\qquad\qquad (14\text{-}14)$$
$$R_3 = \frac{a^2 + b^2 + d^2 - c^2}{2bd}$$

Equation (14-13) is Freudenstein's equation, which relates the crank angles and the lengths of the links for a four-bar linkage.

Very few functions can be generated exactly by a four-bar linkage. Shaffer and Cochin[1] have derived an equation which they call the *compatibility equation*, which can be used to determine whether a function can be generated exactly over a given range.

Freudenstein's equation can be used to design a four-bar linkage that will generate a given function accurately at a finite number of points (precision points) and will approximate the function between these points (Fig. 14-15). That is, the designed linkage will be compatible with Eq. (14-13) only at the precision points. The amount by which the generated function dif-

[1] B. W. Shaffer and I. Cochin, Synthesis of the Quadric Chain When the Position of Two Members Is Prescribed, *Trans. ASME*, October, 1954.

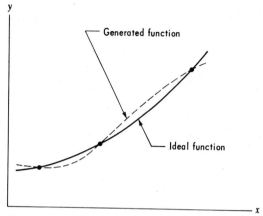

Fig. 14-15

fers from the ideal function between the precision points depends upon the distance between the points and the nature of the ideal function. The methods available for obtaining four and five precision points are complicated. Only three points will be considered here.

EXAMPLE 14-2   Determine the proportions of a four-bar linkage to generate $y = x^{1.8}$ when x varies from 1 to 5.   Let $\phi_s = 30°$, $\Delta\phi = 90°$, $\psi_s = 90°$, and $\Delta\psi = 90°$.   Let there be precision points at $x = 1$, 3, and 5.

Solution   In Table 14-1 the values of x and y which are listed were substituted into Eqs. (14-1) and (14-3) to obtain the corresponding values of ϕ and ψ.   Writing the Freudenstein equation for each of the three positions of the mechanism gives

$$R_1 \cos 30° - R_2 \cos 90° + R_3 = \cos(-60°)$$
$$R_1 \cos 75° - R_2 \cos 122.72° + R_3 = \cos(-47.72°)$$
$$R_1 \cos 120° - R_2 \cos 180° + R_3 = \cos(-60°)$$

Table 14-1

x	y	ϕ, deg	ψ, deg
1	1	30	90
3	7.2247	75	122.72
5	18.1195	120	180

or

$$0.8660\,R_1 + R_3 = 0.5000$$

$$0.2588R_1 + 0.5405R_2 + R_3 = 0.6727$$

$$-0.5000R_1 + R_2 + R_3 = 0.5000$$

From these equations

$$R_1 = 1.3180 \qquad R_2 = 1.8004 \qquad R_3 = -0.6415$$

Next, we choose the length of one link and solve for the others using Eqs. (14-14). Let $a = 1.000$. Then $b = 0.555$, $c = 1.557$, and $d = 0.759$. The resulting mechanism is shown in Fig. 14-16.

In problem solutions a negative result for length b or d indicates that the link extends opposite to that shown in Fig. 14-14.

14-7 ROLLING CURVES

Rolling curves can be designed to satisfy a given functional relationship.[1] An example will be used to illustrate the method for determining the contours of the rolling bodies.   In Fig. 14-19 bodies 2 and 3 rotate about centers O_2 and O_3, and their angular displacements are expressed by the variables ϕ and ψ.   Suppose it is desired to design two rolling curves to satisfy the relationship $y = 1.5x^2$ from $x = 1$ to $x = 5$.   The function is plotted in Fig. 14-17.   Input x and output y are represented by their analogs ϕ and ψ in Fig. 14-18.   Let it be desired that the range for ϕ and ψ be 120°.   In order to have a linear relationship between x and ϕ and between y and ψ, we will

[1] H. E. Golber, Rollcurve Gears, *Trans. ASME*, vol. 61, p. 223.

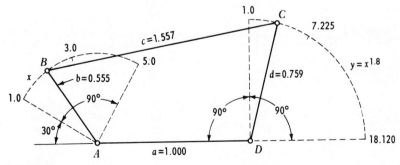

Fig. 14-16

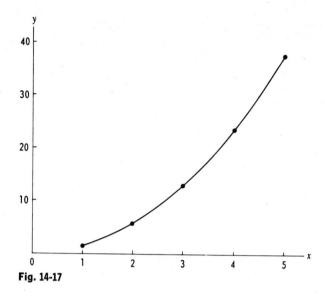

Fig. 14-17

substitute the given data into Eqs. (14-1) and (14-3).   Then

$$\phi = 0° + \frac{120° - 0°}{5 - 1}(x - 1) = 30(x - 1) \tag{14-15}$$

$$\psi = 0° + \frac{120° - 0°}{37.5 - 1.5}(y - 1.5) = 10\tfrac{3}{3}(y - 1.5) \tag{14-16}$$

In Table 14-2 assumed values of x are shown, and the corresponding values of y were computed.   These values were then substituted in Eqs. (14-15) and (14-16) in order to obtain the corresponding values of ϕ and ψ.   Next, the values of ϕ and ψ were laid off as shown in Fig. 14-19.

In Chap. 2, where rolling contact is discussed, it is shown that in order

Table 14-2

POSITION	x	y	ϕ, deg	ψ, deg	$\dfrac{d\psi}{d\phi}$	r	R
0	1	1.5	0	0	0.33	0.75	0.25
1	2	6	30	15	0.67	0.60	0.40
2	3	13.5	60	40	1.00	0.50	0.50
3	4	24	90	75	1.33	0.43	0.57
4	5	37.5	120	120	1.67	0.38	0.62

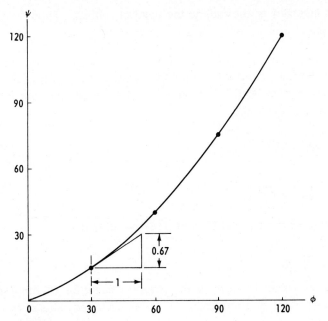

Fig. 14-18

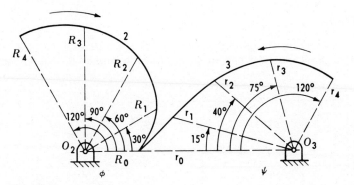

Fig. 14-19

for bodies 2 and 3 in Fig. 14-19 to roll on one another, their contours must be such that the point of contact at all times lies on the line of centers O_2O_3. Hence in Fig. 14-19

$$R + r = O_2O_3 \qquad (14\text{-}17)$$

Figure 14-18 shows ψ plotted as a function of ϕ. The slope of this curve represents the angular-velocity ratio for bodies 2 and 3, which we know varies inversely as the radii to the point of contact.   Thus

$$\frac{d\psi}{d\phi} = \frac{R}{r} \tag{14-18}$$

In order to obtain an expression for ψ as a function of ϕ, we can solve Eqs. (14-15) and (14-16) for x and y and substitute the results into our original equation $y = 1.5x^2$ to obtain

$$\psi = 5\left(\frac{\phi}{30} + 1\right)^2 - 5 \tag{14-19}$$

Then

$$\frac{d\psi}{d\phi} = \frac{1}{3}\left(\frac{\phi}{30} + 1\right) \tag{14-20}$$

By eliminating R from Eqs. (14-17) and (14-18) we obtain

$$r = \frac{O_2O_3}{1 + d\psi/d\phi} \tag{14-21}$$

In Table 14-2 the values of $d\psi/d\phi$ were computed using Eq. (14-20), and the values of r were obtained from Eq. (14-21) for an assumed value of $O_2O_3 = 1$ in.   The resulting curves are shown in Fig. 14-19.

PROBLEMS

14-1   Design a four-bar linkage (Fig. P14-1) to perform the following function. B_1C_1 are two points on a link and are to be moved to position B_2C_2 in 0.7 sec and

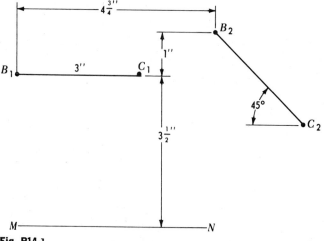

Fig. P14-1

returned in 0.5 sec. The two positions are the extreme positions of the link. The drawing may be made on $8\frac{1}{2} \times 11$ in. paper with point B_1 located 3 in. from the top and 1 in. from the left edge. Let a line from C_2 to the rocker-arm pivot D represent the rocker arm. The coupler is to be pinned to the rocker arm at a point on line C_2D. This point is to be $3\frac{1}{2}$ in. from D. The driving crank is to pivot about a point on line MN and the crank is to be 1 in. in length. Find the rpm of the driving crank and indicate on the drawing its direction of rotation.

14-2 Use the overlay method discussed in Sec. 14-4 to design a four-bar linkage to generate $y = e^x$ from $x = -1$ to $x = +1$. First complete the table below using Eqs. (14-1) and (14-3) to determine the values of ϕ and ψ. Let the driving crank have a length of 1.5 in. and let the coupler length be 4 in. Make a full-size drawing of the final mechanism showing it in its starting position. Draw it in dashed lines for the other positions.

POSITION	x	y	ROTATION FROM START, deg DRIVER ϕ CW; DRIVEN ψ CW	
0	-1		0	0
1	-0.5			
2	0			
3	$+0.5$			
4	$+1$		100	100

14-3 Use Freudenstein's method to design a four-bar linkage to generate $y = \sin x$, when x varies from 0 to 90°, with precision points at $x = 0$, 60, and 90°. Let $\phi_s = 300°$, $\Delta\phi = 90°$, $\psi_s = 250°$, and $\Delta\psi = 90°$. Let the ground link a be 1.000 in. Make a full-size drawing of the mechanism showing it in its starting position. Draw it in dashed lines for the other positions.

14-4 Use Freudenstein's method to design a four-bar linkage to generate $y = 1/x$, when x varies from 1 to 2, with precision points at 1, $1\frac{1}{2}$, and 2. Let $\phi_s = 30°$, $\Delta\phi = 100°$, $\psi_s = 120°$, and $\Delta\psi = 90°$. Let the ground link a be 1.000 in. Using a scale of 1 in. = 0.5 in., make a drawing of the mechanism showing it in its starting position. Draw it in dashed lines for the other positions.

14-5 Rolling curves are to be used to mechanize the function $y = \ln x$ from $x = 1$ to $x = 2$. Let ϕ vary from 0 to 100° and ψ from 0 to 150°. Use 0.2 increments in x. Make a table similar to Table 14-2. Draw the rolling curves using a center distance $= 3$ in.

15

ANALOG
COMPUTING MECHANISMS[1]

15-1 INTRODUCTION

The increased demand for automatic controls and the trend toward automa-
tion have brought about a continual development in computer mechanisms.
Computer mechanisms are of two types, digital computers and analog
computers.

Digital computing machines receive input quantities in numerical
form, usually on a keyboard, and perform arithmetical operations of addition, subtraction, multiplication, and division by repetitive addition in counting devices. Integration is done by summation, and converging series are
substituted for trigonometric functions. The results are finally presented
in numerical form to the operator. The adding machine, the electric desk
calculator, and the electronic digital computer are examples of digital computers. Many digital computers receive information from punched cards
or tapes. A characteristic of these machines is their production of numer-

[1] The author is indebted to two articles, Computing Mechanisms, *Prod. Eng.*, March
and April, 1956, and Analog Computing Mechanisms, G. W. Michalec, *Mach. Des.*,
March, 1959, for the approach to most of the mechanisms discussed in this chapter.

ical results by calculations in discrete steps which involve appreciable time delays. The cumulative effect of these delays can be considerable if the computational work is complex. Digital computers which receive information from punched or magnetic tapes, disks, or drums are being used in factories for the automatic operation of machine tools and assembling of parts.

Analog computers deal with magnitudes instead of purely mathematical values. They give instantaneous solutions to specific problems or continuous solutions as the input quantities vary. These devices may be mechanical, electrical, hydraulic, or pneumatic. Usually they are mechanical or electrical. Analog computers can be designed to carry out the usual operations of algebra and calculus, i.e., addition, subtraction, multiplication, division, integration, and resolution of vectors. Some examples of simple analog computers are the slide rule, planimeter, and speedometer. Most important, analog computers can be used to generate functions of one or more independent variables. Examples of this type are those used in connection with gunsights, bombsights, and automatic pilots.

15-2 ADDITION AND SUBTRACTION

Mechanisms which perform addition or subtraction are usually based on the differential principle. Two basic types are the slide-and-link differential and the bevel or spur gear differential. These mechanisms solve the

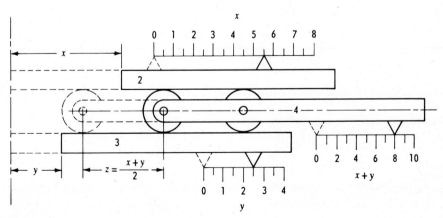

Fig. 15-1

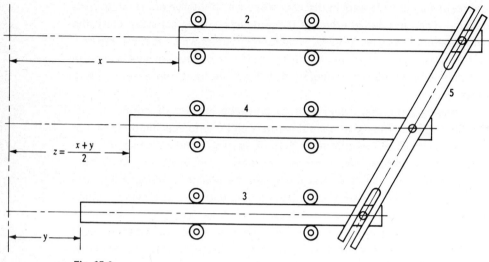

Fig. 15-2

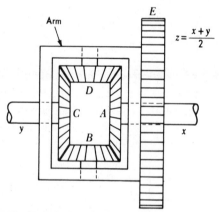

Fig. 15-3

basic equation

$$z = \frac{x + y}{2} \tag{15-1}$$

where z is the output and x and y are the inputs.   In Figs. 15-1 and 15-2 bars 2 and 3 move horizontally on rollers.   This results in a horizontal motion of bar 4 whose displacement is given by Eq. (15-1).   When two numbers are to be added, bar 2 is moved to one of the numbers and bar 3 is moved to the other; the sum is then given by bar 4.   The scales of 2 and 3 are equal, and the scale of 4 is twice as large as that for 2 and 3.   Subtraction can be performed by extending the scales to the left.

Probably the most commonly used mechanism for addition and subtraction is the bevel gear differential shown in Fig. 15-3.   Gear E is integral with the spider or arm and gears A, B, C, and D can all rotate relative to the arm.   The angular displacement of gears A and C are the inputs x and y, and the angular displacement of the arm or gear E is output z.   An explanation of the operation of this device is given in Sec. 13-6, where it is shown that it satisfies Eq. (15-1).   The operation of the spur gear differential, Fig. 15-4, is similar to the bevel gear differential.   Gear differentials are compact and have unlimited angular-displacement capacity.

15-3 MULTIPLIERS

The simplest multiplication to mechanize is multiplication of a variable by a constant; that is, $y = cx$.   This can be done with a gear ratio equal to c for angular motion and by means of a lever system such as shown in Fig. 15-5 for linear motion.   In the figure $y = cx$, where $c = b/a$.

Multiplication of two variables can be performed by a slide type of

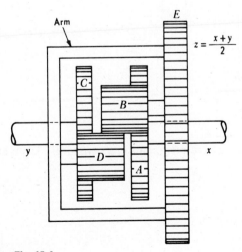

$$z = \frac{x+y}{2}$$

Fig. 15-4

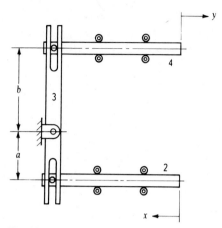

Fig. 15-5

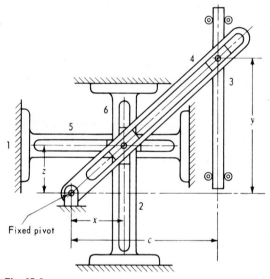

Fig. 15-6

multiplier as shown in Fig. 15-6, where distance c is a constant. From similar triangles

$$\frac{z}{x} = \frac{y}{c}$$

Thus

$$z = \frac{xy}{c} \qquad\qquad (15\text{-}2)$$

15-4   TRIGONOMETRIC FUNCTIONS

The mechanism in Fig. 15-7 consists of two Scotch yokes and generates the sine and cosine functions over the entire range of θ from 0 to 360°.

Another mechanism for generating the sine and cosine of an angle utilizes a planetary gear train, Fig. 15-8.   The pinion of radius $r = R/2$ rolls on a fixed internal gear having a radius R.   A fixed point C on the circumference of the pinion will move along a diameter of the gear with simple harmonic motion.

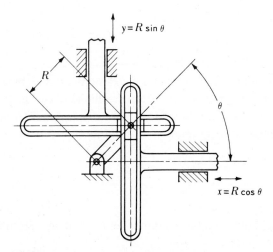

Fig. 15-7

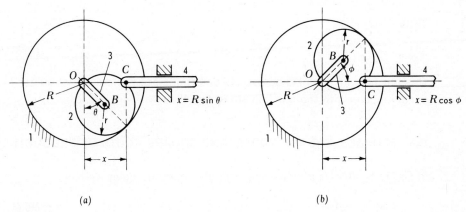

(a)   (b)

Fig. 15-8

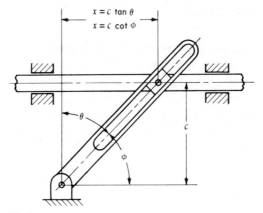

Fig. 15-9

Since the tangent and cotangent functions extend to infinity, they can be generated only over a limited range. The mechanism in Fig. 15-9 is commonly used to generate these functions. The displacement x gives either the tangent or cotangent, depending on whether θ or ϕ is the input angle.

15-5   INVERTER

The mechanism in Fig. 15-10 gives the reciprocal of a number. From similar triangles

$$\frac{x}{a} = \frac{b}{y}$$

Thus

$$x = \frac{ab}{y} \tag{15-3}$$

This mechanism can also be used to determine sec θ and csc θ when used in conjunction with the mechanism in Fig. 15-7.

15-6   SQUARES, SQUARE ROOTS, AND SQUARE ROOTS OF PRODUCTS

In Fig. 15-11, BDC is a right angle. From similar triangles we have

$$\frac{z}{x} = \frac{y}{z} \qquad z^2 = xy \qquad z = \sqrt{xy} \tag{15-4}$$

Link 4 is pivoted to link 7 and x, y, and z are variables, each starting at point O. If x and y are inputs, then z will be the square root of their product. If y is made constant by fixing point C at $y = 1$, then $z^2 = x$ or $z = \sqrt{x}$. If y is held constant but not equal to unity, then a scale factor must be introduced.

15-7   SLOT–TYPE FUNCTION GENERATOR

Figure 15-12 shows a slot type of function generator used to give $y = f(x)$. Such an arrangement is suitable for functions of one variable.

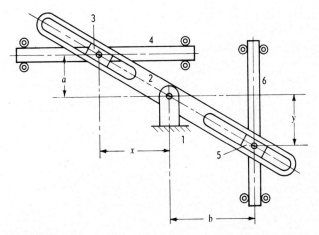

Fig. 15-10

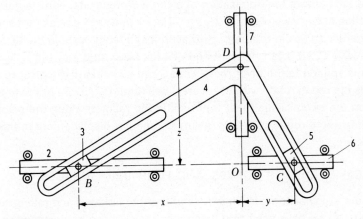

Fig. 15-11

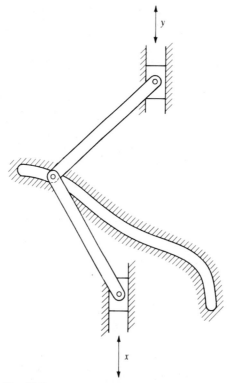

Fig. 15-12

15-8 FUNCTIONS OF TWO VARIABLES

Figure 15-13 shows the application of a three-dimensional cam to generate a function of two variables, $z = f(x,y)$. Block 2 moves in guides (not shown) which prevent it from rotating. However, it can reciprocate in the direction of the lead screw axis. The left end of 2 fits freely into a slot in follower 3, which is splined to the pivot rod. Thus 3 can slide along the pivot rod but causes the rod to rotate with it. A steel ball fitted in the end of the follower rolls on the cam. Thus the x input shifts the follower along the pivot rod and the y input rotates the cam, giving an angular displacement to the output gear z.

15-9 CAMS, ROLLING CURVES, AND FOUR–BAR LINKAGES

From our study of cams in Chap. 10 and four-bar linkages and rolling curves in Chap. 14, we have seen that these mechanisms can be designed to generate a wide variety of functions. Though the four-bar linkage is difficult to design, its cost of manufacture is less than for most of the other types of computing devices we have discussed. Four-bar linkages are relatively simple in construction and can be made with high precision.

15-10   MULTIPLICATION OF COMPLEX FUNCTIONS

Multiplication of complex functions can be accomplished by controlling by means of cams the motions of the sliding bars of the multiplier in Fig. 15-6. Division of two variables can be done by feeding one of the variables through an inverter and then multiplying by the other. The mechanism in Fig. 15-14a solves the equation

$$z = f(y)x^2$$

The operation is shown in the schematic diagram in Fig. 15-14b. Figure 15-14c shows schematically the generation of the function

$$y = \frac{\cos \theta}{x}$$

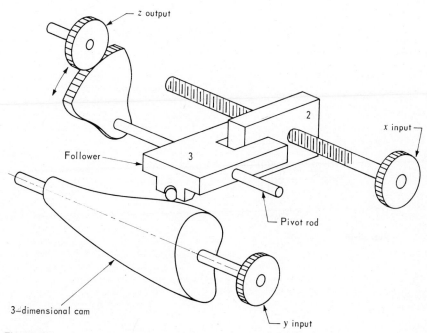

Fig. 15-13

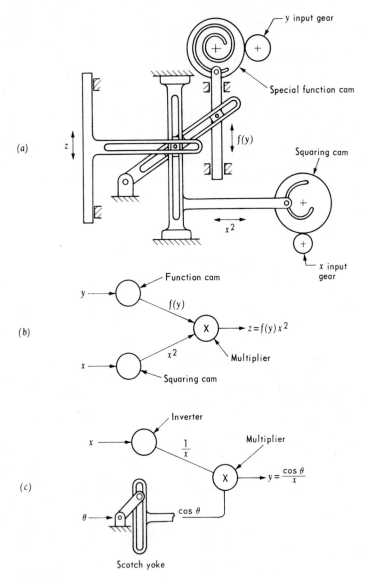

(a)

(b)

(c)

Fig. 15-14

A complex computer consists of a number of computer linkages in combination. Schematic diagrams as illustrated in Fig. 15-14*b* and *c* are an aid in the initial planning of a complex computer.

15-11 INTEGRATOR

The integrator shown in Fig. 15-15*a* is basically a variable-speed drive. Positioned between the disk and roller are two balls, which provide pure rolling in all directions. The disk is rotated by means of the x input while the y input varies the distance y from the center of the disk to the friction balls.

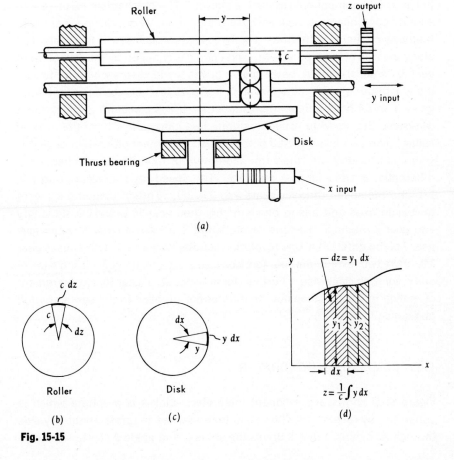

(a)

(b) Roller

(c) Disk

(d) $dz = y_1\, dx$

$$z = \frac{1}{c}\int y\, dx$$

Fig. 15-15

Rotation of the disk an infinitesimal amount dx produces rolling along equal arc lengths on the disk and roller. Thus

$$y \, dx = c \, dz$$

or

$$dz = \frac{1}{c} y \, dx$$

For a given number of revolutions x, the total number of z revolutions will equal the integral of $(1/c)y \, dx$, where y varies as called for in the problem.

15-12 · COMPONENT RESOLVER

In Fig. 15-16 a component resolver is shown. The mechanism resolves into x and y components a vector whose magnitude z and direction θ are continuously changing with time. Figure 15-16a is a plan view. The vector lies along the axis of the lead screw. Direction θ is changed by rotation of input gear 3. A second input, bevel gear 4, drives the lead screw in order to vary the magnitude z. From the elevation in Fig. 15-16b it can be seen that input gears 3 and 4 have independent motions. If z is constant, a change in θ will cause the screw to rotate, resulting in a change in z from the desired value. This can be corrected if, for a given increment of rotation of gear 3, gear 4 is provided with the same increment of rotation. A compensating differential is used for this purpose. The speed ratio for gears 6 to 7 is 1:1, and the speed ratio for gears 8 to 9 is 2:1. Thus if magnitude z were to remain fixed and θ is to be changing, then gear 10 would be stationary and gear 3 rotating. For one revolution of 3 clockwise, gear 6 would give gear 7 of the differential one revolution counterclockwise. This causes gear 8 to make two revolutions counterclockwise. As a result gears 9 and 4 make one revolution clockwise. Further, for any motion of gear 10, any increment of motion in 3 is by means of the differential supplied in the same amount and sense to gear 4.

15-13 COMPONENT INTEGRATOR

Figure 15-17 shows a component integrator. Link 2 is a sphere which is supported by rollers (not shown) so that it is free to rotate about any axis through its center. Disk 3 drives the sphere, and angle θ can be varied as

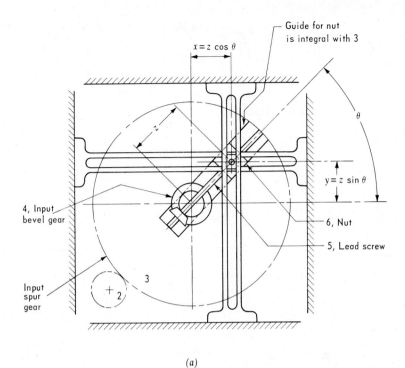

(a)

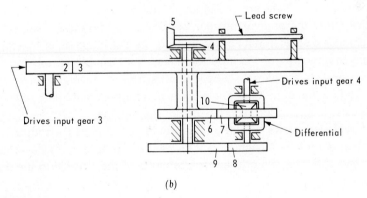

(b)

Fig. 15-16

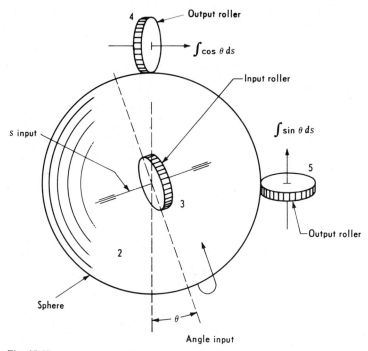

Fig. 15-17

disk 3 rotates. There are two input variables, the angular displacement s of disk 3 and the angular position θ. Disks 4 and 5 have fixed axes and the angular displacements of these disks are the integrals of the sine and cosine components of the input motions. An example of how the device can be used is as follows. Suppose a particle is moving in an x-y plane, and its direction of motion (angle θ, which a tangent to its path makes with the x axis) is known for each position of the particle. Then the distance it has traveled along its path of motion would be fed in as the s input and its direction of motion at each instant as θ input. Output disks 4 and 5 would then give the integrals of the displacements of the particle in the x and y directions.

15-14 ACCURACY

There are two types of errors in mechanical computing mechanisms: (1) kinematic errors, which result from an approximation in the generation of the function, and (2) fabrication errors, which result from manufacturing tolerances and the clearances used for satisfactory operation of the moving parts.   All of the mechanisms discussed in this chapter, along with cams and rolling curves which were treated in earlier chapters, are theoretically correct and thus have no kinematic errors.   However, the four-bar linkage which was discussed in Chap. 14 cannot be designed to generate most functions exactly at all points over a given range, and hence it produces kinematic errors.   The ball disk integrator may have an error due to slip, which is characteristic of friction devices.   All of the mechanisms will have fabrication errors, and these should be kept to the minimum which is economically feasible.

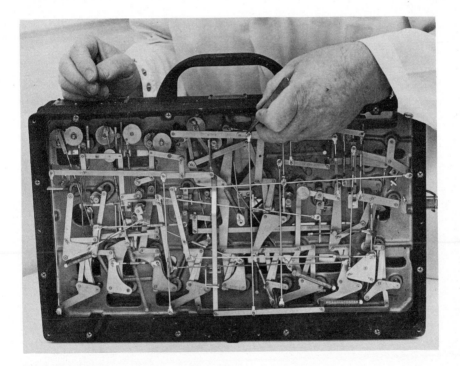

Fig. 15-18. Typical linkage computer. *(General Precision, Inc.)*

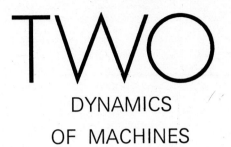

TWO

DYNAMICS
OF MACHINES

16

STATIC FORCES
IN MACHINES

16-1 INTRODUCTION

Forces in machines arise from various sources. There are forces due to weights of parts, forces of assembly, forces from applied loads, and forces from energy transmitted. There are also friction forces, inertia forces, spring forces, impact forces, and forces due to change in temperature. All these forces must be considered in the final design of a machine so that the parts can be proportioned so that they will not fail. In this book we shall consider the effect of all these forces except those due to assembly, change in temperature, and impact. For a consideration of these latter forces the student is referred to books on machine design.

In the analysis of static forces acting on machine members, inertia forces (caused by acceleration) are neglected. If the inertia forces are taken into account, the analysis is called a *dynamic-force analysis*. Often the weights of the parts of a machine are small compared with the other static forces which are present, and then they are neglected in the static-force analysis. In this chapter we shall consider only static forces.

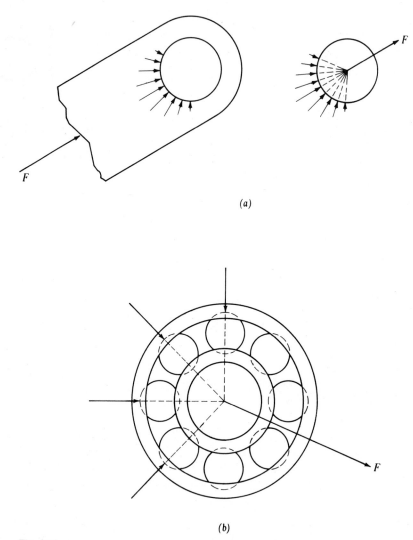

(a)

(b)

Fig. 16-1

16-2 TRANSMISSION OF FORCES IN A MACHINE

In Chap. 1 a machine was defined as a mechanism which transmits forces. Forces in machines are transmitted through the various pairs which connect the links and through the links themselves. The force transmitted from one link to another is normal to the surfaces in contact if friction is not considered. In well-lubricated machines the friction forces are small in comparison with the other forces present and can be disregarded.

A turning pair is illustrated by a journal bearing or a pin joint. In Fig. 16-1a the journal or pin fits into a hole in the link. The forces on the pin act normal to its surface, and thus their resultant force F passes through the center of the pin. Ball and roller bearings (Fig. 16-1b) are also examples of turning pairs. Since the forces in the figure act normal to the surfaces in contact, their resultant force F acts through the center of the bearing.

A lower sliding pair is represented by any link which slides on another and where there is contact over an area. A piston or slider (Fig. 16-2) is an example. Suppose P and Q are applied forces. Then the resultant of

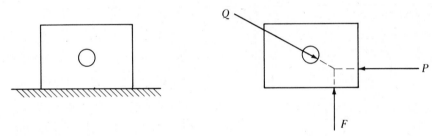

Fig. 16-2

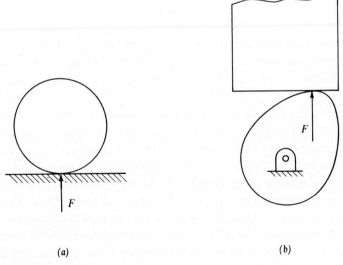

(a)   (b)

Fig. 16-3

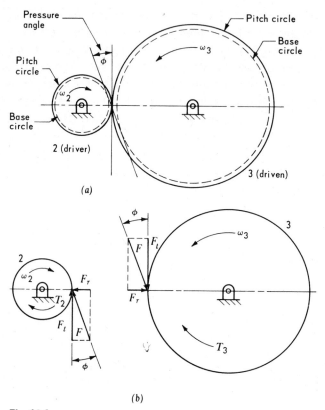

(a)

(b)

Fig. 16-4

the forces acting on the lower surface of the slider is F.   It acts normal to the contacting surfaces and must pass through the point of intersection of forces P and Q.   If force P acts through the center of the pin, then force F will also.

Higher pairs (rolling pairs and sliding pairs) have point or line contact.   If friction is neglected, the force transmitted through the pair is normal to the surfaces at their point of contact.   Rolling contact exists in a ball or roller bearing (Fig. 16-1b) or for the wheel rolling on a plane shown in Fig. 16-3a.   Higher pairing with sliding contact exists between a disk cam and flat-faced follower (Fig. 16-3b) and between the teeth of spur gears as illustrated in Fig. 16-4.   In higher pairs point or line contact does not exist if force is transmitted.   Because of the force the members deform, and then instead of contact at a point or along a line there is contact on a very small area.   However, the force can be assumed to act at a point.   In Fig.

16-4, if tooth friction is disregarded, the force transmitted between the gears is F as shown. This force passes through the pitch point and is directed along the pressure line, which is the common normal to the contacting teeth. F_t and F_r are the tangential and radial components of force F. T_2 is the driving torque applied to gear 2 by its shaft, and T_3 is the resisting torque exerted on gear 3 by its shaft.

16-3 SLIDER–CRANK MECHANISM

Figure 16-5a shows a slider-crank mechanism. A force P is applied to the piston and may be assumed to be the result of gas pressure. In order to maintain equilibrium, a moment or torque T_2 must be exerted on crank 2 by the shaft at O_2.

   The procedure which we will follow in all problems of force analysis will be to draw free-body diagrams of each of the links. In some instances we shall consider several links combined as a free body. In a free-body

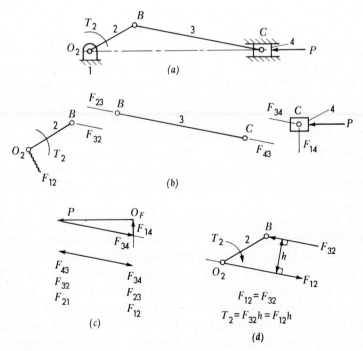

Fig. 16-5

diagram the body is isolated, and all the external forces and moments acting on the body must be shown.   If the number of unknowns is not greater than three, the problem can be solved by the equations of equilibrium.   If there are more than three unknowns on a single body, then additional information must be obtained from a consideration of the equilibrium of the other links before the problem can be solved.   The importance of isolating members and drawing their free-body diagrams cannot be overemphasized.

Figure 16-5b shows each body isolated with the known quantities of the forces indicated.   At a pin connection there can be no moment.   Thus at each end of link 3 there can be only a force.   Whenever a body is in equilibrium under the action of only two forces, the forces must be equal in magnitude, opposite in sense, and colinear.   In mechanics, such a body is known as a *two-force member*. F_{23} denotes the force body 2 exerts on body 3, and F_{43} is the force body 4 exerts on body 3.   We shall use a solid line without an arrowhead to indicate that the direction of a force is known but its magnitude is unknown.   Whether link 3 is in tension or compression cannot be determined from an analysis of link 3 itself.

Link 4 has three forces acting on it.   Force P is known in magnitude and direction and thus is shown in Fig. 16-5b as a solid line with an arrowhead. F_{34} is the force link 3 exerts on 4, and because it must be opposite to F_{43}, its direction is known.   However, its magnitude is unknown. F_{14} is the force link 1 exerts on link 4.   It is directed normal to the surfaces of contact and is unknown in magnitude.   There are only two unknowns in the free-body diagram for link 4.   They are the magnitudes of F_{34} and of F_{14}.

Link 2 has four unknowns: force F_{32} exerted by link 3 on link 2, known in direction but unknown in magnitude, force F_{12} exerted by link 1 on link 2, unknown in magnitude and direction, and the unknown moment T_2 exerted on crank 2 by the shaft.   A wavy line placed at O_2 indicates that we do not know the magnitude or direction of the force F_{12}, which acts through this point.

Link 4, which has only two unknowns, is analyzed first.   The two unknown magnitudes can be found by laying out a force polygon as shown in Fig. 16-5c.   Point O_F is any conveniently located point used as a pole for the force polygon, and force P is laid off using any convenient scale.   The magnitudes of F_{34} and F_{14} can then be scaled from the polygon.   Note that F_{34} and F_{14} must have the sense indicated because the sum of the forces acting on body 4 must be zero for equilibrium.   That is, when the vectors are added they must bring us back to the pole.   Since F_{34} is found from the polygon to act downward to the right, F_{43} must act upward to the left.

Thus F_{23} acts downward to the right, and we find that link 3 in this case is in compression.

From Fig. 16-5d we note that F_{12} must be equal and opposite to F_{32} to balance forces on link 2. However, the two equal, opposite, and parallel forces produce a couple which can only be balanced by another couple. The balancing couple T_2 is equal to $F_{32}h$; it is clockwise and is the torque which the shaft exerts on crank 2.

16-4 FOUR–BAR LINKAGE

The four-link mechanism in Fig. 16-6a has acting on it two forces, P and Q, as shown. A moment T_2 must be applied to link 2 to maintain equilibrium. Suppose we wish to find the forces at the various pins and the magnitude of the moment T_2.

We can first try to solve the problem by considering links 2, 3, and 4 combined as a free body. We will then have five unknowns: the magnitude and direction of F_{12}, the magnitude and direction of F_{14}, and the magnitude of T_2. Since there are only three equations of equilibrium, we cannot obtain a solution.

Next, we can isolate each link as a free body, as shown in Fig. 16-6b. If link 2 is considered, there are five unknowns, and if link 3 is considered, there are four unknowns; if we consider link 4 there are four unknowns to be found. Thus each link cannot be analyzed by itself. Note that if links 3 and 4 are considered, there are only six unknowns, because F_{43} is known to be equal and opposite to F_{34}. Since there are six equations of equilibrium, three for each link, we can obtain a solution. Hence we shall consider links 3 and 4 as shown in Fig. 16-6c, where F_{34} is broken into components $F_{34}{}^{N4}$ and $F_{34}{}^{T4}$, which are parallel and perpendicular, respectively, to O_4C. The magnitude of $F_{34}{}^{T4}$ is found by taking moments about O_4. That is

$$F_{34}{}^{T4} = \frac{Pa}{O_4C}$$

On link 3 the reactions at C are equal and opposite to those at C on link 4. The magnitude of $F_{43}{}^{T4}$ is known because the magnitude of $F_{34}{}^{T4}$ was found from the analysis of link 4. Next, if we examine link 3, we notice that there are three unknowns: magnitude and direction of F_{23} and mag-

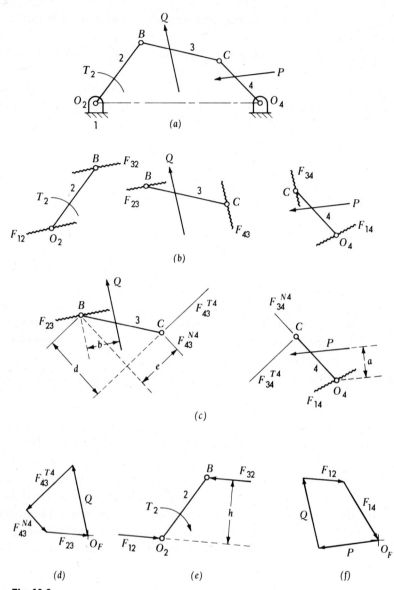

Fig. 16-6

nitude of $F_{43}{}^{N4}$. The magnitude of $F_{43}{}^{N4}$ can be found by taking moments about point B:

$$Qb - F_{43}{}^{T4}d + F_{43}{}^{N4}e = 0$$

Then

$$F_{43}{}^{N4} = \frac{F_{43}{}^{T4}d - Qb}{e}$$

Note that in the moment equation we assumed $F_{43}{}^{N4}$ to be directed upward to the left. Thus if the result for $F_{43}{}^{N4}$ comes out positive, it is directed as assumed. If the result comes out negative, then $F_{43}{}^{N4}$ is directed opposite to what was assumed and acts downward to the right. The force polygon for link 3, Fig. 16-6d, is drawn next in order to obtain the magnitude and direction of F_{23}.

In Fig. 16-6e, F_{32} must be equal and opposite to F_{23} found in Fig. 16-6d. Then F_{12} must be equal and opposite to F_{32}. Taking moments about O_2 we obtain T_2, the torque which the shaft at O_2 exerts on link 2, as follows:

$$T_2 = F_{32}h$$

F_{14} is obtained from the force polygon for bodies 2, 3, and 4, taken as a whole system as shown in Fig. 16-6f.

16-5   SHAPER MECHANISM

In the shaper mechanism, Fig. 16-7a, crank O_3B is integral with gear 3, and gear 2 is the driver. Suppose we want to find the force F_{23} acting on the gear teeth in order to overcome the known resistance Q, and the reactions at O_2, O_3, and O_5 are to be determined. We can begin by considering slider 7 as a free body as shown in Fig. 16-7b. The directions of F_{67} and F_{17} are known and their magnitudes can be found from a force polygon as shown. In Fig. 16-7c the free-body diagram for link 6 is drawn. F_{76} is equal and opposite to F_{67}, and because 6 is a two-force body, F_{56} must be equal and opposite to F_{76}.

The free-body diagram for link 5 is shown in Fig. 16-7d where F_{65} is equal and opposite to F_{56}. F_{45} is directed perpendicular to link 5 but its magnitude is unknown. F_{15} is unknown in magnitude and direction. From a scale drawing we can scale off the values of b and e. Then by taking moments about O_5, the magnitude of F_{45} can be computed. Next, from a force polygon in Fig. 16-7e, the magnitude and direction of F_{15} can be deter-

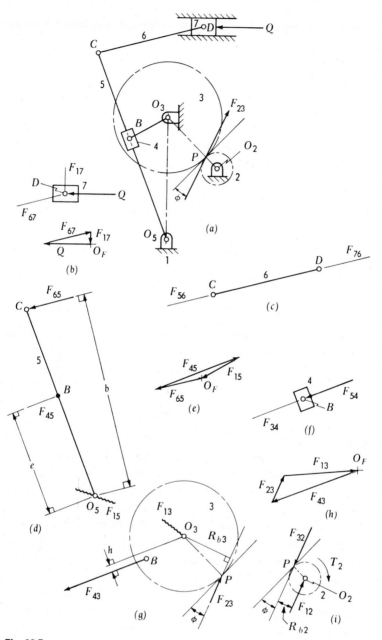

Fig. 16-7

mined.   The free-body diagram for slider 4 is shown in Fig. 16-7f, where F_{54} is equal and opposite to F_{45}. F_{34} must be equal and opposite to F_{54}.

The free-body diagram for link 3 appears in Fig. 16-7g, where F_{43} is equal and opposite to F_{34}, and F_{13} is unknown in magnitude and direction. The values of the moment arm h and the radius R_{b3} of the base circle can be scaled from a drawing where ϕ is the pressure angle.   By taking moments about O_3, the magnitude of F_{23} can be computed.   Next, the magnitude and direction of F_{13} can be found from a force polygon as shown in Fig. 16-7h. Finally, as shown in Fig. 16-7i, F_{32} is equal and opposite to F_{23}, and F_{12} is equal and opposite to F_{32}. T_2, the torque which the pinion shaft exerts on the pinion, equals $F_{32} \times R_{b2}$ and is clockwise.

PROBLEMS

In the following problems, friction forces are to be neglected unless otherwise stated.

16-1   In Fig. P16-1, $P = 200$ lb.   (a) Draw the mechanism half-size and find the torque in inch-pounds which the shaft at O_2 exerts on the crank.   Force scale: 1 in. = 100 lb.   (b) Determine the forces which bodies 2 and 4 exert on the frame.

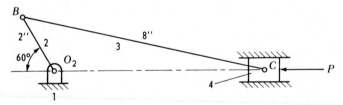

Fig. P16-1

16-2   In the linkage in Fig. P2-20, the shaft at O_2 exerts a torque of 5 in.-lb cw on link 2. Also, there is a 10-lb force acting vertically downward on link 3 midway between B and C.   Draw the mechanism full-size and use a force scale of 1 in. = 5 lb. Determine the resisting torque which the shaft at O_4 exerts on link 4 and find the forces exerted on the frame at O_2 and O_4.

16-3   In the mechanism in Fig. P2-22, the torque in the camshaft is 10 in.-lb.   Draw the mechanism full-size and determine the resisting torque in the shaft of the follower and the forces exerted on the frame at O_2 and O_4 by bodies 2 and 4 if friction between bodies 2 and 4 is neglected.

16-4   In the linkage in Fig. P6-12, disregard velocity V_E, but assume there is a force of 100 lb applied to slider 6 and that it acts vertically downward through point E. Draw the mechanism half-size and use a force scale of 1 in. = 100 lb.   Determine the resisting torque which the shaft at O_2 exerts on 2 and find all the forces exerted on the frame.

16-5   Figure P16-5 shows an inversion of the slider-crank mechanism.   Crank 1 is the frame or stationary member.   Link 4 rotates about point O_4 in the frame. A shaft having its axis perpendicular to the paper and located at O_4 is attached

to link 4. The piston diameter is 2.5 in. and the gas pressure p is 50 psig.
Draw the mechanism half-size and use a force scale of 1 in. = 100 lb. Determine the resisting torque T_4 which the shaft exerts on link 4. Also, find the
forces exerted on the frame.

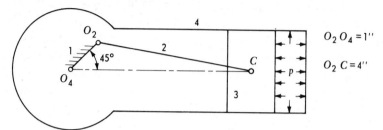

Fig. P16-5

16-6 In Fig. 3-21, the distance from O_2 to line O_4D is 4 in. Crank 2 is at 30° to the
horizontal. $O_2B = 0.75$ in., $BC = 3$ in., $O_4C = CD = 3$ in. Draw the mechanism half-size and use a force scale of 1 in. = 25 lb. Determine the force F
if the shaft at O_2 applies a torque of 25 in.-lb cw to crank 2. Also, find all the
forces exerted on the frame.

16-7 Figure P16-7 shows a Rockwood or pivoted motor belt drive. F_1 and F_2 are the
belt forces and W is the weight of the motor and base. By shifting the pivot
point O_2, the moment arm c can be changed. By increasing c, both forces F_1
and F_2 are increased, permitting more power to be transmitted. Let $F_1 = 300$
lb, $F_2 = 100$ lb, and $W = 150$ lb. $a = 3$ in., $b = 10$ in. Compute the value of c.
Draw a diagram of the mechanism one-sixth size, use a force scale of 1 in. = 100
lb, and determine the magnitude and direction of the force F_{21} which the base
exerts on the frame at O_2.

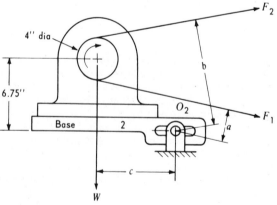

Fig. P16-7

16-8 The garage-type automobile jack shown in Fig. P16-8 is actuated by hydraulic
pressure in the cylinder. Lay out links 2, 3, 4, and 5 to a scale of 1 in. = 4 in.
and use a force scale of 1 in. = 1,000 lb. Determine (*a*) the force exerted by
the cylinder to raise a weight $W = 1,000$ lb neglecting friction and (*b*) the force
in link 2.

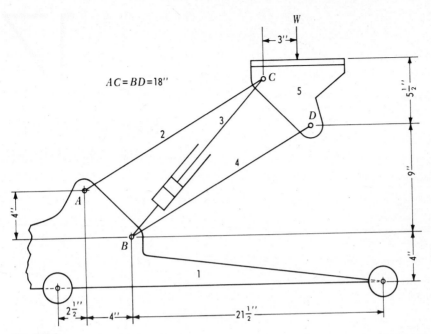

Fig. P16-8

16-9 An elevator brake is shown schematically in Fig. P16-9. The coefficient of friction for the brake shoes is 0.3. For an operating force $P = 100$ lb, determine (a) the braking torque for clockwise rotation of the drum and (b) the braking torque for counterclockwise rotation of the drum.

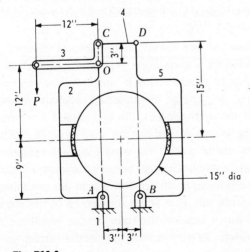

Fig. P16-9

17

INERTIA FORCES
IN MACHINES

17-1 INTRODUCTION

In the preceding chapter it was explained that all the forces acting on a machine are considered static forces except those due to acceleration. Further, we saw how the static forces are transmitted through the links of a mechanism.

Forces due to acceleration are called *inertia forces* or *dynamic forces*. In order to analyze inertia forces, a knowledge of accelerations is essential, and that is why the analysis of accelerations in mechanisms was covered in detail in earlier chapters. In general, the links of a mechanism are subjected to both static forces and inertia forces. In high-speed machines the accelerations and resulting inertia forces can be very large in relation to the static forces which do useful work. For example, in a reciprocating engine such as an automobile engine, at high speeds the inertia forces can be greater than the force produced on the piston due to gas pressure. In a gas turbine the inertia force due to a small unbalance of the rotor can produce, on the bearings which support the rotor, forces which are many times the weight of the rotor. In such cases the inertia forces must be considered in the design of the machine.   In slow-speed machinery the inertia forces may be negligible.

In this chapter we shall study how the inertia forces along with the static forces are transmitted through the links of a mechanism and what effect they have on the frame.   Such a combined-force analysis is called a *complete-force analysis*.

17-2   EQUATIONS OF MOTION

If we know the acceleration of one point in a body, and if the angular velocity ω and acceleration α of the body are known, then the acceleration of any other point in the body can be found.   For example, in Fig. 17-1a, let us assume that A_B, the acceleration of point B, is known.   Then the acceleration A_C of any point C is

$$A_C = A_B + A_{C/B}^n + A_{C/B}^t \tag{17-1}$$

where the magnitude of $A_{C/B}^n$ is $(BC)\omega^2$ and the magnitude of $A_{C/B}^t$ is $(BC)\alpha$. The solution for A_C is shown in Fig. 17-1b.

The displacement of a body can be considered as a linear displacement of some point in the body plus an angular displacement of the body about this point.   The same concept holds for the velocity of a body.   It follows that the acceleration of a body can be considered as a linear acceleration of some point in the body plus an angular acceleration of the body about this point.   It is convenient to take the center of gravity as this point.

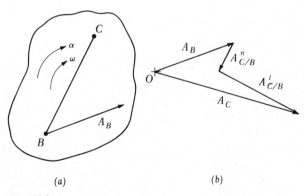

(a)   (b)

Fig. 17-1

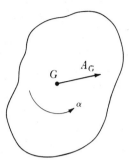

Fig. 17-2

Consider the body in Fig. 17-2.   Let G be the center of gravity of the body and A_G the acceleration of this point.   Also, suppose that the angular acceleration α for the body is known in magnitude and has the sense shown. We want to find what force, torque, or force and torque must be applied to the body to produce A_G and α.

The body of Fig. 17-2 is shown again in Fig. 17-3a.   From mechanics we know that a force F applied at G and acting in the direction of A_G will produce this linear acceleration; that is

$$F = MA_G \qquad\qquad (17\text{-}2)$$

where M is the mass of the body.   This is Newton's equation for linear motion.   In order to produce the angular acceleration α, a torque T must be applied to the body in the same sense as α such that

$$T = I\alpha \qquad\qquad (17\text{-}3)$$

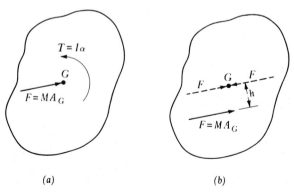

(a)   (b)

Fig. 17-3

where I is the mass moment of inertia of the body about an axis which passes through G and is perpendicular to the plane of rotation; i.e., the axis is perpendicular to the paper. The latter equation is Newton's law for angular motion. In general, a body having plane motion may be subjected to more than a single force and a single torque. Then in Fig. 17-3a, F is the resultant of all forces acting, and T is the resultant of all the torques acting on the body. Hence, F is called the *resultant force* and T the *resultant torque.*

In mechanics the terms *torque, moment,* and *couple* are often used synonymously. An important principle of mechanics is that a force and a couple can be replaced by a single force. The body in Fig. 17-3a appears again in Fig. 17-3b, where the three forces shown are equal in magnitude. The force shown in dashed lines acting upward to the right fills the purpose of the force F in Fig. 17-3a. The dashed force acting downward to the left along with the force F shown solid comprise a couple Fh, which must equal T in Fig. 17-3a in both magnitude and sense. Hence

$$Fh = T = I\alpha$$

or

$$h = \frac{I\alpha}{F} \tag{17-4}$$

In Fig. 17-3b, the two equal and opposite forces at G cancel, and thus the single force F at a distance h from the center of gravity replaces the force F and the couple (or torque) T in Fig. 17-3a.

17-3 INERTIA FORCE AND INERTIA TORQUE

In discussing equations of motion, we saw that when a body is acted upon by a system of forces, the acceleration of the center of gravity is given by the equation

$$F = MA_G \tag{17-5}$$

where F is the resultant force. Also, when a body is acted upon by a system of torques, the angular acceleration of the body is given by the equation

$$T = I\alpha \tag{17-6}$$

where T is the resultant torque.

The *inertia force* is defined as the reversed resultant force and the *inertia torque* is defined as the reversed resultant torque. Along with the resultant force acting on the body, if we assume the inertia force to be acting with the same magnitude and line of action but opposite in sense, then the acceleration of the center of gravity will be zero. Also, along with the resultant torque acting on the body, if we assume the inertia torque to be acting with the same magnitude but opposite in sense, then the angular acceleration of the body will be zero. Thus, by adding the inertia force and inertia torque to a body which is acted upon by a resultant force and a resultant torque, the body is brought to equilibrium. This is known as D'Alembert's principle and aids in the solution of problems in dynamics by permitting them to be solved as problems in statics.

17-4   INERTIA FORCES ON A FOUR–BAR LINKAGE

As an example to illustrate inertia forces acting on a mechanism, let us consider the four-bar linkage in Fig. 17-4a, where the magnitude of ω_2 is assumed known and constant. Points G_2, G_3, and G_4 denote the centers of gravity of links 2, 3, and 4. In our analysis we shall determine the torque which the shaft at O_2 must exert on link 2 to give the desired motion.

The acceleration polygon must be constructed in order to find the linear accelerations of points G_2, G_3, and G_4. From the magnitude and sense of the tangential components of acceleration on the polygon, the magnitude and sense of α_3 and α_4 can be determined. Their sense was found to be as shown in Fig. 17-4a. It is important at this time to indicate in Fig. 17-4a the sense of the angular acceleration of each link so that later the sense of the resultant torque and inertia torque on each link will be indicated correctly.

Link 2 is shown in Fig. 17-4c, where A_{G_2} is the acceleration of the center of gravity. The resultant force F_2, equal to $M_2 A_{G_2}$, where M_2 is the mass of link 2, has the same sense and line of action as A_{G_2}. The inertia force f_2 is equal and opposite to F_2. Note: f will be used to denote inertia force.

In Fig. 17-4d, link 3 is shown with the acceleration of the center of gravity G_3 indicated as A_{G_3}. The resultant force F_3, equal to $M_3 A_{G_3}$, where M_3 is the mass of link 3, has the same sense and line of action as A_{G_3}. f_3 is the inertia force. In order to produce α_3, there must be a resultant torque T_3 equal to $I_3\alpha_3$ having the same sense as α_3 and where I_3 is the mass moment of inertia of link 3 about an axis perpendicular to the paper and

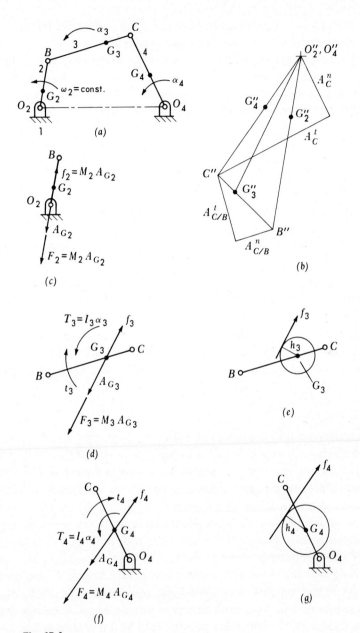

Fig. 17-4

passing through G_3. Equal and opposite to T_3 is the inertia torque t_3 as shown. Note that t will be used to denote inertia torque.

Link 3 is again shown in Fig. 17-4e, where the inertia force f_3 and inertia couple t_3 have been replaced by a single force f_3 as shown. The magnitude, direction, and sense of f_3 must be the same as in Fig. 17-4d, but the line of action is displaced from G_3 by an amount h_3 such that

$$f_3 h_3 = t_3$$

or

$$h_3 = \frac{t_3}{f_3} = \frac{I_3 \alpha_3}{M_3 A_{G_3}}$$

In Fig. 17-4e, f_3 can be easily located by drawing a circle of radius h_3 with its center at G_3. Note that f_3 is drawn tangent to the left side of the circle rather than the right because f_3 must produce a torque about G_3 in the same sense as t_3.

Link 4 is shown in Fig. 17-4f where f_4, the inertia force, is equal and opposite F_4. T_4, the resultant torque, equals $I_4 \alpha_4$ where I_4 is the mass moment of inertia of link 4 about an axis perpendicular to the paper and passing through G_4. The inertia torque t_4 is equal and opposite T_4. Link 4 appears again in Fig. 17-4g, where the inertia force f_4 and inertia couple t_4 have been replaced by a single force f_4 as shown. Since $f_4 h_4$ must equal t_4,

$$h_4 = \frac{t_4}{f_4} = \frac{I_4 \alpha_4}{M_4 A_{G_4}}$$

We shall next find the forces at each pin connection and the torque which the shaft at O_2 exerts on link 2 to give the prescribed motion. The free-body diagrams of links 2, 3, and 4 are shown in Fig. 17-5a to c. The inertia forces are treated as known external forces, and each link is in equilibrium under the action of the inertia forces and the unknown reactions. The determination of these reactions is then the same as for a static analysis as explained in Chap. 16. (It is recommended that the student review that chapter.)

Starting with link 4, we can take moments about point O_4 and determine $F_{34}{}^{T4}$. Then on link 3, $F_{43}{}^{T4}$ will be equal and opposite to $F_{34}{}^{T4}$. For equilibrium of link 3 the sum of the moments about B equals zero. This determines $F_{43}{}^{N4}$. The force polygon for link 3 is shown in Fig. 17-5d and is drawn next, in order to determine F_{23}.

Link 2 appears in Fig. 17-5e. Here F_{32} is equal and opposite to F_{23} found in Fig. 17-5d. Then F_{12} must be equal and opposite to $f_2 \nleftrightarrow F_{32}$.

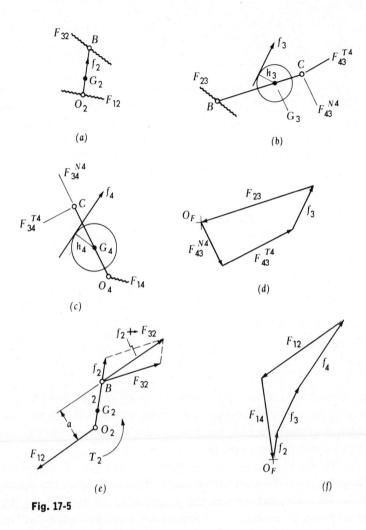

Fig. 17-5

Taking moments about O_2 we obtain T_2, the torque which the shaft at O_2 exerts on link 2, as follows:

$$T_2 = (f_2 + F_{32})a$$

Force F_{14} is obtained from the force polygon for bodies 2, 3, and 4 taken as a whole system as shown in Fig. 17-5f.

Shaking force is defined as the resultant of all the forces acting on the frame of a mechanism due to inertia forces only. A consideration of this

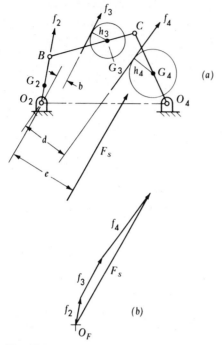

Fig. 17-6

force is important because the frame must be strong enough to withstand it. The shaking force may set up troublesome vibrations in the frame, and if the machine is placed in a building, this force will be transmitted to the floor and may have disturbing effects.

The mechanism of Fig. 17-4 is shown again in Fig. 17-6a. The shaking force is always equal to the sum of the inertia forces acting on the mechanism. If we are interested only in the magnitude and direction of the shaking force, it can be determined from the force polygon shown in Fig. 17-6b, where F_s denotes shaking force. However, if we are interested also in the location of the shaking force, we can determine this by taking moments about any convenient point in the plane of the inertia forces. Since the shaking force is the resultant of the inertia forces, it must produce the same moment. If we take moments about point O_2, then

$$F_s e = f_3 b + f_4 d$$

or

$$e = \frac{f_3 b + f_4 d}{F_s}$$

17-5 COMBINED STATIC– AND INERTIA–FORCE ANALYSIS

We have discussed static forces and inertia forces in mechanisms as separate analyses. There is no need to separate them; we can combine them in a single analysis. The total forces acting on the links of a mechanism are the sums of the static and inertia forces.

Slider-crank mechanism

In Fig 17-7a let P, the force on the piston due to gas pressure, be known and also let ω_2 be known and assumed constant. Points G_2, G_3, and G_4 are the centers of gravity of links 2, 3, and 4. Suppose we wish to find the torque which crank 2 exerts on the crankshaft. The magnitude, direction, and location of the shaking force are also to be determined.

The velocity and acceleration polygons are constructed first; the latter appears in Fig. 17-7b. Links 3 and 4 combined as a free body are shown in Fig. 17-7c. In all our problems, unless otherwise specified, we shall assume that the weights of the links are small compared with the inertia forces, and thus the gravity force acting downward through the center of gravity of each link will be omitted. The inertia force f_3, its moment arm relative to G_3, and inertia force f_4 are determined as in the previous example. The unknowns are the magnitude and direction of F_{23} and the magnitude of F_{14}. By taking moments about B, the magnitude of F_{14} is found as follows:

$$F_{14}a + f_3b + f_4d - Pd = 0$$

or

$$F_{14} = \frac{Pd - f_3b - f_4d}{a}$$

Force F_{23} can then be found by a summation of forces on bodies 3 and 4 taken together as a free body. The force polygon appears in Fig. 17-7d.

The free-body diagram for link 2 is shown in Fig. 17-7e, where F_{12} must be equal and opposite to $f_2 \dplus F_{32}$ for equilibrium. T_2 is the torque which the shaft at O_2 must exert on crank 2 for equilibrium; thus

$$T_2 = (f_2 \dplus F_{32})e$$

The torque which the crank exerts on the crankshaft is equal to T_2 but opposite in sense.

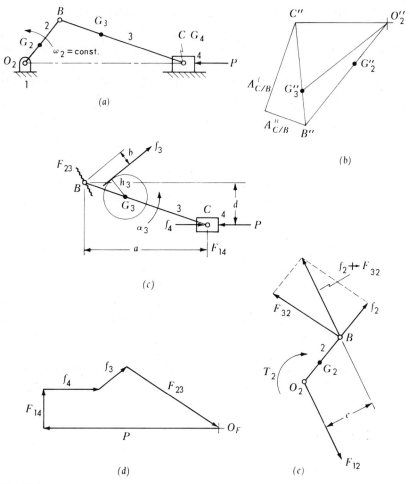

Fig. 17-7

Shaper mechanism

Let us consider the mechanism in Fig. 17-8a where link 2 is the driver and rotates with a known constant angular velocity ω_2. The motion is opposed by a known force P acting on slider 6 as shown. We wish to find the forces at all the pairs or joints and the torque which the shaft at O_2 must exert on crank 2 to drive the mechanism. It will be assumed that the forces due to the weights of the links are negligible. The acceleration polygon appears in Fig. 17-8b. In the remainder of the figure each link is shown isolated and the magnitude, direction, and location of the inertia force on each link are determined as in the previous examples.

The force analysis is started with link 6, which is shown in Fig. 17-8c. The unknowns are the magnitude of F_{16} and the magnitude and direction of F_{56}. The horizontal component of F_{56} is $F_{56}{}^H$, and its magnitude can be found from the summation of horizontal forces on link 6.

In Fig. 17-8d, force $F_{65}{}^H$ is equal and opposite to $F_{56}{}^H$. The magnitude of $F_{65}{}^V$ can be found by summation of moments about C. Then from a force polygon for link 5, the magnitude and direction of F_{45} are found. Next, in Fig. 17-8e, F_{54} is known and is equal and opposite to F_{45}. There are four unknowns in Fig. 17-8e: the magnitude and direction of F_{14} and the magnitude and location of F_{34}. Since there are only three equations of equilibrium, we cannot determine these forces by considering this link alone. If we consider link 3 shown in Fig. 17-8f, we note that here there are also four unknowns: F_{23} is unknown in magnitude and direction and F_{43} is perpendicular to link 4 but is unknown in magnitude and location. However, for the combination of links 3 and 4 we have only six unknowns, and they can be analyzed in combination. From the free body of link 3 we see that F_{23} causes no torque about the center of gravity B_3, and thus F_{43} must be of such a magnitude as to balance the forces, and its line of action must be displaced from B_3 enough to balance the inertia torque. F_{43} can then be resolved into a force passing through B_3 and a pure torque about B_3 sufficient to balance the inertia torque. This is shown in Fig. 17-8g. The equal and opposite force and torque on link 4 as shown in Fig. 17-8h makes link 4 a free body with three unknowns. The magnitude of F_{34} can be found by setting the sum of the moments about O_4 equal to zero. F_{14} can then be found from a force polygon for link 4.

We replaced F_{43} in Fig. 17-8f with the force F_{43} and T_{43}, which are shown in Fig. 17-8g. Thus in Fig. 17-8f

$$F_{43}a = T_{43} = I_3\alpha_3$$

and thus the moment arm a can be determined as follows:

$$a = \frac{I_3\alpha_3}{F_{43}}$$

This locates the actual line of action of F_{43}. In the free-body diagram of link 3, F_{43} will be equal and opposite to F_{34}, which was determined earlier. F_{23} can now be determined from a force polygon for link 3.

The free-body diagram of link 2 is shown in Fig. 17-8i. F_{32} is equal

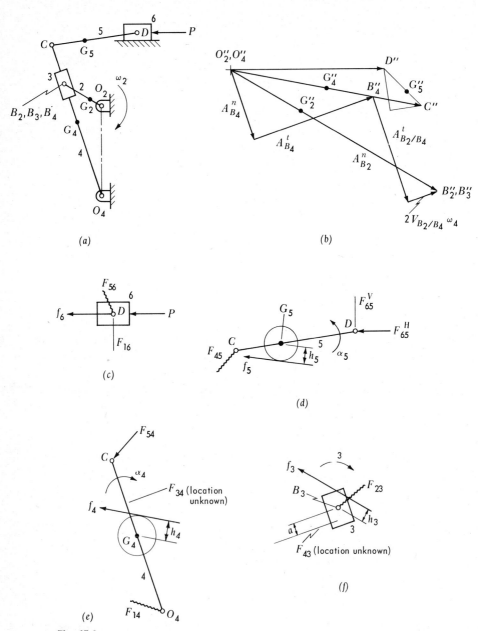

Fig. 17-8

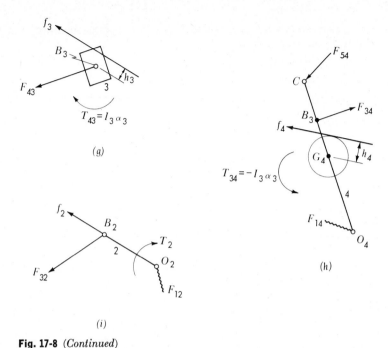

(g)

(h)

(i)

Fig. 17-8 (*Continued*)

and opposite to F_{23}, and F_{12} can be determined from a force polygon for link 2. Finally, by summing moments about O_2, the torque T_2, which the shaft at O_2 exerts on crank 2, can be determined.

Cam mechanism

At high speed the contact force for a cam and follower can be high and cause serious wear. A typical disk cam with radially reciprocating follower is shown in Fig. 17-9. The various forces are as follows:

P = force exerted on follower by whatever other body above (not shown) it actuates

f = inertia force of follower

W = weight of follower

F_s = spring force exerted on follower

F_1, F_2 = normal forces exerted by frame on follower

N = normal force exerted by cam on follower

a = follower overhang

b = distance between bearing surfaces (for a single bearing, b is the length of the bearing)

d = diameter of follower stem

ϕ = pressure angle

μ = coefficient of friction between follower and its guide

Forces P, f, W, and F_s act along the centerline of the follower, and we

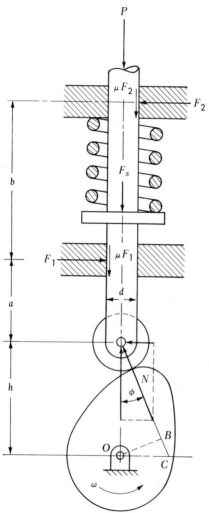

Fig. 17-9

shall let F denote their sum. Then

$$F = P + f + W + F_s \tag{17-7}$$

Summation of vertical forces on the follower gives

$$N \cos \phi = F + \mu(F_1 + F_2) \tag{17-8}$$

In the horizontal direction

$$F_1 = F_2 + N \sin \phi \tag{17-9}$$

Summing moments about the point of application of F_1 gives

$$F_2 (b - \mu d) = Na \sin \phi + \frac{d}{2} (F - N \cos \phi) \tag{17-10}$$

Eliminating F_1 and F_2 from the last three equations, we obtain

$$N = \frac{Fb}{b \cos \phi - (2\mu a + \mu b - \mu^2 d) \sin \phi} \tag{17-11}$$

This equation gives the normal force on the cam for any cam position in which the velocity of the follower is upward. It is to be noted, however, that ϕ and F vary with the angular position of the cam. The friction forces have a considerable effect on N when the pressure angle becomes large. N becomes infinite when the denominator of Eq. (17-11) becomes zero. Hence the limiting value for ϕ is found as follows:

$$b \cos \phi_m - (2\mu a + \mu b - \mu^2 d) \sin \phi_m = 0$$

or

$$\tan \phi_m = \frac{b}{\mu(2a + b - \mu d)} \tag{17-12}$$

The inertia force depends on the mass and acceleration of the follower and will reverse in direction when the acceleration reverses direction. The spring force must at all times be sufficient to keep the follower in contact with the cam. If the spring force is not sufficiently large, the follower will move away from the cam at high cam speeds. This is called *jump* and when the follower returns and strikes the cam, impact loading occurs, which can cause vibrations. The spring should have a preload so as to ensure contact even when the follower is in its lowest position. The spring force is usually supplied by a helical spring having a constant rate; thus the spring force will be proportional to displacement. In designing the spring, a curve of the external force P plus weight W of the follower plus the inertia force f should be plotted versus cam position.   From this curve the spring force at the critical positions can be determined.

The torque T required to drive the cam can be determined from a consideration of Fig. 17-9; thus

$$T = N(OB) \tag{17-13}$$

Since point C is the instant center, the velocity of the follower is

$$V = (OC)\omega \tag{17-14}$$

Equations (17-11) to (17-14) are valid also for a cam having an offset follower.   It can be seen that Eqs. (17-11), (17-13), and (17-14) are applicable also to a cam having a flat-faced reciprocating follower if friction between cam and follower is neglected.

Planetary gear train

It is desired to find the tooth forces on the various gears of the planetary train shown in Fig. 17-10a.   Gear 2 is the driver and rotates clockwise at n_2 rpm, and T_2 is the driving torque in the shaft of gear 2.   The arm (link 3) is the driven member and rotates at n_3 rpm. T_3 is the resisting torque which exists in the driven shaft (link 3).   Spur gears having involute teeth with a pressure angle of ϕ degrees are used.

The free-body diagrams of each of links 2, 3, and 4, as shown in the figure, aid in determining the forces required for static equilibrium.   The inertia forces are zero for links 1, 2, and 3 because the accelerations of the mass centers for these members are zero; the inertia torques are also zero because the train operates at constant angular velocity.   However, there are inertia forces acting on the planet gears because of the normal acceleration of their mass centers.

The free-body diagram for gear 2 appears in Fig. 17-10b, where D_2 is the pitch diameter and the two forces F_{42} are the tooth forces.   Since hp, the horsepower transmitted, and n_2, the rpm of gear 2, are known, the torque T_2 in inch-pounds which the driving shaft exerts on gear 2 can be found as follows:

$$T_2 = \frac{\text{hp } (63{,}000)}{n_2}$$

Since two couples in equilibrium act on gear 2,

$$F_{42}D_2 \cos \phi = T_2$$

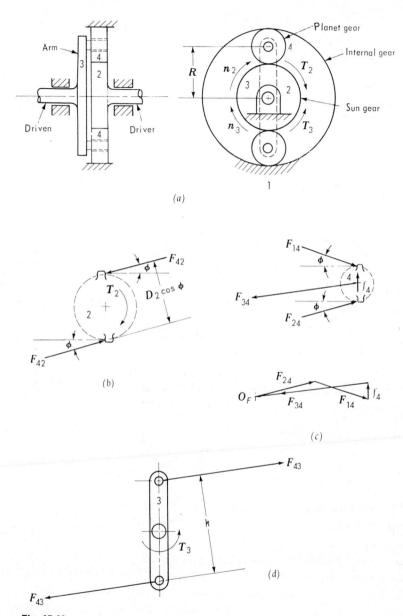

Fig. 17-10

or

$$F_{42} = \frac{T_2}{D_2 \cos \phi}$$

In Fig. 17-10c the free-body diagram for the planet gear 4 is shown. Here F_{24} is equal and opposite to F_{42}. Since the sum of the moments about the center of gear 4 must be zero, F_{14} must be equal in magnitude to F_{24}. The acceleration of the mass center of this gear is

$$A_{G_4} = R\omega_3{}^2$$

where ω_3 is the angular velocity of link 3 and R is the distance shown in Fig. 17-10a. The inertia force on gear 4 is then

$$f_4 = M_4 A_{G_4}$$

and is directed as shown in Fig. 17-10c. The only unknown force in Fig. 17-10c is F_{34}, which acts through the center of gear 4. It can be determined from the force polygon as shown.

From the free-body diagram of link 3 in Fig. 17-10d, the resisting torque T_3 which the driven shaft exerts on the arm can be determined. For equilibrium of moments about the axis of the arm,

$$T_3 = F_{43}h$$

T_3 (in.-lb) could also be obtained from hp, the horsepower transmitted, and n_3, the rpm of the arm, as follows:

$$T_3 = \frac{\text{hp } (63,000)}{n_3}$$

17-6 DETERMINATION OF CENTER OF GRAVITY AND MOMENTS OF INERTIA

In order to analzye the inertia forces acting on the links of a mechanism, the location of the center of gravity for each link must be known. The *center of gravity* of a body is that point through which the weight of the body acts regardless of the position of the body. Center of gravity is also called *center of mass*. The mass center of a system of particles we will call G. Its x, y, and z coordinates x_G, y_G, and z_G from an arbitrary origin are given by

$$Mx_G = \sum m_i x_i \qquad My_G = \sum m_i y_i \qquad Mz_G = \sum m_i z_i$$

in which m_i represents any mass point, and x_i, y_i, and z_i are its coordinates. The total mass of the system of particles is M.

Many machine elements have two axes of symmetry in their plane of motion, and the center of gravity of such bodies lies at the intersection of these axes. Examples are flywheels, pulleys, gears, and sliders. Engineering handbooks give equations for locating the center of gravity for bodies having common geometrical shapes. And often the location of the center of gravity for an irregular-shaped body can be conveniently computed by considering the body to be composed of a number of segments, each in the form of one of these common shapes. The procedure for doing this is explained in elementary mechanics textbooks.

One experimental method which can be used for locating the center of gravity is to suspend the body at some point about which it is free to rotate and to draw a vertical line on the body through the point of suspension. Then by suspending the body at a different point, and by again drawing a vertical line on the body through the point of suspension, the intersection of these lines will locate the center of gravity. It should be mentioned that for some odd-shaped bodies the intersection of these lines may not intersect at a point on the physical body. This is because the body can be considered as a system of masses, and the center of gravity of a system of masses may not lie within one of the masses.

Another method for experimentally determining the location of the center of gravity, which is convenient to use for some members, is to support the member on two scales as shown in Fig. 17-11. The total distributed weight of the member is W, and it acts through the center of gravity G. The scale readings give the values of the reaction forces F_1 and F_2. If we sum

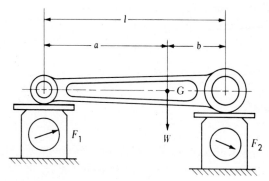

Fig. 17-11

moments about the left support, then

$$Wa = F_2 l$$

and since W is the sum of the scale readings,

$$a = \frac{F_2 l}{F_1 + F_2}$$

The moment of inertia I of a body about a given axis is defined as

$$I = \sum m_i r_i^2$$

where m_i is any mass point in the body and r is the distance from the axis to the mass point. Ordinarily the moment of inertia is desired about an axis passing through the center of gravity. If we call its value I, then the moment of inertia about some parallel axis is

$$I_o = I + Md^2$$

where M is the total mass of the body and d is the distance between the axes. The latter is known as the *parallel-axis theorem;* elementary mechanics textbooks explain how we can use this theorem to find the moment of inertia of some composite body about any desired axis, if we know the moment of inertia of each segment of the body about an axis through the center of gravity of the segment.

An experimental method for determining the moment of inertia of a body is illustrated in Fig. 17-12. The body is supported on a knife-edge at any point O, other than the center of gravity. If the body is displaced a small angle θ and released, it will oscillate about point O, and by observing the time required for a given number of oscillations, we can ultimately determine the moment of inertia about the center of gravity G.

The relation between the moment about point O and the angular acceleration α is

$$T_o = I_o \alpha \qquad\qquad (17\text{-}15)$$

or

$$-Wr \sin \theta = I_o \frac{d^2\theta}{dt^2}$$

where $r \sin \theta$ is the moment arm of the force W, and the negative sign is used because the torque is opposite in direction to the angle θ. If θ is small, the sine of the angle is approximately equal to the angle in radians,

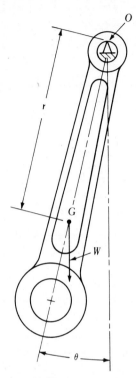

Fig. 17-12

and the above equation can be rewritten

$$-Wr\theta = I_o \frac{d^2\theta}{dt^2}$$

or

$$\frac{d^2\theta}{dt^2} + \frac{Wr}{I_o}\theta = 0$$

which is the differential equation of motion. This is a linear differential equation of the second order, and the general solution is

$$\theta = A \sin\sqrt{\frac{Wr}{I_o}}t + B \cos\sqrt{\frac{Wr}{I_o}}t$$

The boundary conditions are that when $t = 0$, $\theta = \theta_{max}$, and $d\theta/dt = 0$; that is, the angular speed is zero. Solving for the constants of integration, we find that $A = 0$ and $B = \theta_{\max}$. Thus

$$\theta = \theta_{\max} \cos \sqrt{\frac{Wr}{I_o}}\, t$$

The function is a cosine wave which goes through a complete cycle when

$$\sqrt{\frac{Wr}{I_o}}\, t = 2\pi$$

Solving for t, we find the time required for one cycle, which is called the period T.

$$T = 2\pi \sqrt{\frac{I_o}{Wr}} \tag{17-16}$$

Solving this for I_o, we obtain

$$I_o = Wr \left(\frac{T}{2\pi}\right)^2 \tag{17-17}$$

The moment of inertia I about the center of gravity may then be found by the parallel-axis theorem; thus

$$I = I_o - Mr^2$$

$$= Wr \left[\left(\frac{T}{2\pi}\right)^2 - \frac{r}{g} \right] \tag{17-18}$$

Accuracy in the determination of I depends on the accuracy in determining T and r. If the two terms in the brackets are of the same order of magnitude, then a small error in either one of them will make for a large error in I. In order to make this error small, it is desirable to make T large and r small. This can be done for the rod in Fig. 17-12 by suspending it from the end opposite that which is shown.

The moment of inertia of a part may also be determined by supporting it on a pendulum consisting of a lightweight table suspended by chords as shown in Fig. 17-13. To determine the moment of inertia of the part about an axis G-G through its center of gravity, it is placed on the table so that the axis G-G is directly below and parallel to the pivot axis O-O. The period for small oscillations can be observed by counting the number of oscillations which occur in several minutes. The moment of inertia of the part about its center-of-gravity axis G-G can be determined as follows. Let

W_p = weight of part
W_t = weight of table

r_p = distance from O to center of gravity of part
r_t = distance from O to center of gravity of table
I_{po} = moment of inertia of part about axis O-O
I_{to} = moment of inertia of table about axis O-O
T = period of table with part
T_t = period of table alone

Equation (17-16) gives the period of the table with the part on it, where I_o is the sum of the moments of inertia of the table and the part about axis O-O, and W is the total weight of the table and the part.   Thus

$$T = 2\pi \sqrt{\frac{I_{po} + I_{to}}{(W_p + W_t)r}} \qquad (17\text{-}19)$$

where r is the distance from O-O to the center of gravity of the part and table combined.   Elementary texts on mechanics explain that r may be found by taking static moments about O-O as follows:

$$(W_p + W_t)r = W_p r_p + W_t r_t$$

or

$$r = \frac{W_p r_p + W_t r_t}{W_p + W_t} \qquad (17\text{-}20)$$

Substitution of Eq. (17-20) into Eq. (17-19) gives

$$\frac{T}{2\pi} = \sqrt{\frac{I_{po} + I_{to}}{W_p r_p + W_t r_t}} \qquad (17\text{-}21)$$

Solving for I_{po} we obtain

$$I_{po} = \left(\frac{T}{2\pi}\right)^2 (W_p r_p + W_t r_t) - I_{to} \qquad (17\text{-}22)$$

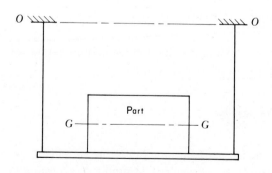
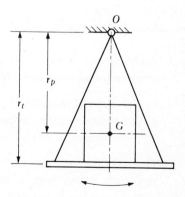

Fig. 17-13

But from Eq. (17-17)

$$I_{to} = W_t r_t \left(\frac{T_t}{2\pi}\right)^2$$

Substitution of the latter into Eq. (17-22) yields

$$I_{po} = \left(\frac{T}{2\pi}\right)^2 W_p r_p + \frac{W_t r_t}{4\pi^2}(T^2 - T_t^2) \tag{17-23}$$

However, by the parallel-axis theorem

$$I_{po} = I_p + \frac{W_p}{g} r_p{}^2 \tag{17-24}$$

where I_p is the moment of inertia of the part about its center-of-gravity axis G-G. Substitution of Eq. (17-24) into Eq. (17-23) gives

$$I_p = W_p r_p \left[\left(\frac{T}{2\pi}\right)^2 - \frac{r_p}{g}\right] + \frac{W_t r_t}{4\pi^2}(T^2 - T_t^2) \tag{17-25}$$

Since the *moment of inertia* of a body about a given axis is defined as the sum of its mass particles multiplied by the square of the distance from the axis to the particle, it is often convenient to express it as

$$I = Mk^2 \tag{17-26}$$

where M is the total mass of the body and k is a constant called the *radius of gyration*. In other words, if all the mass of a body were considered to be concentrated at a distance k from the axis under consideration, then the moment of inertia for such a system would be the same as that of the body.

17-7   DYNAMICALLY EQUIVALENT SYSTEMS

We have seen from our study of a body under the influence of a system of external forces that its acceleration is dependent upon:

1. The mass of the body
2. The location of the center of gravity
3. The moment of inertia

The solution to problems in dynamics can sometimes be simplified by replacing a link with a dynamically equivalent system. A *dynamically equivalent system* is defined as a group of bodies rigidly connected which

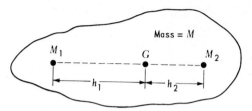

Fig. 17-14

will be given the same accelerations as the actual link or body when acted upon by the same forces.

The body in Fig. 17-14 has a mass M and a moment of inertia I about the center of gravity G. Although there is no limit on the number of masses which may be used in an equivalent system, the simplest system will be composed of two masses. Since the equivalent system must be a rigid body, we shall consider it to consist of two point masses, M_1 and M_2, connected by a weightless rod as shown. For the two point masses to be dynamically equivalent to the original link, the following equations must be satisfied:

$$M_1 + M_2 = M$$
$$M_1 h_1 = M_2 h_2 \tag{17-27}$$
$$M_1 h_1{}^2 + M_2 h_2{}^2 = I$$

The first of Eqs. (17-27) must be satisfied because

$$\sum F = MA_G$$

that is, if the acceleration A_G of the center of gravity is to be the same for the link and its equivalent system, they must have the same total mass M. For the acceleration of the center of gravity of the two systems to be the same, it must have the same location, and thus the second equation must be satisfied. Further, because

$$\sum T = I\alpha$$

and since α is to be the same for both systems, I for the equivalent system must be the same as that for the link. This requirement is stated by the third equation.

There are four unknowns: M_1, M_2, h_1, and h_2. Any one of these may be assumed, and then the other three are determined from the three Eqs.

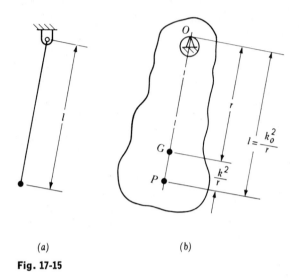

(a)   (b)

Fig. 17-15

(17-27). We shall find that the dynamically equivalent system is a useful concept in our study of balancing of the slider-crank mechanism in Chap. 20.

17-8 CENTER OF PERCUSSION

Another concept which is useful in the study of dynamics is the idea of center of percussion. We will develop the idea of center of percussion by means of simple and compound pendulums. A *simple pendulum* consists of a concentrated mass at the end of a rod of negligible weight as shown in Fig. 17-15a. A body (Fig. 17-15b) which oscillates in the manner of a simple pendulum but whose mass is distributed and not concentrated as in a simple pendulum is called a *compound pendulum*. The mass of a compound pendulum may be considered concentrated at a single point such that the period of oscillation remains unchanged; this point is then known as the *center of percussion*. The period of the simple pendulum is

$$T = 2\pi \sqrt{\frac{l}{g}} \tag{17-28}$$

Earlier in this chapter we found the period of a compound pendulum to be expressed by Eq. (17-16); that is,

$$T = 2\pi \sqrt{\frac{I_o}{Wr}} = 2\pi \sqrt{\frac{Mk_o{}^2}{Wr}} = 2\pi \sqrt{\frac{k_o{}^2}{gr}} \tag{17-29}$$

where W and M are the weight and mass of the pendulum and I_o and k_o are the moment of inertia and radius of gyration about the point of suspension O. If we equate the period for the simple pendulum to that of the compound pendulum, we find

$$l = \frac{k_o^2}{r} \qquad\qquad (17\text{-}30)$$

Thus if all the mass of the compound pendulum in Fig. 17-15b were concentrated at point P, then its period would be unchanged. Point P is called the center of percussion relative to point O. We cannot speak merely of the center of percussion of a body but must always refer it to some other point on the body. If the body in Fig. 17-15b were supported at some other point, the center of percussion would be at some other point than P.

We shall next determine the distance from the center of gravity to the center of percussion. By the parallel-axis theorem

$$I_o = I + Mr^2$$

where I is the moment of inertia about the center of gravity. If we let k be the radius of gyration about the center of gravity, then

$$Mk_o^2 = Mk^2 + Mr^2$$

and

$$k_o^2 = k^2 + r^2$$

Substituting the latter into Eq. (17-30), we have

$$l = \frac{k^2}{r} + r$$

and we see that the distance from the center of gravity to the center of percussion is k^2/r.

If a pendulum is given an acceleration α about its point of suspension, it can be shown that the inertia force and inertia torque can be replaced by a single force at the center of percussion. Consider the pendulum in Fig. 17-16a to have an acceleration α as shown. Then the inertia force f acts through G and is opposite in sense to $A_G = r\alpha$. Also, there will be an inertia torque t which is opposite in sense to α. The force f and torque t are again shown in Fig. 17-16b, where the torque is represented by the couple fh. Then

$$t = fh$$

or

$$I\alpha = MA_Gh$$

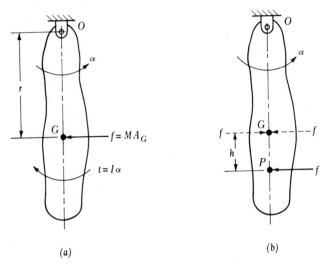

(a)   (b)

Fig. 17-16

and

$$h = \frac{I\alpha}{MA_G} = \frac{Mk^2\alpha}{Mr\alpha} = \frac{k^2}{r} \tag{17-31}$$

Thus we find that h comes out equal to the distance GP, which we found earlier. Since the two forces at G are equal and opposite, they cancel, and we find that the only inertia effect can be represented by a single force f at the center of percussion. This means that if a pendulum is struck by a force which is perpendicular to the line OG, there will be no reaction force at the point of suspension if the blow passes through the center of percussion. An example of this is the stinging effect produced on the hands when batting a ball if the ball strikes the bat other than at the center of percussion.

PROBLEMS

17-1 The slider-crank mechanism of a single-cylinder diesel engine is shown in Fig. P17-1. A gas force $P = 4{,}000$ lb acts to the left through piston pin C. The crank rotates counterclockwise at a constant speed of 1,800 rpm. Draw the mechanism to a scale of 1 in. = 2 in. Use a velocity scale of 1 in. = 25 fps, an acceleration scale of 1 in. = 1,875 fps², and a force scale of 1 in. = 1,000 lb. Make a combined static- and inertia-force analysis. (a) Determine the forces F_{14} and F_{12} and the torque T_2 in foot-pounds exerted by the crankshaft on the crank for equilibrium. (b) Determine the magnitude and direction of the shaking force and its location from point O_2.

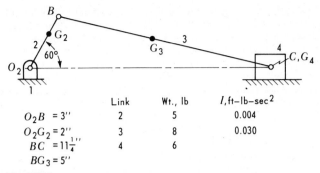

	Link	Wt., lb	I, ft-lb-sec^2
O_2B = 3''	2	5	0.004
O_2G_2 = 2''	3	8	0.030
BC = $11\frac{1}{4}''$	4	6	
BG_3 = 5''			

Fig. P17-1

17-2 A slider-crank mechanism used in a compressor is shown in Fig. P17-2. The flywheel serves as the crank. An external torque applied to the crankshaft drives the flywheel at a nearly constant speed of 600 rpm ccw. Air pressure acting to the left on the piston at the instant is 30 psig. Draw the mechanism, using a scale of 1 in. = 5 in. Use a velocity scale of 1 in. = 10 fps, an acceleration scale of 1 in. = 500 fps², and a force scale of 1 in. = 300 lb. Make a combined static- and inertia-force analysis and determine the necessary torque T_2 in magnitude in foot-pounds and sense which must be exerted on the crank by the crankshaft. (*b*) Determine the magnitude and direction of the shaking force and its location from point O_2.

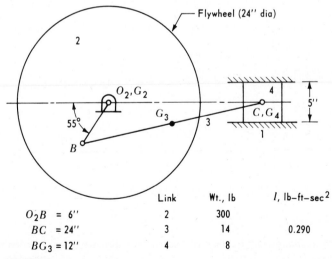

	Link	Wt., lb	I, lb-ft-sec^2
O_2B = 6''	2	300	
BC = 24''	3	14	0.290
BG_3 = 12''	4	8	

Fig. P17-2

17-3 Consider the cam in Fig. P6-8. The cam rotates at a constant speed of 15 rad/sec ccw. The load on the follower exerted by a link above (not shown) which it actuates is 5 lb. The weight of the follower is 2 lb. The spring has a spring rate of 20 lb/in. and exerts a force of 4 lb on the follower when the latter is in its lowest position. Draw the mechanism full-size with the cam in the position shown. The velocity and acceleration of the follower for this position have been found to be 0.72 fps and 36.8 fps² and are both directed upward. Let the cam and follower system be represented by Fig. 17-9, $a = 1\frac{3}{32}$ in., $b = 2$ in., $d = \frac{3}{8}$ in., and $\mu = 0.1$. Determine (a) the normal force exerted on the cam, (b) the limiting value for the pressure angle, and (c) the torque in inch-pounds that the camshaft exerts on the cam.

17-4 In the planetary gear train in Fig. P17-4 the gears have 20° involute teeth and a diametral pitch of 8. The arm is the driver and rotates counterclockwise at 2,000 rpm when viewed from the right end. A is a fixed gear and there are two planet gears B, each weighing 0.5 lb. The internal gear C is the driven member. Fifty horsepower is transmitted; assume 100 percent efficiency. (a) Determine the rpm and direction of rotation of gear C when viewed from the right end. (b) Determine all the forces on each of the gears. Use a force scale of 1 in. = 100 lb. Determine also the torque in inch-pounds which the driving shaft exerts on the arm, the torque that the driven shaft exerts on gear C, and the resisting torque which the frame exerts on gear A.

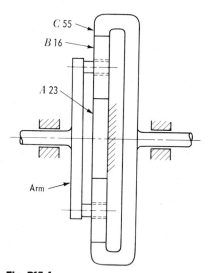

C 55

B 16

A 23

Arm

Fig. P17-4

17-5 In order to determine the moment of inertia of a flywheel, it was suspended on a knife-edge passing through the bore in the flywheel for accommodating the shaft. The bore was 2 in. and thus the distance from the knife-edge support to the center of gravity was 1 in. and the flywheel made 100 oscillations in 198.5 sec. Determine the mass moment of inertia in pound-inches-second squared of the flywheel about an axis through its center of gravity if the flywheel weighs 88 lb.

17-6   The connecting rod shown in Fig. 17-12 was suspended on a knife-edge as shown
and allowed to oscillate through a small angle.   It completed 200 cycles in 189
sec.   The distance between centers of the bearings at each end of the rod is
10 in. and distance r in the figure is 7 in.   The rod weighs 2.25 lb.   (a) Determine the moment of inertia of the rod in pound-inches-second squared about
its center of gravity.   (b) Determine W_1, W_2, and h_2 of the two-mass-equivalent
system for the connecting rod if W_1 is located at the center of the upper bearing.
The bore at the upper end of the rod is 0.75 in. in diameter.   (c) Locate the
center of percussion of the connecting rod about the center of the upper bearing.

18

FLYWHEELS

18-1 INTRODUCTION

A *flywheel* is a rotating mass which is used as an energy reservoir in a machine. The kinetic energy of a rotating body is $\frac{1}{2} I\omega^2$, where I is the mass moment of inertia about the axis of rotation and ω is the angular velocity. If the speed of the machine is increased, energy is stored up in the flywheel, and if the speed decreases, energy is given up by the flywheel.

There are two types of machines which benefit from the use of a flywheel. The first type is illustrated by an electric generator which is driven by an internal-combustion engine. Consider a single-cylinder four-stroke-cycle internal-combustion engine. The torque delivered to the generator varies considerably because there is one power stroke only once every two revolutions. The voltage output of the generator is a function of the speed, and a change in the voltage would cause a flicker in the lights. A flywheel is used in this instance to ensure a fairly uniform velocity and torque to the generator.

The second type of machine which benefits from a flywheel is illustrated by a punch press. The punching process requires a large amount of power in spurts, and if a flywheel were not used, all this power would have to be supplied by the motor, and a large motor would be required. By using a flywheel, a much smaller motor can be used. This is because energy from the motor is being stored in the flywheel during the interval between the actual punching operations and is available for use when the actual punching occurs.

18-2 COEFFICIENT OF FLUCTUATION

The coefficient of fluctuation is the permissible variation in speed and is defined as

$$C = \frac{\omega_1 - \omega_2}{\omega}$$

where ω_1 = maximum angular speed of flywheel
ω_2 = minimum angular speed of flywheel
ω = average angular speed of flywheel

Also

$$C = \frac{V_1 - V_2}{V}$$

where V_1 = maximum speed of given point on flywheel
V_2 = minimum speed of same point on flywheel
V = average speed of same point on flywheel

Values of this coefficient that have been found satisfactory in practice vary from about 0.2 for crushing machinery to 0.002 for electric generators. Recommended values for various types of machinery can be found in engineering handbooks and machine-design textbooks.

18-3 WEIGHT OF FLYWHEEL FOR A GIVEN COEFFICIENT OF FLUCTUATION IN SPEED

Consider the flywheel in Fig. 18-1 and let it be assumed that the angular velocity varies where V_1 is the maximum rim speed, V_2 is the minimum rim speed, and V is the average rim speed. Then

$$V = \frac{V_1 + V_2}{2}$$

or

$$2V = V_1 + V_2 \tag{18-1}$$

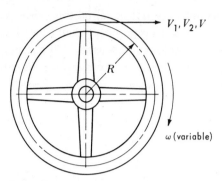

Fig. 18-1

Since the coefficient of fluctuation in speed is

$$C = \frac{V_1 - V_2}{V}$$

then

$$CV = V_1 - V_2 \qquad\qquad (18\text{-}2)$$

Suppose that the entire weight of the flywheel were concentrated at the mean rim radius R and let this weight be called W. Then the kinetic energy at speeds V_1 and V_2 would be

$$KE_1 = \frac{1}{2}\frac{W}{g}V_1^2 \qquad \text{and} \qquad KE_2 = \frac{1}{2}\frac{W}{g}V_2^2$$

If we let E represent the change in kinetic energy, then

$$E = \frac{1}{2}\frac{W}{g}(V_1^2 - V_2^2) \qquad\qquad (18\text{-}3)$$

Multiplication of Eqs. (18-1) and (18-2) by one another gives

$$2CV^2 = V_1^2 - V_2^2 \qquad\qquad (18\text{-}4)$$

and substitution of Eq. (18-4) into (18-3) yields

$$E = \frac{W}{g}CV^2 \qquad\qquad (18\text{-}5)$$

In an actual flywheel not all the weight is concentrated at the rim as we have assumed, but the flywheel is usually designed so that most of its weight is in the rim so that the kinetic energy will be greater for a given angular speed. In the equations above, W is the effective weight of the flywheel at the rim and is equal to the weight of the rim plus the effect of the arms and hub. For a flywheel with arms, the actual weight of the rim is about 90 percent of the effective weight W. Since the stresses in the rim and arms are due to centrifugal forces which are a function of the speed, the rim speed V is usually limited to 6,000 fpm for cast iron and 8,000 fpm for steel. The density of cast iron is 0.256 lb/in³ and of steel is 0.283 lb/in³.

EXAMPLE   As an example of flywheel analysis we will determine the size of flywheel required to be used on a punch press. A diagram of the press is shown in Fig. 18-2, where the slider of a slider-crank mechanism serves as the punch to force a slug of metal in the plate through the hole in the die. The press is to punch 30 holes per minute, and thus the time between punching operations is 2 sec. We will assume that the actual punching takes place ⅙ of the interval between punches, or the actual punching takes place in ⅙ (2) = ⅓ sec. A hole ¾ in. in diameter is to be punched in a plate of SAE 1025 steel ½ in. in thickness. The driving motor runs at 900 rpm with a velocity reduction through gears to give the 30 punching operations per minute.

We shall first calculate the energy necessary for punching. The maximum force required for shearing the material from the plate can be expressed as

$$P = \pi dts$$

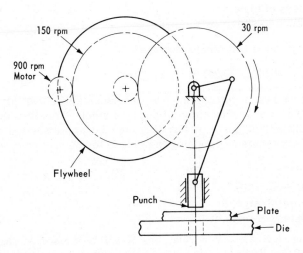

Fig. 18-2

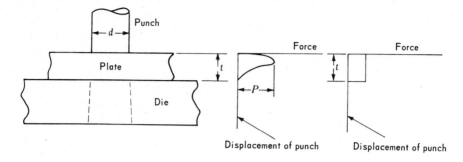

(a)

(b)

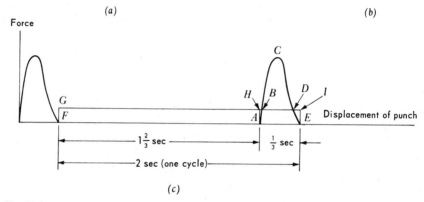

(c)

Fig. 18-3

where d = diameter of hole, in.
t = thickness of plate, in.
s = resistance to shear, psi†

Then the maximum force can be expressed as

$$P = \pi \, (\tfrac{3}{4})(\tfrac{1}{2}) \; 45{,}000 = 53{,}000 \text{ lb}$$

Figure 18-3 shows a typical force-vs.-displacement curve for the punching of a hole in a ductile material such as steel. The area under the force-vs.-displacement curve in Fig. 18-3a can be approximated as a triangle, and the work done in punching the hole is then

$$W_k = \tfrac{1}{2}Pt$$

where W_k = work done, ft-lb
P = maximum force, lb
t = thickness of plate, ft

† Machinery's Handbook (The Industrial Press, New York, N.Y.) gives for s a value of 0.75 the ultimate stress in tension or thus approximately 45,000 psi.

Thus

$$W_k = \tfrac{1}{2}(53,000)\ \tfrac{1}{2}(\tfrac{1}{12}) = 1,100 \text{ ft-lb}$$

Analysis without flywheel

The average power required during the punching operation, assuming the force-deflection curve is rectangular as shown in Fig. 18-3b, is

$$\frac{W_k}{\text{Time}} = \frac{1,100}{\tfrac{1}{3}} = 3,300 \text{ ft-lb/sec}$$

and the corresponding horsepower = 3,300/550 = 6. However, since P in Fig. 18-3a is twice as large as that in Fig. 18-3b, the maximum instantaneous horsepower would actually be approximately 12.

Analysis with flywheel

If a flywheel were used, the motor horsepower could be reduced considerably. Figure 18-3c shows that the time between punching operations is 2 sec and that the time required for a punching operation is $\tfrac{1}{3}$ sec. The work done per cycle was found earlier to be 1,100 ft-lb; it is represented by the area $ABCDE$ in the figure and must be supplied in $\tfrac{1}{3}$ sec. If a flywheel is used, this same amount of energy must be supplied per cycle and is represented by the area $FGIE$. Thus with a flywheel the 1,100 ft-lb of energy is supplied in 2 sec, indicating that $1,100/2(550) = 1$-hp motor is required. During the $\tfrac{1}{3}$-sec punching interval, the motor supplies the energy represented by area $AHIE$, which is 183 ft-lb, but the energy required for punching is represented by area $ABCDE$, which is 1,100 ft-lb. Thus the energy that has to be taken from the flywheel is $1,100 - 183 = 917$ ft-lb. Thus we see that the energy supplied by the motor during the actual punching operation is $183/1,100 = \tfrac{1}{6}$ the total energy required during the total cycle.

To determine the weight and size of the flywheel, we shall assume a mean rim diameter of 3 ft and then the average rim speed will be

$$V = \frac{\pi D n}{60} = \frac{\pi(3)150}{60} = 23.6 \text{ fps}$$

If we assume a permissible coefficient of speed fluctuation of 0.10, then from Eqs. (18-1) and (18-2)

$$47.2 = V_1 + V_2 \quad \text{and} \quad 2.36 = V_1 - V_2$$

and

$$V_1 = 24.8 \text{ fps} \quad \text{and} \quad V_2 = 22.4 \text{ fps}$$

As a check,

$$C = \frac{V_1 - V_2}{V} = \frac{24.8 - 22.4}{23.6} = 0.10$$

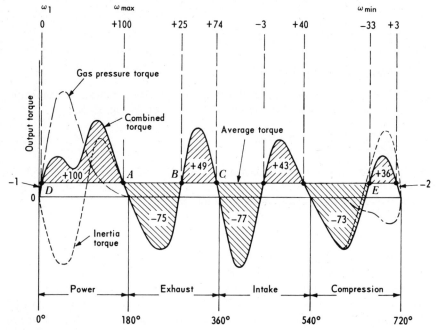

Fig. 18-4

From Eq. (18-5)

$$W = \frac{Eg}{CV^2} = \frac{917(32.2)}{0.1(23.6)^2} = 530 \text{ lb}$$

Assuming a flywheel with arms, the actual weight of the rim would be about 90 percent of W, or 0.90 (530) = 477 lb.

18-4 FLYWHEEL FOR AN INTERNAL–COMBUSTION ENGINE

As another example of the use of a flywheel let us consider the flywheel required for a single-cylinder four-stroke-cycle engine. In Fig. 18-4 the output torque is plotted versus crank position. Four strokes of the piston occur during a cycle, and there is a power stroke only once every two revolutions. The combined-output torque is the sum of the torque due to gas pressure and the inertia torque. The gas pressure on the piston can be obtained from an indicator diagram, which gives the gas pressure as a function of piston position. The output torque due to gas pressure can be found from a static-force analysis and is plotted versus the angular position of the crank. The inertia-output torque is obtained from an inertia-force analysis.

The areas lying above the zero-torque line represent positive work done by the engine, and the areas below represent negative work. The average torque is obtained by taking the algebraic sum of these areas and dividing it by the crank displacement of 720°. If the machine which is driven by the engine offers a resisting torque equal to the average torque, some means must be provided to level out the combined-torque curve. The purpose of the flywheel is to give a uniform-output torque. The crosshatched areas above the average-torque line represent excess energy which increases the speed and stores energy in the flywheel, and the crosshatched areas below the average-torque line represent a deficiency in energy which reduces the speed and removes energy from the flywheel. The relative magnitudes of the crosshatched areas in the figure are given by the numbers in the areas as shown.

In order to determine the energy which must be stored in the flywheel, we must find the positions at which the crank speed is a maximum and a minimum. At first we are likely to believe that the speed would be a minimum at the beginning of the power stroke and a maximum at the end of the power stroke. However, from an examination of Fig. 18-4 we shall see that these are not the correct positions. If we start at point A and go to point B, we note that the area under the combined-torque curve is negative, indicating that there is a deficiency of energy, and thus the speed decreases. Then as we go from B to C, we note that the area is positive, indicating an excess of energy, and thus the speed increases. Hence the maximum and minimum speeds occur somewhere at points where the combined-torque curve intersects the average-torque curve.

In order to find the points of maximum and minimum speed we shall arbitrarily start at D, the first point where the combined-torque curve crosses the average curve and call the speed here ω_1. At this point the kinetic energy is E_1, and when we proceed to point A, because of the excess energy of 100, the energy at A is $E_1 + 100$. Then as we proceed to point B, there is a deficiency in energy of 75 units, and thus when we get to B the energy has been reduced to $E_1 + 25$. At the top of the figure the relative-energy level is indicated for each intersection, and we note that the maximum speed occurs at point A, which is the point where the energy is a maximum, and the minimum speed occurs at point E, where the energy is a minimum. Then the maximum energy change is given by the algebraic sum of the shaded areas between these points; this is the energy to be used in Eq. (18-5).

PROBLEMS

18-1   The crankshaft of a crusher rotates at 60 rpm and the force-vs.-displacement diagram is similar to Fig. 18-3c.   The average power input is 10 hp.   The actual crushing operation requires $\frac{1}{8}$ sec and occurs once each revolution of the crankshaft.   (a) Determine the energy in foot-pounds which must be stored in the flywheel.   (b) Determine the weight of flywheel rim required if the flywheel is mounted on the crankshaft.   Assume that the arms and hub contribute 10 percent of the effective weight at rim.   The mean diameter of the rim is to be 72 in.   Assume a coefficient of speed fluctuation of 0.2 and that the average rim speed occurs at 60 rpm.

18-2   A $\frac{5}{8}$-in.-diameter hole is to be punched in a plate of SAE 1025 steel $\frac{3}{4}$ in. thick. Twenty holes per minute are to be punched, and the actual punching takes place in one-fifth the interval between punches.   The driving motor runs at 1,200 rpm and is geared to a countershaft which runs at 160 rpm and upon which the flywheel is mounted.   The countershaft in turn is geared to the crankshaft of the press in order to give the 20 punching operations per minute.   The resistance to shear for the plate may be taken as 45,000 psi.   (a) Make a sketch similar to Fig. 18-2 and label the speeds of the shafts.   Also make a sketch similar to Fig. 18-3c.   Find the horsepower required for the motor if no flywheel is used.   (b) Find the horsepower required for the motor assuming a flywheel is used.   (c) Determine the weight of flywheel rim required assuming that 90 percent of the effective weight at the rim is due to the rim alone.   The average speed at the mean diameter of the rim is 4,000 fpm and the coefficient of speed fluctuation is 0.10.   (d) If the total weight of the flywheel is approximately 1.25 times the weight of the rim, find the approximate total weight of the flywheel. (e) If the flywheel rim has a square cross section, determine the necessary dimensions.   The rim is made of cast iron which weighs 0.256 lb/in³.   (f) If the flywheel were made from a solid, circular plate of cast iron, determine the necessary flywheel weight for an average peripheral speed of 4,000 fpm.   The permissible coefficient of speed fluctuation is 0.10, as used above.

18-3   A four-stroke-cycle single-cylinder engine is to be analyzed in order to determine the flywheel requirements.   The gas-pressure-torque, inertia-torque, and combined-torque curves have been plotted and are similar to those in Fig. 18-4. The shaded areas lying above and below the average-torque curve have been measured with a planimeter in order to locate the points where ω is a maximum and a minimum.   These points were found to be 485° apart.   The shaded areas between points A and E were found to be -1.20 in.², $+0.79$ in.², -1.23 in.², $+0.69$ in.², and -1.17 in.².   The scale used for ordinates on the diagram was 1 in. $= 100$ ft-lb.   The full 720° length of the diagram was drawn 7.20 in.   The engine has a 4-in. stroke.   Find the equivalent weight of flywheel rim.   The permissible coefficient of speed fluctuation is to be 0.05, the engine runs at 1,000 rpm, and the mean diameter of the flywheel rim is to be 24 in.   Hint: First find how many foot-pounds of work or energy are represented by each square inch of area on the torque-vs.-displacement diagram.   Recall that work or energy equals torque times angular displacement in radians.

19

BALANCING
ROTATING MASSES

19-1 INTRODUCTION

In Chap. 17 we studied inertia forces in various mechanisms. The effect of inertia forces in setting up shaking forces on the structure was also discussed. The question is what can be done about the shaking forces.

It is possible to balance wholly or in part the inertia forces in the system by introducing additional masses which serve to counteract the original forces. This procedure is applied to two different kinds of problems. The first is a system of rotating masses, as illustrated by an automobile wheel or an automobile crankshaft, and the second is a system of reciprocating masses, as illustrated by a slider-crank mechanism. Balancing of rotating masses is discussed in this chapter and the balancing of reciprocating masses is discussed in Chap. 20.

19-2 SINGLE ROTATING MASS

To illustrate the principles involved, we shall begin by considering Fig. 19-1, where the shaft is assumed to be rotating at a constant speed ω and there is a single concentrated mass of weight W at radius R. Let W_e be the

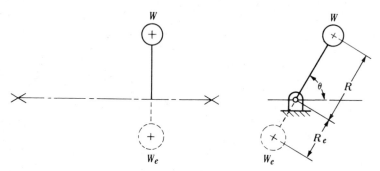

Fig. 19-1

weight which must be added at some radius R_e in order to produce equilibrium.

Static balance will be produced if the sum of the moments of the weights about the axis of rotation is zero:

$$-WR \cos \theta + W_e R_e \cos \theta = 0$$

or

$$W_e R_e = WR \tag{19-1}$$

If the value of R_e is arbitrarily chosen, then the value of W_e can be found by Eq. (19-1). When static balance exists, the shaft will have no tendency to rotate in its bearings regardless of the position to which it is rotated.

Dynamic balance requires that the sum of the inertia forces in Fig. 19-1 be zero; thus

$$\frac{W}{g} R\omega^2 - \frac{W_e}{g} R_e\omega^2 = 0$$

or

$$W_e R_e = WR \tag{19-2}$$

From Eqs. (19-1) and (19-2) we see that static and dynamic balance will be attained if we make

$$W_e R_e = WR$$

19-3   SEVERAL ROTATING MASSES IN THE SAME TRANSVERSE PLANE

In Fig. 19-2a and b, W_1, W_2, and W_3 are concentrated weights all lying in the same plane of rotation. W_e represents the weight which must be added at some radius R_e and angular position θ_e in order to produce equilibrium. For static balance the sum of the moments due to the original weights and the added weight W_e about the axis of rotation must be zero; thus

$$\sum WR \cos \theta + W_e R_e \cos \theta_e = 0 \tag{19-3}$$

For dynamic balance the inertia forces must be in equilibrium, and hence the sum of their horizontal components must be zero; thus

$$\sum \frac{W}{g} R\omega^2 \cos \theta + \frac{W_e}{g} R_e\omega^2 \cos \theta_e = 0 \tag{19-4}$$

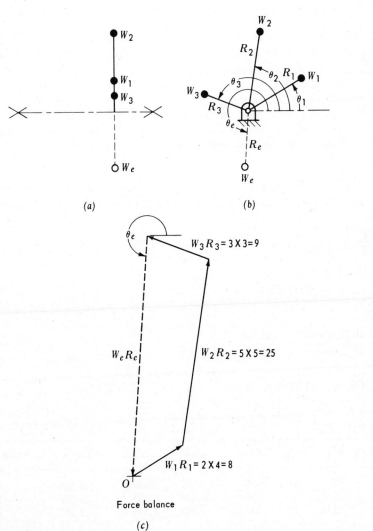

(a)

(b)

Force balance

(c)

Fig. 19-2

and the sum of their vertical components must be zero; thus

$$\sum \frac{W}{g} R\omega^2 \sin\theta + \frac{W_e}{g} R_e\omega^2 \sin\theta_e = 0 \qquad (19\text{-}5)$$

If we divide Eqs. (19-4) and (19-5) by ω^2/g, we obtain

$$\sum W_e R \cos\theta + W_e R_e \cos\theta_e = 0$$
$$\sum W_e R \sin\theta + W_e R_e \sin\theta_e = 0 \qquad (19\text{-}6)$$

Thus Eqs. (19-6) are the conditions required for dynamic balance. From an examination of Eqs. (19-3) and (19-6) we see that static and dynamic balance will be attained if Eqs. (19-6) are satisfied.

EXAMPLE 19-1   For the rotor in Fig. 19-2 the values of the weights W_1, W_2, and W_3, their radii of rotation R_1, R_2, and R_3, and their angular positions θ_1, θ_2, and θ_3 are known.   It is desired to find the weight W_e at a 3.5-in. radius and its angular position θ_e required to balance the rotor statically and dynamically.

Mathematical Solution   A mathematical solution of Eqs. (19-6) to find W_e and θ_e can be most easily done by listing the quantities as shown in Table 19-1.

From Eqs. (19-6),

$$2.818 + W_e R_e \cos\theta_e = 0$$
$$31.703 + W_e R_e \sin\theta_e = 0 \qquad (19\text{-}7)$$

Then

$$\frac{W_e R_e \sin\theta_e}{W_e R_e \cos\theta_e} = \frac{-31.703}{-2.818} \qquad (19\text{-}8)$$

or

$$\tan\theta_e = 11.25$$

Table 19-1

NUMBER	W, lb	R, in.	θ, deg	cos θ	sin θ	WR cos θ	WR sin θ
1	2	4	30	0.866	0.500	6.928	4.000
2	5	5	80	0.174	0.985	4.350	24.625
3	3	3	160	−0.940	0.342	−8.460	3.078
						$\sum = 2.818$	$\sum = 31.703$

It is important to retain the signs in Eq. (19-8) in order that the proper quadrant for θ_e can be determined. From Eq. (19-8) we see that sin θ_e is negative and cos θ_e is negative. Hence θ_e lies in the third quadrant, and

$$\theta_e = \tan^{-1} 11.25 = 264.9°$$

From Eqs. (19-7),

$$W_e = \frac{-31.703}{R_e \sin \theta_e} = \frac{-31.703}{3.5(-0.9960)} = 9.09 \text{ lb}$$

Graphical Solution Equations (19-6) can be solved graphically for the values of W_e and θ_e by laying out the WR vectors to scale as shown in Fig. 19-2c. Since the WR vectors represent inertia forces, they act radially outward and must be drawn parallel to the corresponding radii in Fig. 19-2b. The vector W_eR_e as shown is required to close the polygon and produce equilibrium. The magnitude of W_eR_e, when scaled, was found to be 32 units. Hence

$$W_e = \frac{32}{R_e} = \frac{32}{3.5} = 9.15 \text{ lb}$$

θ_e in Fig. 19-2c was measured with a protractor and found to be 266°.

19-4 SEVERAL ROTATING MASSES LYING IN SEVERAL TRANSVERSE PLANES

In Fig. 19-3, a rotor having two equal concentrated weights W_1 and W_2 lying in different transverse planes C and D is shown. It is evident that the static forces are in balance and also that the dynamic forces F_1 and F_2 are

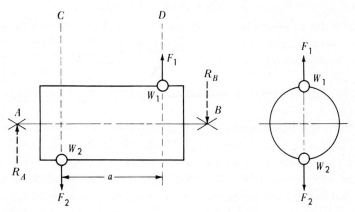

Fig. 19-3

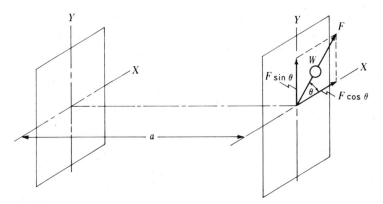

Fig. 19-4

equal and thus in balance.   However, F_1 and F_2 produce an unbalanced couple equal to $F_1 a$, which will set up reactive forces R_A and R_B at the bearings A and B.   The purpose of balancing any rotating equipment is to eliminate or to reduce as much as possible the forces transmitted to the bearings in order to eliminate vibration.   Thus we see that in balancing a rotating system we must balance not only forces but also moments.

We shall first define the moment of a force about a plane.   In Fig. 19-4, consider two parallel planes distance a apart.   Let F be a force in the right plane which makes an angle θ with the X axis.   This could be an inertia force acting on a rotating weight W.   The moment of the force F about the left plane is defined as the moment of the vertical component of F about the X axis, which is $Fa \sin \theta$, plus the moment of the horizontal component of F about the Y axis, which is $Fa \cos \theta$.

In order to illustrate the method for balancing a system of masses lying in several transverse planes, let us consider the system of weights W_1, W_2, and W_3 in Fig. 19-5.   The procedure is as follows:

1. Choose any two transverse reference planes A and B as shown in the figure.
2. Let the distance in the axial direction from plane A to weights W_1, W_2, W_3, etc., be a_1, a_2, a_3, etc., respectively.   Distances to the right of plane A are considered positive $(+)$ and distances to the left are negative $(-)$.
3. Since inertia force is $F = (W/g) R\omega^2$, the F forces are proportional to WR. We can balance moments about plane A by adding a weight W_B in plane B so that the sum of the moments about the X axis is zero and the

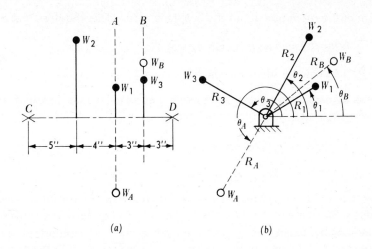

(a)   (b)

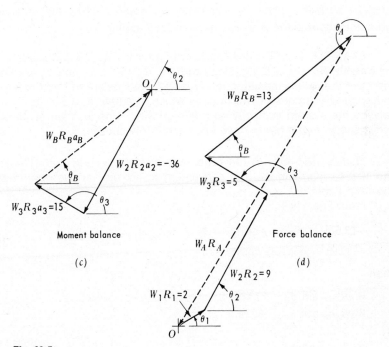

(c)   (d)

Fig. 19-5

sum of the moments about the Y axis is zero. This requires that:

$$\sum WRa \sin \theta + W_B R_{Ba_B} \sin \theta_B = 0$$

$$\sum WRa \cos \theta + W_B R_{Ba_B} \cos \theta_B = 0$$

(19-9)

4. Next, we can add a weight W_A in plane A so as to balance all forces in the X direction and the Y direction, that is,

$$\sum WR \cos \theta + W_A R_A \cos \theta_A = 0$$

$$\sum WR \sin \theta + W_A R_A \sin \theta_A = 0$$

(19-10)

If Eqs. (19-9) and (19-10) are satisfied, then the system will be in dynamic balance. Since the first of Eqs. (19-10) is similar to Eq. (19-3), the system will also be in static balance if it is in dynamic balance. It is important to note that in step 4 when we add the weight W_A in order to balance the forces, it must be added in plane A; otherwise we would destroy the balance of moments about plane A which we produced earlier.

EXAMPLE 19-2   Let us consider the rotor in Fig. 19-5 and assume that the weights W_1, W_2, and W_3, along with R_1, R_2, R_3, θ_1, θ_2, and θ_3, are known and have the values indicated in Table 19-2. We desire to find the weights and angular positions of two masses which, if added at a 3-in. radius, will balance the system statically and dynamically. The location of planes A and B shown in the figure have been arbitrarily chosen.

Mathematical Solution   In order to satisfy Eqs. (19-9) and (19-10) and find W_B, θ_B, W_A, and θ_A, it is convenient to list the various quantities as shown in Table 19-2.
From Eqs. (19-9),

$$-23.676 + W_B R_{Ba_B} \sin \theta_B = 0$$

$$-30.99 + W_B R_{Ba_B} \cos \theta_B = 0$$

(19-9')

Then

$$\frac{W_B R_{Ba_B} \sin \theta_B}{W_B R_{Ba_B} \cos \theta_B} = \frac{23.676}{30.99}$$

or

$$\tan \theta_B = 0.765$$

From Eqs. (19-9') we see that $\sin \theta_B$ is positive and $\cos \theta_B$ is positive. Hence θ_B lies in the first quadrant, and

$$\theta_B = \tan^{-1} 0.765 = 37.4°$$

Table 19-2

NUMBER	W, lb	R, in.	θ, deg	a, in.	cos θ	sin θ
1	1	2	30	0	0.866	0.500
2	3	3	60	−4	0.500	0.866
3	2	2.5	150	3	−0.866	0.500

NUMBER	$WR \cos \theta$	$WR \sin \theta$	$WRa \cos \theta$	$WRa \sin \theta$
1	1.732	1.000	0	0
2	4.500	7.794	−18.00	−31.176
3	−4.330	2.500	−12.99	7.500
	$\sum = 1.902$	$\sum = 11.294$	$\sum = -30.99$	$\sum = -23.676$

From Eqs. (19-9′)

$$W_B = \frac{23.676}{R_B a_B \sin \theta_B} = \frac{23.676}{3 \times 3 \times 0.6074} = 4.33 \text{ lb}$$

From Eqs. (19-10)

$$1.902 + (4.33 \times 3 \times 0.794) + W_A R_A \cos \theta_A = 0$$

$$11.294 + (4.33 \times 3 \times 0.607) + W_A R_A \sin \theta_A = 0 \tag{19-10′}$$

Then

$$\frac{W_A R_A \sin \theta_A}{W_A R_A \cos \theta_A} = \frac{-19.194}{-12.202} = 1.57$$

or

$$\tan \theta_A = 1.57$$

From Eqs. (19-10′) we see that $\sin \theta_A$ is negative and $\cos \theta_A$ is negative. Hence θ_A lies in the third quadrant, and

$$\theta_A = \tan^{-1} 1.57 = 237.5°$$

From Eqs. (19-10′)

$$W_A = \frac{-12.202}{R_A \cos \theta_A} = \frac{-12.202}{3(-0.5373)} = 7.58 \text{ lb}$$

Graphical solution   A listing of the various quantities appears in Table 19-3.   Equations (19-9) can be solved graphically for the values of W_B and θ_B by laying out the WRa vectors using any convenient scale.   This is shown in Fig. 19-5c.   It is important to note that these moment vectors have the same sense as the inertia forces acting on the weights in Fig. 19-5a; that is,

Table 19-3

NUMBER	θ, deg	W, lb	R, in.	a, in.	WR	WRa
1	30	1	2	0	2	0
2	60	3	3	−4	9	−36
3	150	2	2.5	3	5	15

they are directed radially outward unless the moment is negative.   In Fig. 19-5c, $W_2R_2a_2$ would be drawn at an angle θ_2 with the horizontal and would be directed upward to the right from the pole O if its value were positive, but since it is negative it is directed opposite.   The vector $W_BR_Ba_B$ shown dotted is required for equilibrium and when scaled was found to be 39 units.   Then

$$W_B = \frac{39}{R_Ba_B} = \frac{39}{3 \times 3} = 4.33 \text{ lb}$$

In Fig. 19-5c, θ_B was found to measure 38°.   Since in Fig. 19-5a we see that a_B is positive, then θ_B is as shown and not $(38 + 180)$°.

Next, Eqs. (19-10) can be solved for the values of W_A and θ_A by laying out the WR vectors, using any convenient scale as shown in Fig. 19-5d. Since these vectors represent the inertia forces, they are always positive in value and are directed radially outward from the corresponding weights in Fig. 19-5b.   Vector W_AR_A shown dotted is required for equilibrium and was found to scale 22.7 units.   Then

$$W_A = \frac{22.7}{R_A} = \frac{22.7}{3} = 7.57 \text{ lb}$$

In Fig. 19-5d, θ_A was found to measure 238°.

19-5   GENERAL COMMENTS ON BALANCING

As we have seen earlier, the necessary counterbalance was determined as the product of a weight and a radius.   If we arbitrarily choose one, the other can be determined.   In practice the radius is chosen as some convenient value and the weight is left to be determined.   The weight is either added by attaching a weight to the rotor, or an equivalent effect is produced by removing weight by drilling a hole in the rotor at a position 180° away and in the same transverse plane.   If counterweights are added, they should be placed at the maximum radius available in order that they will be a minimum. Likewise when weights must be added in two planes, these planes should be selected as far apart as possible so as to reduce the amount of counterweights needed.

Although any rotor may be balanced by adding a weight in each of two transverse planes, such a method leaves a bending moment in the shaft. For this reason it is often desirable to balance each unbalance in the plane of the unbalance. For example, in an automobile crankshaft each crank produces an unbalance, and frequently the crankshaft is balanced by adding a counterbalance opposite each crank.

19-6 BALANCING MACHINES

In spite of all the care that may be taken in the design and manufacture of a rotating part, whether the part is completely machined, cast, or forged, or if it is assembled from various parts as in the case of the armature of an

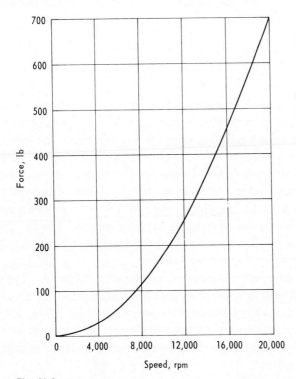

Fig. 19-6

electric motor, it is uncommon for it to run smoothly, particularly if the operating speed is high.   Variations in dimensions due to machining, non-homogeneity of the material, variations in methods of assembly, and eccentricity of bearing surfaces all contribute to offsetting the center of gravity from the axis of rotation.

The curve in Fig. 19-6 emphasizes the effect of a small amount of unbalance at high speeds.   The curve shows centrifugal force produced by an inch-ounce of unbalance at various angular speeds.   (An inch-ounce is defined as one ounce of weight at one inch from the axis of rotation.) The centrifugal force due to 1 in.-oz at 1,000 rpm is 1.76 lb.   At 10,000 rpm it is 176 lb.   That is, it increases as the square of the speed.   It is evident that the centrifugal force produced on a large rotor can be very large, even if the center of gravity of the rotor is displaced only a small amount from the axis of rotation, and consequently large shaking forces will be produced on the structure.   For example, consider the rotor of an aircraft gas turbine weighing 400 lb which operates at 16,000 rpm and suppose the center of gravity is 0.001 in. from the axis of rotation.   The 6.4 in.-oz of unbalance would cause a centrifugal force

$$F = MR\omega^2$$

$$= \frac{400}{32.2} \times \frac{0.001}{12} \left(\frac{2\pi \times 16,000}{60} \right)^2 = 2,900 \text{ lb}$$

Such a force could cause considerable damage to the machine.   Since it is usually impossible to manufacture the rotor of a machine so that the center of gravity will lie within 0.001 in. from the axis of rotation, the part must be balanced after manufacture, and the balancing is done experimentally.

The type of balancing required depends on the type of machine, its size, speed, and economics.   In general, static balance is sufficient for members of short axial length which rotate at low or moderate speeds, because the unbalanced moment due to dynamic effects is small.   Examples of members of this type are automobile wheels, airplane propellers, and narrow fans.

The trial-and-error method may be used for static balance.   The part is set on level ways, and temporary weights are placed upon the part until the required balancing weight and position are determined.   Then either a permanent weight or weights are attached to the part or the proper amount of weight may be removed at a position diametrically opposite in order to produce the same effect.

Figure 19-7 shows diagrammatically a machine for statically balancing

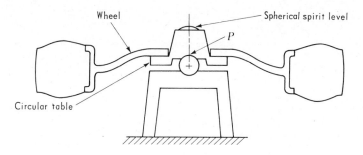

Wheel   Spherical spirit level

P

Circular table

Fig. 19-7

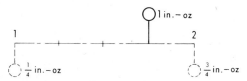

1 in.−oz

1   2

$\frac{1}{4}$ in.−oz $\frac{3}{4}$ in.−oz

Fig. 19-8

automobile wheels.   The apparatus consists of a circular table which is mounted at point P on a sphere so that it is free to tilt in any direction.   The wheel is placed on the table.   The center of gravity of the wheel, tire, and table combined lies below point P, and the table contains a spherical spirit level.   The position of the bubble in the spirit level indicates the direction of the light side where lead weights are then fastened to the rim to achieve balance.

The next type of balancing machine we shall consider is one used to balance rotors which are not of short axial length and hence where the moments due to dynamic effects must be balanced.   Such a rotor can be balanced completely, i.e., statically and dynamically, by adding masses in two reference planes such that the moments about each of these planes will be balanced.   The following is a proof.   Let us consider the rotor axis as divided into a number of equal segments as shown in Fig. 19-8 and con-sider two reference planes 1 and 2 perpendicular to the axis.   Suppose that the only unbalance can be represented by 1 in.-oz three units to the right of plane 1.   Then in order to balance moments about plane 1 we can add $\frac{3}{4}$ in.-oz in plane 2, that is, $3(1) = \frac{3}{4}(4)$.   Next, in order to balance moments about plane 2 we can add $\frac{1}{4}$ in.-oz in plane 1, that is, $1(1) = \frac{1}{4}(4)$.   As a result of balancing moments about planes 1 and 2, we have moments in balance about all other planes, and further, the forces are in balance.

Many balancing machines are based on the principle of balancing

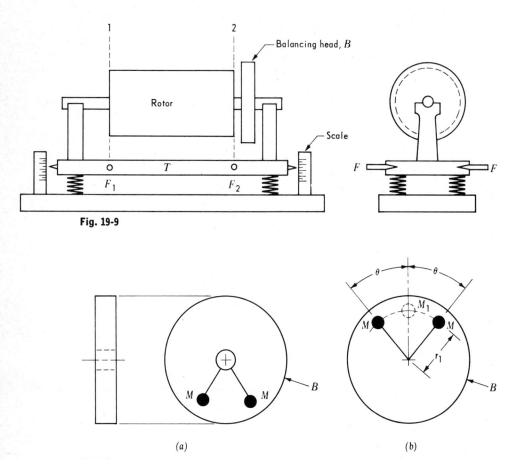

Fig. 19-9

Fig. 19-10

moments about two planes as has been just discussed.   One type uses
what is called a *balancing head*.   A diagram of the machine is shown in Fig.
19-9.   Planes 1 and 2 are reference planes in which masses are to be added
in order to balance the rotor.   Table T is mounted on springs and can be
made to pivot about pivots F_1 or F_2.

A *balancing head*, Fig. 19-10a, is a device which is solidly coupled to the
rotor to be tested and contains two arms with masses M.   These arms
rotate with the rotor, provided that the operator does not interfere.   There
is a motor and a system of gears within disk B that allow the operator to
rotate the masses relative to B.   Power is supplied through slip rings.   The
operator has before him two buttons.   If he presses the first, the two arms

rotate in the same direction with a fixed angle between them; if he presses the second, the arms rotate in opposite directions.   In each case the arms make one revolution relative to B in about 5 sec.

Since the two arms form the only unbalance in B, it is possible for the operator to change the magnitude as well as the direction of the unbalance.

The procedure for finding the weights to be added to the rotor and their angular positions is as follows.   The table is allowed to pivot about pivots F_1, and the rotor, driven by a belt or flexible shaft, is brought to a speed in which it is in resonance with the springs.   By pressing button 1, which rotates the arms in the same direction, a maximum and a minimum amplitude are observed once each 5 sec.   After removing his finger from button 1 at the minimum amplitude, the operator makes the arms rotate against each other by pressing button 2.   During this operation the line which bisects the angle between the arms does not move relative to B, and hence the direction of the additional unbalance does not change, but the magnitude varies from two masses (when the arms coincide) to zero (when they are 180° apart).   Button 2 is released when the vibration is reduced to zero, the rotor is stopped, and from the position of the arms, the magnitude and direction of the correction are readily determined.   In Fig. 19-10b, let M_1 be the equivalent mass lying on the bisector.   Since centrifugal force is proportional to mass times radius,

$$2Mr_1 \cos \theta = M_1 r_1$$

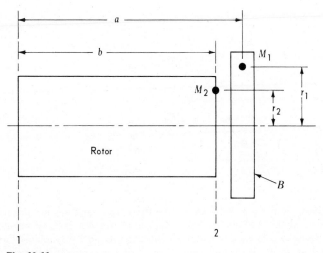

Fig. 19-11

or

$$M_1 = 2M \cos \theta$$

Now let us consider Fig. 19-11 in order to find the mass M_2 required to be added in plane 2 (at any convenient radius r_2) which will be equivalent to M_1. Taking moments of the centrifugal forces about plane 1 we have

$$M_1 r_1 a = M_2 r_2 b$$

or

$$M_2 = M_1 \frac{r_1 a}{r_2 b}$$

Next, the balancing head is placed on the left end of the rotor, pivots F_1 in Fig. 19-9 are removed, and the table is made to pivot about pivots F_2. The entire procedure is then repeated to find the magnitude and location of the mass to be added in plane 1 which will balance moments about plane 2.

PROBLEMS

19-1   The rigid rotor in Fig. P19-1 is to be balanced by the addition of a fourth mass at a 7-in. radius.   Make a drawing of the system using a scale of 1 in. = 10 in. Determine the weight and angular position of the balancing mass (*a*) using the mathematical method and (*b*) using the graphical method with a scale of 1 in. = 20 lb-in. for the WR vectors.   Using dashed lines, show on your drawing the required balancing mass in its actual position.

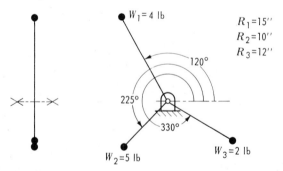

Fig. P19-1

19-2   Make a drawing of Fig. P19-2.   Determine the weights and angular positions of two masses which, if added at a 2-in. radius in planes *A* and *B*, will balance the rotor.   Begin by balancing moments about plane *A*.   Solve graphically using a scale of 1 in. = 100 WRa units and a scale of 1 in. = 50 WR units.   Using

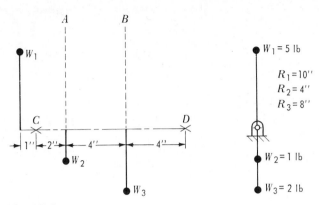

Fig. P19-2

dashed lines draw in the weights to be added, showing them in their true positions.

19-3   Same as Prob. 19-2 except solve by the mathematical method.

19-4   Make a drawing of Fig. P19-2.   Determine the bearing reactions at C and D for a shaft speed of 1,000 rpm, if the system of three masses as shown were left unbalanced.   Neglect the reactions due to static loads.   Solve graphically using for moment vectors a scale of 1 in. = 50 WRa units and for force vectors a scale of 1 in. = 25 WR units.   Begin by taking moments about a plane through C.   On your drawing of the rotor show the bearing reactions and indicate the values of their magnitude and angular positions.

19-5   Same as Prob. 19-4 except solve by the mathematical method.

19-6   Consider the rotor in Fig. 19-5 and let W_1, W_2, W_3, R_1, R_2, R_3, θ_1, θ_2, and θ_3 have the values given in Table 19-2.   Make a drawing of Fig. 19-5 using a scale of 1 in. = 5 in.   Determine the bearing reactions at C and D for a shaft speed of 500 rpm if the rotor were left unbalanced.   Neglect the reactions due to static loads.   Solve graphically and begin by taking moments about a plane through C.   Use a scale of 1 in. = 30 WRa units and a scale of 1 in. = 3 WR units.   On your drawing of the rotor show the bearing reactions and indicate the values of their magnitudes and angular positions.

19-7   Same as Prob. 19-6 except solve by the mathematical method.

20

BALANCING
RECIPROCATING MASSES

20-1 INTRODUCTION

In Chap. 17 we found that the shaking force is the resultant of all the forces acting on the frame of a mechanism due to inertia forces only. Thus if the resultant of all the forces due to inertia effects acting on the frame is zero, there is no shaking force; however, there may nevertheless be a shaking couple present. Balancing a mechanism consists of eliminating the shaking force and shaking couple. In some instances we can accomplish both. We shall discover that in most mechanisms, by adding appropriate balancing weights, we can reduce the shaking force and shaking couple, but it is usually not practical to provide means to completely eliminate them.

20-2 FOUR–BAR LINKAGE

We shall begin by considering the four-bar linkage in Fig. 20-1. Crank 2 rotates with constant angular velocity, rocker 4 oscillates, and coupler 3 has a combined motion of translation and rotation. The mechanism O_2BCO_4

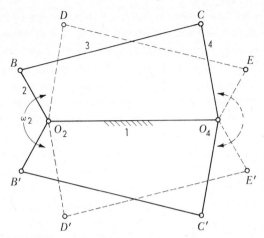

Fig. 20-1

can be balanced by introducing a mechanism which produces opposite effects. If we employ the mechanism $O_2B'C'O_4$, which is the mirror image of the original mechanism, and if we make it move in the opposite sense, this will balance the vertical shaking forces and the shaking moments due to the angular accelerations of links 3 and 4. Unbalanced horizontal shaking forces would remain, however, and in order to balance these, we could further employ the mechanism O_2DEO_4 and $O_2D'E'O_4$, which is the mirror image of the first combination and which moves in the opposite sense. This method of balancing is complicated, and instead, a method which provides partial balance is usually used.

Partial balance of the four-bar linkage can be obtained as shown in Fig. 20-2. The student should review the concept of a dynamically equivalent system presented in Chap. 17. There it is explained that in order to obtain a dynamically equivalent system we must satisfy Eqs. (17-27). In these equations M can be written as W/g, and the equations then become Eqs. (20-1); thus

$$W_1 + W_2 = W$$
$$W_1h_1 = W_2h_2 \tag{20-1}$$
$$W_1h_1{}^2 + W_2h_2{}^2 = Ig$$

These equations state that two members are dynamically equivalent if they have the same weight, the same center of gravity, and the same moment of inertia. Here, in treating the four-bar linkage, we shall balance only the shaking forces, and hence we do not need to fulfill the requirement of equal moment of inertia. The equivalent coupler shown in Fig. 20-2 satisfies the first two conditions and replaces the original coupler. Link 2 and weight W_1 can be balanced by W_3 to bring the center of gravity to the center of rotation O_2. Similarly, link 4 and weight W_2 can be balanced by W_4. This will eliminate the horizontal and vertical shaking forces, but there will remain an unbalanced, variable shaking couple due to the angular accelerations of links 3 and 4.

The shaking couple cannot be eliminated completely by simple means in most mechanisms, but it can be reduced by keeping the moment of inertia of the links to a minimum. It is neither practical nor necessary to obtain complete balance. After providing reasonable balance, troublesome vibrations can be avoided by using other means. Some of these are avoiding resonance, the use of dynamic-vibration absorbers, and the use of vibration-isolation mountings.

20-3 SLIDER–CRANK MECHANISM

Balancing the slider-crank mechanism is accomplished in a manner similar to that used for the mechanism in the previous section. Since the slider-crank mechanism is so widely used in such machines as internal-combustion engines and compressors, considerable work has been done on balancing these mechanisms.

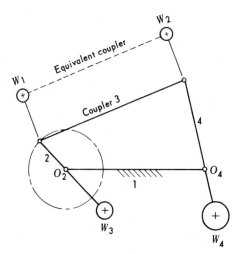

Fig. 20-2

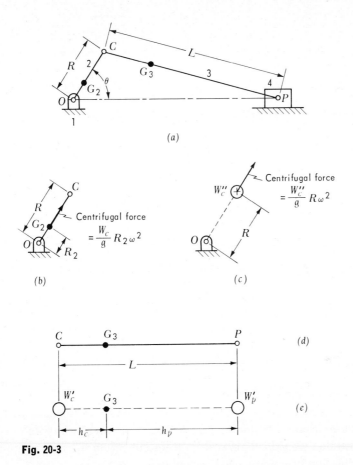

Fig. 20-3

A slider-crank mechanism is shown in Fig. 20-3a, and we shall assume that the crank rotates with constant angular velocity ω.   Points C and P are the crankpin and piston pin, respectively, and G_2 and G_3 are the centers of gravity of links 2 and 3.   In Fig. 20-3b the crank appears again.   The sum of the weights of the crank and crankpin is Wc, and the centrifugal force acts outward along line G_2C.   In Fig. 20-3c the crank and crankpin have been replaced by a concentrated weight W_c'' located at C such that the centrifugal forces in Fig. 20-3b and c will be equal.   Thus

$$\frac{W_c''}{g} R\omega^2 = \frac{W_c}{g} R_2\omega^2$$

or

$$W_c'' = \frac{R_2}{R} W_c \qquad\qquad (20\text{-}2)$$

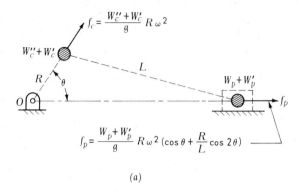

$$f_c = \frac{W_c'' + W_c'}{g} R\omega^2$$

$$f_p = \frac{W_p + W_p'}{g} R\omega^2 \left(\cos\theta + \frac{R}{L}\cos 2\theta\right)$$

(a)

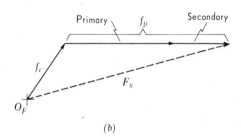

(b)

Fig. 20-4

The connecting rod of weight W in Fig. 20-3d can be replaced by the dynamically equivalent system shown in Fig. 20-3e, which consists of the two concentrated weights W_c' and W_p'. If we substitute the various quantities into the first two of Eqs. (20-1), we obtain

$$W_c' + W_p' = W$$

$$W_c'h_c = W_p'h_p$$

(20-3)

The location of the center of gravity G_3 of the connecting rod can be found experimentally by the method described in Sec. 17-6. Then lengths h_c and h_p are known. Substitution of their values into the equations above then gives us the values of W_c' and W_p'. By taking h_c and h_p as the distances from G_3 to points C and P as we have done, the third of Eqs. (20-1) will be very nearly satisfied. Thus the equivalent system in Fig. 20-3e will have the same inertia forces as the actual connecting rod but its inertia torque will be slightly different.

Figure 20-4a shows the slider-crank mechanism replaced by a dynamically equivalent system consisting of only two concentrated weights. At the crankpin we have the equivalent weight of crank W_c'' plus the equivalent weight W_c' of the connecting rod.   At the piston pin we have W_p, which is the sum of the weights of the piston and piston pin.   Also at this point we have the equivalent weight W_p' of the connecting rod.   Since the concentrated weight at the crankpin has pure rotation and that at the piston pin has rectilinear translation, the analysis of the inertia forces on the mechanism is greatly simplified.

In Fig. 20-4a, f_c and f_p are the inertia forces and are equal to the mass times the acceleration.   In Sec. 9-2 we found the acceleration of the piston to be expressed by Eq. (9-10) where $n = L/R$.   The piston acceleration can then be written

$$A = -R\omega^2 \left(\cos \theta + \frac{R}{L} \cos 2\theta \right) \tag{20-4}$$

where ω is the angular velocity of the crank.   The other quantities in this equation are indicated in Fig. 20-3a.   A positive result for A in Eq. (20-4) indicates that the piston acceleration is directed away from point O and vice versa.   The inertia force at the piston pin is always directed opposite to the acceleration and is

$$f_p = \underbrace{\frac{W_p + W_p'}{g} R\omega^2 \cos \theta}_{\text{Primary}} + \underbrace{\frac{W_p + W_p'}{g} R\omega^2 \frac{R}{L} \cos 2\theta}_{\text{Secondary}} \tag{20-5}$$

A positive result for f_p in Eq. (20-5) indicates that the inertia force is directed away from point O and vice versa.   The first portion of the equation is a function of the angle θ and is called the *primary inertia force*, and the second, because it is a function of 2θ, is called the *secondary inertia force*.

Shaking force was defined in Sec. 17-4 as the force acting on the frame of a mechanism due to all the inertia forces.   Hence the shaking force F_s is the vector sum of f_c and f_p as shown in Fig. 20-4b.   We note from the figure that both f_c and f_p act through point O for all values of θ.   Hence the shaking force F_s acts on the crankshaft bearing for all positions of the crank.

EXAMPLE 20-1   To illustrate how the shaking force for a single-cylinder engine varies in magnitude and direction and the means used to provide partial balance, let us consider the following example.   An engine has a speed of 1,000 rpm; connecting rod length, 14 in.; stroke, 8 in.; connecting rod weight, 34 lb; weight of reciprocating parts (piston and piston pin), 20 lb; weight of crank and crankpin, 10 lb; distance from crankshaft axis to center of gravity of crank and crankpin combined, 1.5 in.; distance from crankpin center to center of gravity of connecting rod, 4 in.

Angular velocity of crank $\omega = (1,000/60)2\pi = 104.6$ rad/sec. Substitution into Eq. (20-2) gives

$$W_c'' = \frac{1.5}{4}(10) = 3.75 \text{ lb}$$

Substitution into Eqs. (20-3) gives

$$W_c' + W_p' = 34$$
$$W_c'(4) = W_p'(10)$$

from which $W_c' = 24.29$ lb and $W_p' = 9.71$ lb. Thus

Total rotating weight $= W_c'' + W_c' = 3.75 + 24.29 = 28.04$ lb

Total reciprocating weight $= W_p + W_p' = 20 + 9.71 = 29.71$ lb

Centrifugal force on crank is

$$f_c = \frac{W_c'' + W_c'}{g} R\omega^2 = \frac{28.04}{386}(4)(104.6)^2 = 3,220 \text{ lb}$$

Primary inertia force

$$= \frac{W_p + W_p'}{g} R\omega^2 \cos \theta = \frac{29.71}{386}(4)(104.6)^2 \cos \theta = 3,360 \cos \theta \text{ lb}$$

Secondary inertia force $= \dfrac{W_p + W_p'}{g} R\omega^2 \dfrac{R}{L} \cos 2\theta$

$$= \frac{29.71}{386}(4)(104.6)^2 \frac{4}{14} \cos 2\theta = 960 \cos 2\theta \text{ lb}$$

Total inertia force $= f_p = 3,360 \cos \theta + 960 \cos 2\theta$

To visualize the shaking-force action on the crankshaft bearing, it is convenient to plot it on a polar diagram as shown in Fig. 20-5. The inertia forces for various values of θ can be calculated in table form. Positive and negative values resulting from the sign of the cosine function indicate whether the direction is to the right or left.

A circle with a radius of 3,220 units representing the centrifugal force f_c is drawn first with dashed lines, using the pole O as a center. Then for various values of θ the inertia force f_p is laid off horizontally from the corresponding point on the circle. For example, when $\theta = 60°$, $f_p = 1,680 - 480 = 1,200$ lb. Thus where the 60° radial line intersects the f_c circle at A, a horizontal line 1,200 units long is laid off to the right to locate point B. The shaking force F_s, which comes on the bearings for this position, is then represented by a vector directed from O toward B, and its magnitude scales 3,960 lb. This same procedure is then used to locate other points on the resultant egg-shaped curve, which represents the unbalanced shaking force. When

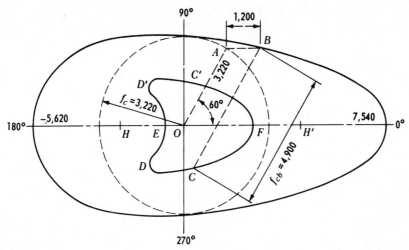

Fig. 20-5

$\theta = 0°$ the shaking force is directed horizontally to the right from point 0 and is $3,220 + 3,360 + 960 = 7,540$ lb. When $\theta = 180°$ the shaking force is directed horizontally to the left from point 0 and is $-3,220 - 3,360 + 960 = -5,620$ lb.

We shall assume that a weight known as a *counterbalance* is placed opposite the crank. The inertia force acting on this weight is known as the *counterbalance force*. By varying the amount of this weight, the distance from its center of gravity to the crankshaft axis, or both, any desired counterbalance force can be obtained. For any engine the counterbalance force should balance the centrifugal force f_c completely and part of the inertia force f_p so as to make the resultant load on the bearings, and hence the foundation, a minimum.

We shall assume that a counterbalance force f_{cb} equal to the centrifugal force plus one-half of the maximum primary inertia force is used and thus

$$f_{cb} = 3,220 + 1,680 = 4,900 \text{ lb}$$

In Fig. 20-5 the counterbalance force is shown acting at the angle θ with the horizontal, starting from the previously found point B in the sample construction. On a line drawn at 60° from the horizontal, a length of 4,900 units is laid off from B to locate point C. A vector directed from 0 toward C then represents the resultant shaking force when the crank angle is 60°. Continuing this construction for the full 360° gives the curve $CDED'C'F$. If the counterbalance force were made to overcome only the centrifugal force f_c and none of the inertia force f_p, then the shaking-force diagram and load on the bearings would fall along the horizontal axis between points HH'.

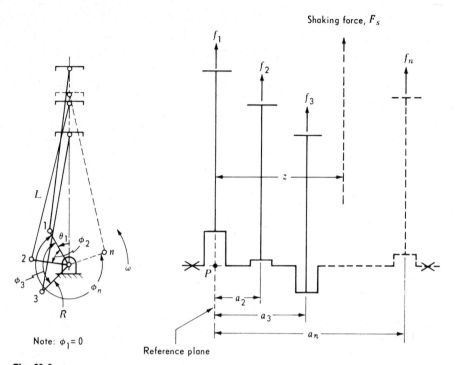

Fig. 20-6

The optimum amount of counterbalance force will be equal to the centrifugal force plus some fraction of the maximum primary inertia force and will result in a shaking-force diagram which will make the maximum horizontal and vertical loads on the bearings approximately equal. For each engine different amounts of counterbalance force can be tried, and diagrams similar to $CDED'C'F$ can be plotted in order to determine the optimum.

20-4   MULTICYLINDER IN–LINE ENGINES

In the previous section it was shown that by providing a counterbalance weight opposite the crankpin, the centrifugal force f_c can be annulled, and if this counterbalance weight were increased, it would only partially annul the inertia force f_p. In our discussion here of multicylinder engines we shall assume that the centrifugal force acting on each crank has been canceled by an appropriate balancing weight opposite each crank. Thus we shall be concerned only with the balancing of the inertia forces f_p acting at the various pistons.   We shall see that it is sometimes possible to balance some or all of these inertia forces by changing the relative angular positions of the cranks.   In addition to balancing these inertia forces, the moments created by these forces axially along the crankshaft must be considered and, if not eliminated, must be reduced as much as possible.

Previously we have denoted the inertia force at the piston as f_p.   In Fig. 20-6 an in-line engine having n cylinders is shown.   The angle crank 1 makes with the vertical is θ_1.   The angles between the cranks are fixed since the crankshaft is rigid.   The angle crank 2 makes with crank 1 is ϕ_2, that between 3 and 1 is ϕ_3, and so on for all the cranks.

Substitution into Eq. (20-5) gives the inertia forces acting at the various pistons, and these will be designated $f_1, f_2, f_3, \ldots, f_n$; thus

$$f_1 = \frac{W_p + W_p'}{g} R\omega^2 \left[\cos(\theta_1 + \phi_1) + \frac{R}{L}\cos 2(\theta_1 + \phi_1) \right]$$

$$f_2 = \frac{W_p + W_p'}{g} R\omega^2 \left[\cos(\theta_1 + \phi_2) + \frac{R}{L}\cos 2(\theta_1 + \phi_2) \right]$$

$$\cdots\cdots\cdots\cdots\cdots\cdots\cdots\cdots\cdots\cdots$$

$$f_n = \frac{W_p + W_p'}{g} R\omega^2 \left[\cos(\theta_1 + \phi_n) + \frac{R}{L}\cos 2(\theta_1 + \phi_n) \right]$$

The algebraic sum of the inertia forces will be the shaking force F_s.   Then

$$F_s = \frac{W_p + W_p'}{g} R\omega^2 \left[\sum_{n=1}^{n=n} \cos(\theta_1 + \phi_n) + \frac{R}{L} \sum_{n=1}^{n=n} \cos 2(\theta_1 + \phi_n) \right] \quad (20\text{-}6)$$

However,

$$\cos(\theta_1 + \phi_1) = \cos\theta_1 \cos\phi_1 - \sin\theta_1 \sin\phi_1$$

$$\cos(\theta_1 + \phi_2) = \cos\theta_1 \cos\phi_2 - \sin\theta_1 \sin\phi_2$$

$$\cdots\cdots\cdots\cdots\cdots\cdots\cdots\cdots\cdots$$

$$\cos(\theta_1 + \phi_n) = \cos\theta_1 \cos\phi_n - \sin\theta_1 \sin\phi_n$$

or

$$\sum_{n=1}^{n=n} \cos(\theta_1 + \phi_n) = \cos\theta_1 \sum_{n=1}^{n=n} \cos\phi_n - \sin\theta_1 \sum_{n=1}^{n=n} \sin\phi_n \quad (20\text{-}7)$$

Similarly, it can be shown that

$$\sum_{n=1}^{n=n} \cos 2(\theta_1 + \phi_n) = \cos 2\theta_1 \sum_{n=1}^{n=n} \cos 2\phi_n - \sin 2\theta_1 \sum_{n=1}^{n=n} \sin 2\phi_n \quad (20\text{-}8)$$

Substituting Eqs. (20-7) and (20-8) into Eq. (20-6), we obtain the following expression for the shaking force:

$$F_s = \frac{W_p + W_p'}{g} R\omega^2 \left[\cos\theta_1 \sum_{n=1}^{n=n} \cos\phi_n - \sin\theta_1 \sum_{n=1}^{n=n} \sin\phi_n \right.$$

$$\left. + \frac{R}{L}\cos 2\theta_1 \sum_{n=1}^{n=n} \cos 2\phi_n - \frac{R}{L}\sin 2\theta_1 \sum_{n=1}^{n=n} \sin 2\phi_n \right] \quad (20\text{-}9)$$

For the inertia forces to be in balance, the resultant inertia force must be zero; that is, $F_s = 0$ for all values of the crank position θ_1. Since in some crankshaft arrangements either the primary forces or secondary forces are balanced while the others are unbalanced, we shall consider these two kinds of forces separately. From Eq. (20-9) we see that the necessary condition for the primary forces to be balanced is

$$\sum_{n=1}^{n=n} \cos\phi_n = 0 \qquad \sum_{n=1}^{n=n} \sin\phi_n = 0 \qquad (20\text{-}10)$$

and the necessary condition for the secondary forces to be balanced is

$$\sum_{n=1}^{n=n} \cos 2\phi_n = 0 \qquad \sum_{n=1}^{n=n} \sin 2\phi_n = 0 \qquad (20\text{-}11)$$

The moment of the shaking force can be found by taking moments of the inertia forces about some point in their plane. In Fig. 20-6, consider a reference plane perpendicular to the crankshaft axis intersecting the axis at point P. Consider moments about point P and let M be the moment of the shaking force. Then

$$M = f_1 a_1 + f_2 a_2 + f_3 a_3 + \cdots + f_n a_n \qquad (20\text{-}12)$$

Equation (20-12) is equivalent to multiplying each term in Eq. (20-9) by its corresponding moment arm. Hence

$$M = \frac{W_p + W_p'}{g} R\omega^2 \left[\cos\theta_1 \sum_{n=1}^{n=n} a_n \cos\phi_n - \sin\theta_1 \sum_{n=1}^{n=n} a_n \sin\phi_n \right.$$

$$\left. + \frac{R}{L}\cos 2\theta_1 \sum_{n=1}^{n=n} a_n \cos 2\phi_n - \frac{R}{L}\sin 2\theta_1 \sum_{n=1}^{n=n} a_n \sin 2\phi_n \right] \quad (20\text{-}13)$$

For the moment of the shaking force to be zero, M in this equation must be zero for all values of the crank position θ_1. Since in some crankshaft arrangements either the primary moments or secondary moments are balanced while the others are unbalanced, we shall consider these two kinds of moments separately.   From Eq. (20-13) we see that the necessary condition for the primary moments to be balanced is

$$\sum_{n=1}^{n=n} a_n \cos \phi_n = 0 \qquad \sum_{n=1}^{n=n} a_n \sin \phi_n = 0 \qquad\qquad (20\text{-}14)$$

and the necessary condition for the secondary moments to be balanced is

$$\sum_{n=1}^{n=n} a_n \cos 2\phi_n = 0 \qquad \sum_{n=1}^{n=n} a_n \sin 2\phi_n = 0 \qquad\qquad (20\text{-}15)$$

The location of the shaking force F_s is denoted by distance z in Fig. 20-6, and since the sum of the moments of the f forces about point P is equal to the moment of the resultant of the forces,

$$z = \frac{M}{F_s} \qquad\qquad (20\text{-}16)$$

where M is computed from Eq. (20-13) and F_s is computed from Eq. (20-9).

20-5 FOUR–CYLINDER IN–LINE ENGINE

The four-cylinder in-line engine will be used to illustrate the application of the above equations in analyzing the balance of engines.   In selecting the crank arrangement for an engine, balance is not the sole criterion.   It is important that the power impulses be applied at equal time intervals in order to provide smooth torque output from the crankshaft.   The sequence in which firing takes place in the cylinders is known as *firing order*.

In four-stroke-cycle engines gases are ignited when the piston is approximately at top dead center, and power is delivered to the crankshaft for 180° of rotation until the piston reaches bottom dead center.   During the next 180° of rotation, the burned gases are exhausted; a fresh charge of combustible mixture is sucked into the cylinder during the next 180° of rotation.   Then the mixture is compressed during the next 180°, after which the cycle is repeated.   A stroke is the movement of the piston from one dead-center position to the next, and we see that power is delivered to the crankshaft during one-half revolution of each two revolutions of the crankshaft.   In our discussion of various engines we shall use charts to study the uniformity of the sequence in which the power strokes occur.   We shall label the power stroke, P; exhaust, E; intake, I; and compression, C.

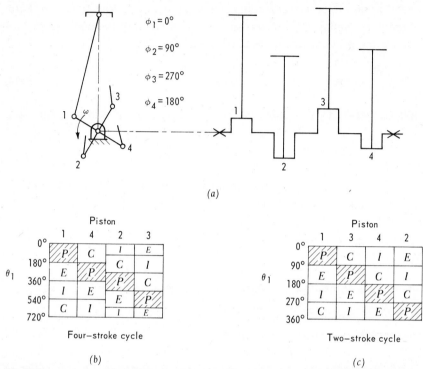

(a)

Piston

	1	4	2	3
0°	P	C	I	E
180°	E	P	C	I
360°	I	E	P	C
540°	C	I	E	P
720°			I	E

θ_1

Four–stroke cycle

(b)

Piston

	1	3	4	2
0°	P	C	I	E
90°	E	P	C	I
180°	I	E	P	C
270°	C	I	E	P
360°				

θ_1

Two–stroke cycle

(c)

Fig. 20-7

For a two-stroke-cycle engine all the events discussed above occur during each revolution of the crankshaft, each occupying approximately one-quarter of a revolution. Power and exhaust occur during one stroke and are followed by intake and compression during the next stroke.

Various crank arrangements can be used for the four-cylinder engine, but we shall consider only two. Any other arrangement can be analyzed by the same methods which we shall use here. Let us first consider the engine with cranks arranged as indicated in Fig. 20-7 where $\phi_1 = 0°$, $\phi_2 = 90°$, $\phi_3 = 270°$, and $\phi_4 = 180°$. The power strokes for the four-stroke-cycle engine and two-stroke-cycle engine are shown in Fig. 20-7b and c, respectively. We see that in the two-stroke-cycle engine, power is applied evenly, while for the four-stroke-cycle, the power strokes for cylinders 4 and 2 overlap for 90° and there is a 90° lapse between the power strokes of pistons 3 and 1.

We have seen that the conditions required for balance of an in-line engine are stated by Eqs. (20-10), (20-11), (20-14), and (20-15). When applying

Table 20-1

CRANK	ϕ	$\cos\phi$	$\sin\phi$	2ϕ	$\cos 2\phi$	$\sin 2\phi$	a	$a\cos\phi$	$a\sin\phi$	$a\cos 2\phi$	$a\sin 2\phi$
1	0°	1	0	0°	1	0	0	0	0	0	0
2	90°	0	1	180°	−1	0	a	0	a	$-a$	0
3	270°	0	−1	540°	−1	0	$2a$	0	$-2a$	$-2a$	0
4	180°	−1	0	360°	1	0	$3a$	$-3a$	0	$3a$	0
Σ		0	0		0	0		$-3a$	$-a$	0	0
		PRIMARY FORCES BALANCED			SECONDARY FORCES BALANCED			PRIMARY MOMENTS UNBALANCED		SECONDARY MOMENTS BALANCED	

these equations to an engine, it is convenient to construct a table. The engine in Fig. 20-7 is analyzed in Table 20-1. We see that the primary and secondary forces and the secondary moments are balanced, but the primary moments are unbalanced. Since the forces are all balanced, the unbalanced primary moment is a pure couple acting to rotate the engine about an axis perpendicular to the axis of the crankshaft. Substituting the summations into the moment equation (20-13), we obtain

$$M = \frac{W_p + W'_p}{g} R\omega^2 \left(-3a\cos\theta_1 + a\sin\theta_1\right)$$

Next, we shall consider the conventional four-cylinder automobile engine, which is shown in Fig. 20-8. The cranks are arranged as follows: $\phi_1 = 0°$, $\phi_2 = 180°$, $\phi_3 = 180°$, and $\phi_4 = 0°$. For the four-stroke-cycle engine power is evenly applied, as shown in Fig. 20-8b, while in the two-stroke-cycle, shown in Fig. 20-8c, there is uneven application of power. Balance of the engine is analyzed in Table 20-2. The primary forces and moments

Table 20-2

CRANK	ϕ	$\cos\phi$	$\sin\phi$	2ϕ	$\cos 2\phi$	$\sin 2\phi$	a	$a\cos\phi$	$a\sin\phi$	$a\cos 2\phi$	$a\sin 2\phi$
1	0°	1	0	0°	1	0	0	0	0	0	0
2	180°	−1	0	360°	1	0	a	$-a$	0	a	0
3	180°	−1	0	360°	1	0	$2a$	$-2a$	0	$2a$	0
4	0°	1	0	0°	1	0	$3a$	$3a$	0	$3a$	0
Σ		0	0		4	0		0	0	$6a$	0
		PRIMARY FORCES BALANCED			SECONDARY FORCES UNBALANCED			PRIMARY MOMENTS BALANCED		SECONDARY MOMENTS UNBALANCED	

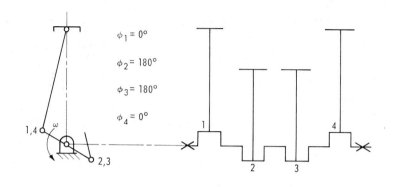

$\phi_1 = 0°$

$\phi_2 = 180°$

$\phi_3 = 180°$

$\phi_4 = 0°$

(a)

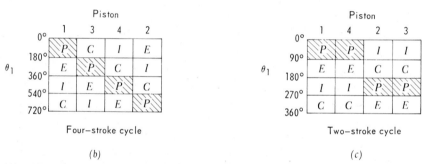

Four-stroke cycle

(b)

Two-stroke cycle

(c)

Fig. 20-8

are balanced, but the secondary forces and moments are unbalanced. Substitution of the summations into Eqs. (20-9) and (20-13) gives

$$F_s = \frac{W_p + W'_p}{g} R\omega^2 \left(4 \frac{R}{L} \cos 2\theta_1 \right)$$

for the shaking force and

$$M = \frac{W_p + W'_p}{g} R\omega^2 \left(6a \frac{R}{L} \cos 2\theta_1 \right)$$

for the moment of the shaking force. By substituting these results into Eq. (20-16), we obtain the location of the shaking force as

$$z = \frac{M}{F_s} = \frac{3}{2} a$$

We see then from Fig. 20-6 that the shaking force is located midway between cylinders 2 and 3.

A common device used to balance the secondary forces for this four-cylinder engine is the Lanchester balancer shown in Fig. 20-9. The balancer consists of two helical gears with eccentric balancing weights as shown. The pitch point for these gears is located on the line of action of the shaking force F_s so that their resulting inertia force F_b balances the shaking force. Since the unbalanced moments in this engine are due to the unbalanced secondary forces, by balancing the secondary forces we also obtain balance of the moments. A crossed helical gear on the crankshaft drives these gears at twice crankshaft speed in order that the balancing forces have the same circular frequency as the unbalanced secondary forces. For an engine where unbalance of primary forces only exists, this type of balancer can be used, but the balancing gears rotate at crankshaft speed.

If an engine has an unbalance which consists of a pure couple, then a balancing device as shown in Fig. 20-10 can be used. Two gears with eccentric balancing weights and arranged as shown produce a pure couple. The idler is used to make the two gears turn in the same direction. The couple caused by the balancing weights is

$$C = \frac{W_b}{g} r\omega^2 b \cos\theta$$

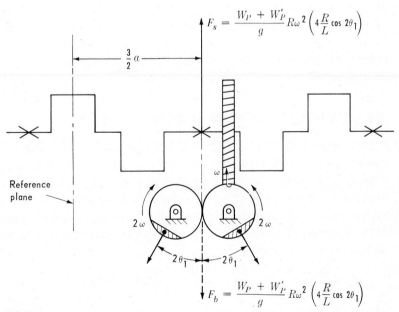

$$F_s = \frac{W_P + W_P'}{g} R\omega^2 \left(4\frac{R}{L}\cos 2\theta_1\right)$$

$$F_b = \frac{W_P + W_P'}{g} R\omega^2 \left(4\frac{R}{L}\cos 2\theta_1\right)$$

Fig. 20-9

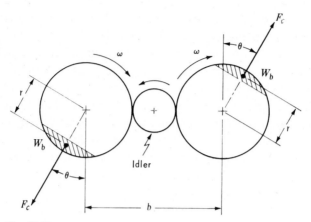

Fig. 20-10

For balancing primary couples, the gears with balancing weights would be driven at crankshaft speed, while for balancing secondary couples, the gears would be driven at twice crankshaft speed.

The methods used here for analyzing the balance of the two four-cylinder engines can also be used for other in-line engines having any number of cylinders. Such an analysis indicates that an in-line six-cylinder engine having a crank arrangement $\phi_1 = \phi_6 = 0°$, $\phi_2 = \phi_5 = 240°$, and $\phi_3 = \phi_4 = 120°$ is perfectly balanced in regard to primary and secondary forces and moments, as is an in-line eight-cylinder engine having a crank arrangement of $\phi_1 = \phi_8 = 0°$, $\phi_2 = \phi_7 = 180°$, $\phi_3 = \phi_6 = 90°$, and $\phi_4 = \phi_5 = 270°$.

20-6   ANALYSIS BY ROTATING VECTORS

A convenient method of analyzing balance of multicylinder engines is by use of rotating vectors. As stated earlier, we shall assume that the centrifugal force acting on each crank has been balanced by a balancing weight opposite the crank. Thus the only unbalance is caused by the inertia force acting on the reciprocating weight, and this force acts along the centerline of the cylinder. Equation (20-5) represents the inertia force acting at one piston and can be rewritten

$$f_p = \frac{W_p + W'_p}{g} R\omega^2 \cos \omega t + \frac{W_p + W'_p}{g} R\omega^2 \frac{R}{L} \cos 2\omega t$$

where θ has been replaced by ωt and t is considered zero when the crank is at top dead center. The primary force can now be represented as the projection on the cylinder of a vector of magnitude $[(W_p + W'_p)/g] R\omega^2$ rotating about its origin with an angular velocity ω, and the secondary force can be represented as the projection of a vector of magnitude $[(W_p + W'_p)/g]$ $R\omega^2 (R/L)$ rotating with a velocity 2ω. The primary and secondary rotating vectors both point in the direction of the crank when the crank is at top dead center. The primary force vector will always point in the direction of the crank since it rotates at the same angular velocity as the crank. The secondary force vector will point in the direction of the crank when the crank is at top dead-center position, but after the crank has rotated 180° and is at bottom dead center, this force vector will point toward top dead-center crank position because this vector rotates at twice the crank velocity.

This vector method will be demonstrated by applying it to the four-cylinder engine in Fig. 20-7, which was analyzed in the previous section by the tabular method. The crank arrangement is shown in Fig. 20-11 where the cylinders are vertical and crank 1 is at top dead-center position. It will be explained that at the instant the crankshaft is in the position shown in Fig. 20-11a, then the vectors in Fig. 20-11b to e will be directed as shown. In Fig. 20-11b the rotating primary-force vectors are shown pointing in the directions of the corresponding cranks. These vectors rotate at the same angular velocity as the crankshaft, and their projections on the vertical are the primary forces. From Fig. 20-11b it can be seen that the sum of these vertical forces will be zero at all times, and thus the primary forces are in balance.

In Figure 20-11c, the secondary-force vectors are shown, and they rotate at twice crankshaft speed. Since the primary and secondary forces point in the direction of the crank when the crank is at top dead center, the secondary-force vector for crank 1 points upward at the instant considered. In Fig. 20-11a we note that cranks 2, 3, and 4 have rotated 90°, 270°, and 180°, respectively, from top dead center. Hence the secondary-force vectors in Fig. 20-11c have rotated 180°, 540°, and 360°, respectively. Note that the magnitudes of these secondary-force vectors are all the same.

The moments of the primary and secondary forces can also be represented as projections of rotating vectors. The moments of these forces about point P depend on their moment arms as shown in Fig. 20-11a. The moments are counterclockwise or clockwise, depending upon whether the force is upward or downward. Therefore the moment of a force can be represented by a rotating vector in the same direction as the force and of magnitude equal to the product of the force and its distance from point P.

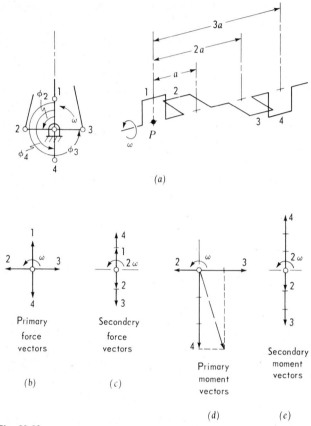

(a)

Primary
force
vectors

(b)

Secondary
force
vectors

(c)

Primary
moment
vectors

(d)

Secondary
moment
vectors

(e)

Fig. 20-11

Figure 20-11*d* shows the rotating primary-moment vectors. Vector 3 is twice as large as vector 2 because cylinder 3 is twice as far from *P* as cylinder 2. Vector 4 is three times as large as vector 2 because cylinder 4 is three times the distance from *P* as cylinder 2. The vertical projections of these rotating vectors give the moments of the primary forces, and we see that they are not in balance. The resultant of the vectors in Fig. 20-11*d* is shown by the dashed vector. The projection of this rotating vector upon the vertical also represents the unbalanced moment of the primary forces. When the projection of this vector on the vertical points upward, the moment is counterclockwise about point *P* in Fig. 20-11*a*. When the vertical projection points downward, the moment is clockwise about *P*. The resultant, as shown in Fig. 20-11d, represents the moment when crank 1 is at top dead center.   The secondary moments are shown in Fig. 20-11e, and we see that they are balanced.

20-7 V–TYPE AND OPPOSED ENGINES

As shown in Fig. 20-12, a V-type engine consists of two in-line engines having a common crankshaft.   The two planes in which the pistons reciprocate intersect at the crankshaft axis and form a V of angle ψ.   In automotive applications V-6, V-8, and V-12 engines in which ψ is either 60° or 90° are common.   In small engines and compressors V-2 and V-4 arrangements are often used.

The angle ψ for a V-type engine is chosen so as to give evenly spaced power strokes.   In a V-type engine a connecting rod from the left bank and one from the right bank are connected to a common crank, or throw as they are sometimes called.   The two rods lie side by side on the crank, and hence do not lie in the same transverse plane, but this small offset is usually neglected in analyzing balance.

In the analysis of a V-type engine each bank of cylinders can be analyzed separately, and then any unbalance of the banks can be combined vectorially.   Let us examine the engine in Fig. 20-12a.   Considering each bank separately, we found previously that a four-cylinder engine with this crank arrangement is balanced except for the primary moments.   Thus for this V-type engine the only unbalance will be primary moments in each bank.

In Fig. 20-12b the primary-force vectors are shown for the crank position where piston number 1 is at top dead center, and in Fig. 20-12c the primary-moment vectors are shown with the resultant moment vector for each bank. The unbalanced moment in the left bank is the projection of the resultant vector in the plane of the left bank, and the unbalanced moment in the right bank is the projection of the resultant vector in the plane of the right bank.   The two resultant vectors have the same magnitude and phase angle ϕ relative to crank 1.   Thus the total unbalance for this engine can be represented by a single vector rotating at crankshaft speed and having a magnitude and phase angle equal to either of the resultant vectors in Fig. 20-12c, since this vector has a component in each bank equal to the moment in that bank.   Since the forces are completely balanced, this vector represents a pure couple.   From Fig. 20-12c we observe that the magnitude of the

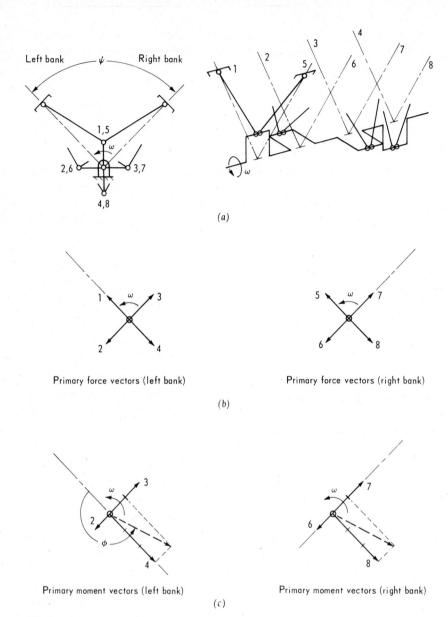

(a)

Primary force vectors (left bank) Primary force vectors (right bank)

(b)

Primary moment vectors (left bank) Primary moment vectors (right bank)

(c)

Fig. 20-12

couple is

$$M = \sqrt{\left(3a\,\frac{W_p + W'_p}{g}\,R\omega^2\right)^2 + \left(a\,\frac{W_p + W'_p}{g}\,R\omega^2\right)^2}$$

$$= \sqrt{10}\,a\left(\frac{W_p + W'_p}{g}\right)R\omega^2$$

and the phase angle as measured counterclockwise from the crank for pistons 1 and 5 is

$$\phi = 180° + \tan^{-1} \tfrac{1}{3} = 198.43°$$

This couple can be represented by the two forces F attached to the crankshaft as shown in Fig. 20-13. These forces rotate with the crankshaft, and they can be balanced by adding two counterweights as shown. The counterweights are appropriately chosen so that the pure couple produced by their inertia forces exactly balances the couple due to the F forces.

The engine shown in Fig. 20-14 is an example of an opposed engine. An *opposed engine* consists of two banks of cylinders or two in-line engines on opposite sides of the crankshaft and lying in a common plane. An opposed engine is a special case of the V-type engine where $\psi = 180°$. The resulting shaking force F_s and shaking couple M lie in the plane in which the pistons reciprocate. Opposed engines of two, four, and six cylinders are common.

The vector method for investigating balance of the engine in Fig. 20-14a is shown in Fig. 20-14b, where the rotating vectors are combined into single

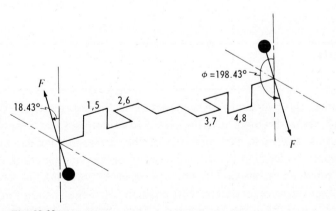

Fig. 20-13

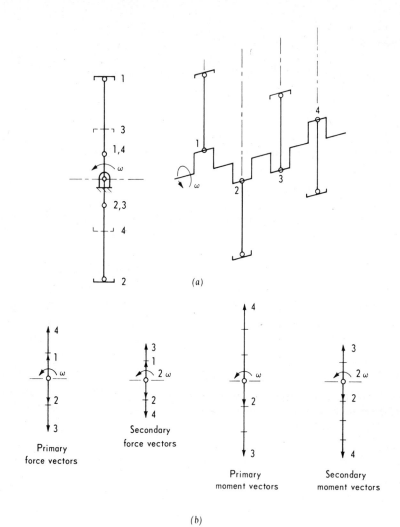

(a)

(b)

Fig. 20-14

diagrams since the two banks of cylinders lie in the same plane. When applying the vector method, it is important to remember that a primary-force vector is always in the direction of the corresponding crank. A secondary-force vector is in the direction of the corresponding crank if the piston is at top dead center, and for other positions of the crank the second-ary-force vector will have rotated from its position at top dead center through twice the angle that the crank has rotated. This engine is balanced except for the secondary moment, which is a pure couple.   Balance can be obtained by adding counterweights which produce a pure couple.   The counterweights must be driven by gears so as to rotate at twice crankshaft speed.

The V-8 engine and opposed engine which were analyzed by the rotating-vector method could have been analyzed using the tabular method for in-line engines by applying it to each bank separately.   However, the rotating-vector method of analysis makes it easier to visualize the unbalance.

20-8   DISCUSSION

Throughout our discussion of engine balance we have assumed that the weights of all the pistons and connecting rods were the same and that the distance between cylinders was uniform.   In applications where this would not be the case, the student should be able to apply the equations presented here to check for unbalance or to vary the length of the vectors if the rotating vector (graphical method) is used.   The advantage of the graphical method over the equation method is that the physical significance of what is going on is not lost.

It is not always practical to balance completely the shaking force or moment in an engine, but the annoyance due to their effects can be reduced by selecting mountings for the engine which are of appropriate stiffness so as to isolate the vibration of the engine from the structure in which it is contained.   The study of mountings to provide for vibration isolation is not within the scope of this book, but is presented in textbooks on mechanical vibrations.

PROBLEMS

20-1   In the mechanism shown in Fig. 9-9, length $OD = 9\frac{1}{4}$ in., $OA = 3\frac{1}{4}$ in., $AB = 6\frac{1}{4}$ in., and $BD = 5$ in.   Link 2 weighs $1\frac{1}{2}$ lb and its center of gravity lies on line OA at $1\frac{1}{4}$ in. from O; link 3 weighs 3 lb and its center of gravity lies on line AB at $2\frac{1}{2}$ in. from A; link 4 weighs $2\frac{1}{4}$ lb and its center of gravity lies on line BD at $1\frac{3}{4}$ in. from D.   Make a drawing of the mechanism using a scale of 1 in. = 4 in., showing all the dimensions stated above.   (a) Link 3 is to be replaced by a two-mass dynamically equivalent system, and the masses are to be located at A and B.   Their weights are to be designated W_A and W_B. Determine W_A and W_B.   (b) Determine the amount of counterbalance in inch-pounds which must be attached to links 2 and 4 in order to eliminate the shaking forces in the mechanism.

20-2   The following data are given for a single-cylinder internal-combustion engine, with the piston moving in the horizontal direction: speed = 1,500 rpm, stroke = 8 in., weight of crank and crankpin = 8 lb, distance from crankshaft axis to center of gravity of crank and crankpin combined = 2.5 in., length of connecting rod = 16 in., weight of connecting rod = 8 lb, distance from center of gravity of connecting rod to crankpin = 4 in., weight of piston and piston pin = 7 lb.   (a) Use a force scale of 1 in. = 2,000 lb and determine graphically the magnitude and direction of the shaking force for a crank angle θ = 150° if no counterbalance were used.   (b) Same as (a) except that a counterbalance weight is used such that the counterbalance force will be equal to the centrifugal force plus 0.6 of the maximum primary inertia force. (c) If the center of gravity of the counterbalance weight in part (b) were 2 in. from the crankshaft axis, determine the weight of counterbalance required.

20-3   (a) Consider the engine in Prob. 20-2.   Make a table as shown below in order to list the values of the primary and secondary inertia forces for each 30° position of the crank.

θ, deg	cos θ	2θ, deg	cos 2θ	PRIMARY f_p, lb	SECONDARY f_p, lb
0		0			
30		60			
60		120			
.		.			
.		.			
.		.			

Using a force scale of 1 in. = 2,000 lb, draw the shaking-force diagrams similar to Fig. 20-5 for the following cases: (1) no counterbalance weight and (2) a counterbalance weight is used such that the counterbalance force will be equal to the centrifugal force plus 0.6 of the maximum primary inertia force. (b) What are the values of the maximum horizontal and maximum vertical shaking forces when scaled from the diagram in (2) of part (a)?

20-4   A single-cylinder engine is shown in Fig. P20-4.   Length of crank = 9 in., weight of crank and crankpin = 9.47 lb, distance from crankshaft to center of gravity of crank and crankpin combined = 4.75 in., length of connecting rod = 36 in., weight of connecting rod = 16 lb, distance from crankpin to center of gravity of connecting rod = 9 in., weight of piston and piston pin = 6 lb.   Gears 1, 2, and 3 are the same size and gear 1 is keyed to the crankshaft. Gears 4 and 5 are half the size of gears 1, 2, and 3.   Note the small clearances between gears 1 and 3 and between 3 and 5.   Design a vibration-damper system to balance the primary and secondary inertia forces.   The engine is to be in balance at all speeds by adding weights W_1, W_2, W_3, W_4, and W_5 in circles a, b, c, d, and e, respectively; that is, there are to be no unbalanced horizontal or vertical forces. W_1 is to balance only the centrifugal force.   Radii of circles b, c, d, and e are 3 in.   Using a scale of 1 in. = 18 in., make a drawing similar to Fig. P20-4.   Show on the drawing the values of the weights W_1, W_2, W_3, W_4, and W_5 and indicate their positions.

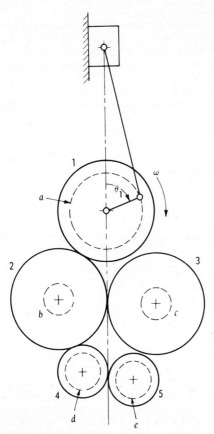

Fig. P20-4

20-5 The crank arrangement for a three-cylinder in-line engine is shown in Fig. P20-5 where $\phi_1 = 0°$, $\phi_2 = 120°$, and $\phi_3 = 240°$. (a) Construct a firing-order chart similar to that shown in Fig. 20-7 if the engine is to be used as a four-stroke-cycle engine and the power impulses are to occur at equal time intervals. (b) Same as part (a) except that the engine is to be used as a two-stroke-cycle engine.

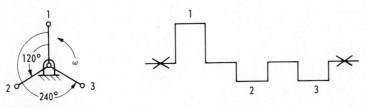

Fig. P20-5

20-6 For the engine in Prob. 20-5 the distance between the centerline of the cylinders is a. (a) Construct a table similar to Table 20-1 and indicate whether the primary and secondary forces or moments are unbalanced. (b) What is the value of the shaking force? (c) Using the results in the table, write an expression for the shaking moment in terms of the position θ_1 of crank 1. (d) Is the shaking moment a pure couple? Why or why not?

20-7 A proposed engine has the crank arrangement shown in Fig. P20-7. The cylinders are equally spaced with a distance a between their centerlines. (a) Construct a table similar to Table 20-1 and indicate whether the primary or secondary forces or moments are unbalanced. (b) Write an expression for the shaking force in terms of the position θ_1 of crank 1. (c) Write an expression for the shaking moment in terms of θ_1. (d) Write an expression for the location of the shaking force in terms of θ_1.

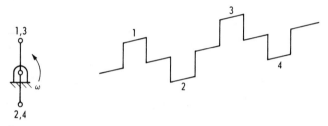

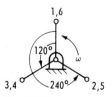

Fig. P20-7

20-8 Consider the engine in Prob. 20-7. (a) State all the possible firing orders starting with cylinder 1 if the power impulses are to occur at equal time intervals and the engine is used as a four-stroke-cycle engine. (b) If the crankshaft as shown were used in a two-stroke-cycle engine, what firing order would exist starting with cylinder 1? (c) How could the angles between cranks be changed to make for a more uniform power input if the crankshaft were to be used in a two-stroke-cycle engine? Make a sketch showing a front-end view of the crankshaft with crank 1 vertical. The crankshaft is to rotate counterclockwise. Number the cranks so that the firing order will be 1, 3, 2, 4.

20-9 An in-line six-cylinder engine (four-cycle), the type used in automobiles, has the cranks arranged as shown in Fig. P20-9, where $\phi_1 = \phi_6 = 0°$, $\phi_2 = \phi_5 = 240°$, and $\phi_3 = \phi_4 = 120°$. Construct a table similar to Table 20-1 to show that this

Fig. P20-9

engine is completely balanced in regard to primary and secondary forces and moments and requires no counterbalance weights other than the counterbalance opposite each crank to balance the weight $W_c' + W_c''$.

20-10 For the engine in Prob. 20-9 power impulses are to be applied to the crankshaft at equal time intervals.   From an examination of Fig. P20-9 it can be seen that one possible firing order is 1, 5, 3, 6, 2, 4.   List other possible firing orders starting with cylinder 1.

20-11 An in-line eight-cylinder engine (four-cycle) has the cranks arranged as shown in Fig. P20-11, where $\phi_1 = \phi_8 = 0°$, $\phi_2 = \phi_7 = 180°$, $\phi_3 = \phi_6 = 90°$, and $\phi_4 = \phi_5 = 270°$.   Construct a table similar to Table 20-1 and indicate whether the primary and secondary forces and moments are in balance.

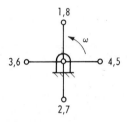

Fig. P20-11

20-12 Consider the engine in Prob. 20-6.   Use the rotating vector method and state whether the primary and secondary forces or moments are balanced.   Let 0.8 in. represent unit length for the primary force and primary-moment vectors and let 0.5 in. represent unit length for the secondary-force and secondary-moment vectors.   If the primary moments are unbalanced, draw in the resultant moment as a dashed line.   Do the same for the secondary moments.

20-13 Use the rotating vector method to analyze the engine in Prob. 20-9.   Let 0.3 in. represent unit length for the primary-force and primary-moment vectors.   Use 0.2 in. to represent unit length for the secondary-force and secondary-moment vectors.

20-14 The two-cylinder opposed-piston engine shown in Fig. P20-14 is running at 4,000 rpm.   The total reciprocating weight at each piston is 3.5 lb, and the two connecting rods (each 6.5 in. long) are joined to a common crank which is 2 in. long.   Assume that $W_c'' + 2W_c'$ (the total equivalent weight at the crankpin) has been balanced by a counterbalance weight opposite the crank.   (a) Use the rotating vector method and draw the rotating primary- and secondary-force vectors in the directions they will assume when the crank is in the position shown in the figure.   Let 1 in. represent unit length for the primary-force vectors and use 0.5 in. to represent unit length for the secondary-force vectors.   State whether the primary- or secondary-force vectors are balanced or unbalanced.   (b) Using Eq. (20-5), compute the values of the inertia forces and the magnitude and direction of their resultant at the instant.   (c) If the inertia forces are to be balanced by adding additional weight opposite the crank and if the center of gravity of this weight is to be 1.5 in. from the crankshaft axis, find the required weight.

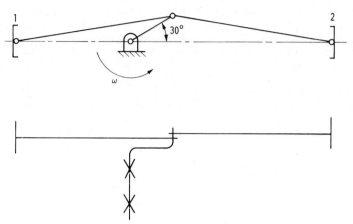

Fig. P20-14

20-15 In the radial engine shown in Fig. P20-15, the three connecting rods are joined to a common crank, and they can be assumed to lie in the same plane. All pistons and connecting rods have the same weight. The equivalent weight of crank at point C is W_c''. The equivalent weights for each connecting rod are W_c' and W_p'. (a) Determine the weight of counterbalance required at a radius r from the crankshaft axis in order to balance the equivalent rotating weights at C. (b) Make a drawing similar to Fig. P20-15 showing piston 1 at top dead center and indicate on the drawing the values of θ_1, θ_2, and θ_3. Analyze the

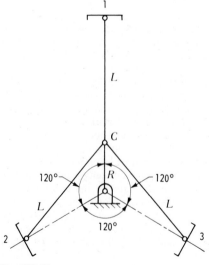

Fig. P20-15

engine using the rotating vector method.   Let 0.5 in. represent unit length for the primary-force vectors and use 0.375 in. for unit length for the secondary-force vectors.   Indicate whether the primary- or secondary-force vectors are balanced.   (*c*) Determine the phase angle ϕ measured counterclockwise from the crank at which a counterbalance weight must be added to balance the inertia forces acting upon the reciprocating weights.   (*d*) Write an equation for the magnitude of the resultant of the primary-force vectors.   (*e*) Write an equation giving the weight of a counterbalance at radius *r* which will balance the inertia forces acting on the reciprocating weights.   (*f*) Write an equation which gives the total of the counterbalance weights found in parts (*a*) and (*e*).

20-16   A proposed V-6 four-stroke-cycle engine is shown in Fig. P20-16.   The firing order is to be 1, 6, 3, 5, 2, 4.   (*a*) Analyze the power distribution by constructing a chart similar to the one in Fig. 20-7*b*.   (*b*) Analyze the balance of the engine using the rotating vector method.   When laying out the vectors, show them in the directions that they will assume when piston 1 is at top dead center.   Let 0.6 in. represent unit length for the primary-force and primary-moment vectors. Use 0.3 in. to represent unit length for the secondary-force and secondary-moment vectors.   State which force and moment vectors are balanced and which are unbalanced.

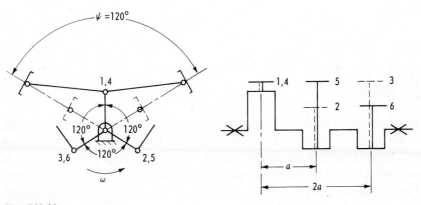

Fig. P20-16

21

GYROSCOPIC EFFECTS

21-1 INTRODUCTION

A body rotating about an axis of symmetry offers a resistance to a change
in the direction of this axis; this is known as the *gyroscopic effect*. Important
applications of the gyroscopic effect are the gyrocompass used on airplanes
and ships and in inertial guidance control systems for missiles and space
travel, where gyroscopes are used to sense the angular motion of a body.

Forces due to gyroscopic effects must sometimes be taken into account in the design of machines. These forces are encountered in the
bearings of an automobile engine as the automobile makes a turn, in ma-
rine turbines as the ship pitches in a heavy sea, and in a jet airplane engine
shaft as the airplane changes direction, to mention a few examples.

21-2 GYROSCOPIC FORCES

Gyroscopic effects can be most easily studied using the principle of angular momentum. We shall begin by reviewing from mechanics that angular velocity is a vector quantity. In Fig. 21-1, axes X and Z lie in plane

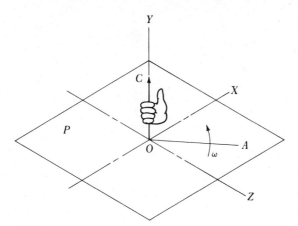

Fig. 21-1

P, and the Y axis is perpendicular to the plane.   Let the line OA rotate in the plane with an angular velocity ω.   Then the angular velocity can be represented by the vector OC, which is directed perpendicular to the plane of rotation P.   In order to determine the sense of the vector, the right-hand rule is used.   The rule is as follows: If the right hand is placed with the fingers pointing in the direction of the angular velocity ω, then the thumb will point in the sense of OC.   Thus an angular velocity directed as shown by the curved arrow is represented by a vector OC whose sense is upward. Similarly, if the angular velocity were directed opposite to that shown by the curved arrow, one would place his fingers in that direction, and then the thumb would point downward, indicating that the sense of vector OC is downward.   The length of vector OC represents the magnitude of the angular velocity to some convenient scale.   Also, angular acceleration and angular momentum have both magnitude and direction and thus are vector quantities and may be represented using the same convention as that used in Fig. 21-1.

In Fig. 21-2a let a particle of mass M have a linear velocity V.   Then the *linear momentum* of the particle is defined as the vector MV which has the same direction and sense as the velocity, and its magnitude is the product of the mass and the magnitude of the velocity vector.

Angular momentum for a particle is defined as the moment of the linear momentum.   In Fig. 21-2b, let the particle M be rotating about point O with angular velocity ω and radius R.   Then the angular momentum H is MVR. As shown in Fig. 21-3, a body of mass M, when rotating, behaves as if all the

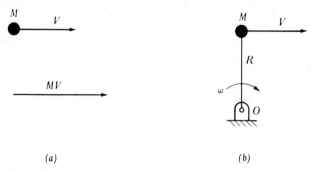

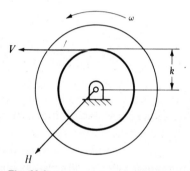

(a)   (b)

Fig. 21-2

Fig. 21-3

mass were concentrated in a ring at a radius k (the radius of gyration) from the axis of rotation. Then the angular momentum of the rotating body is

$$H = MVk$$

and since $V = \omega k$ and $I = Mk^2$, then

$$H = M\omega k^2 = I\omega \tag{21-1}$$

where I is the mass moment of inertia of the body about its axis of rotation and ω is its angular velocity. The angular momentum is represented by the vector H directed along the axis of rotation. The sense of the vector is determined by the right-hand rule. With fingers in the direction of ω, the thumb indicates that the sense of the vector is outward.

If we differentiate Eq. (21-1) with respect to time, we obtain the rate of change of angular momentum as follows:

$$\frac{dH}{dt} = I\frac{d\omega}{dt} = I\alpha$$

where α is the angular acceleration of the body.   However, by Newton's law, a torque T having the same sense as α is required to produce α.   Thus

$$T = \text{rate of change of angular momentum} \qquad (21\text{-}2)$$

Next, let us consider a disk keyed to a shaft which rotates in bearings A and B as shown in Fig. 21-4.   The axis of the shaft is called the *spin axis* S, and ω_s is the angular velocity about the spin axis.   Axes S, P, and T are mutually perpendicular.   The angular momentum is $I\omega_s$ and by the right-hand rule the angular-momentum vector H is directed as shown.   If the shaft is tilted through an angle $\Delta\theta$ about the axis P, the angular-momentum vector changes from H to H'.   Then

$$H' = H \twoheadrightarrow \Delta H$$

where vector ΔH represents the change in angular momentum.   The arc CD is equal to the radius OC times the angle $\Delta\theta$ in radians, and for a small $\Delta\theta$ the chord approximates the arc.   Hence the change in angular momentum is

$$\Delta H = I\omega_s\Delta\theta$$

and the rate of change of angular momentum is

$$\frac{dH}{dt} = \lim_{\Delta t\to 0}\frac{I\omega_s\Delta\theta}{\Delta t} = I\omega_s\frac{d\theta}{dt} = I\omega_s\omega_p \qquad (21\text{-}3)$$

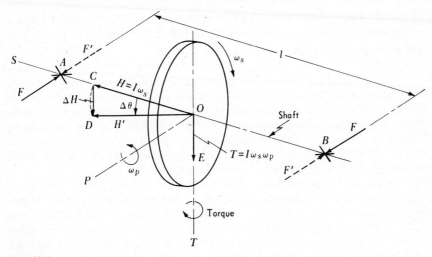

Fig. 21-4

The axis about which the shaft is tilted is called the *axis of precession*, $\Delta\theta$ is the *angle of precession*, and ω_p is the *angular velocity of precession*.

We found earlier that a change in angular momentum requires a torque T. Hence from Eqs. (21-2) and (21-3)

$$T = I\omega_s\omega_p \tag{21-4}$$

where the quantity T in Eq. (21-4) is called the *gyroscopic torque* or *gyroscopic couple*. This required torque is represented by the vector OE in Fig. 21-4, and it must have the same direction and sense as the change in angular momentum ΔH. (Note that as $\Delta\theta$ approaches zero in the limit, vector ΔH becomes parallel to the axis OT.) Then applying the right-hand rule to the torque vector, we see that the torque is clockwise about axis OT when viewed from above and can be indicated as shown by the curved arrow. Further, this torque must be externally supplied and is applied to the shaft in the form of a couple by the bearings which exert the forces F on the shaft. Thus

$$Fl = T$$

The forces F' shown dotted in the figure are equal and opposite to the F forces and represent the forces which the shaft exerts against the bearings.

To summarize, if we have a body rotating about an axis of symmetry OS and we give this axis a change in direction $\Delta\theta$, this latter rotation is called a *precession*. The precession makes for a change in the angular momentum which requires that an external torque T be applied to the system. Similarly, if we have a body rotating about an axis of symmetry OS and we apply an external torque T about an axis OT which is perpendicular to OS, this will produce a precession.

EXAMPLE 21-1   In Fig. 21-5*a*, the propeller shaft of an airplane is shown. The propeller rotates at 2,000 rpm cw when viewed from the rear and is driven by the engine through reduction gears. Suppose the airplane is flying horizontally and is making a turn to the right at 0.2 rad/sec when viewed from above. The propeller weighs 60 lb and has a radius of gyration of 3 ft. It is desired to find the gyroscopic forces which the propeller shaft exerts against bearings A and B which are 6 in. apart.

When working problems, it is helpful to make a sketch similar to Fig. 21-4 labeling all quantities in order to determine their proper sense. The propeller in Fig. 21-5*a* can be represented by the rotating disk in Fig. 21-5*b*. The spin axis S, which represents the axis of the shaft, is drawn first. The angular momentum has the same sense as the angular velocity, and by the right-hand rule the angular-momentum vector H is directed as

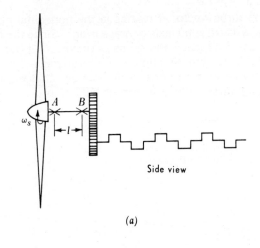

Side view

(a)

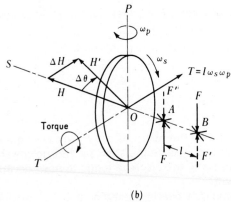

(b)

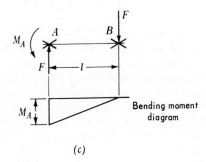

Bending moment
diagram

(c)

Fig. 21-5

shown. As the airplane turns, vector H rotates in the horizontal plane to H', and thus the vertical axis OP is the axis of precession. Since the torque vector T must have the same direction and sense as vector ΔH, it is directed as shown, and this establishes the direction of the torque axis OT. Then applying the right-hand rule to the torque vector, we see that the torque is clockwise when looking inward along the torque axis; thus

$$\omega_s = \frac{2{,}000(2\pi)}{60} = 209 \text{ rad/sec}$$

$$I = \frac{W}{g} k^2 = \frac{60}{32.2} (3)^2 = 16.8 \text{ lb-ft-sec}^2$$

The gyroscopic torque is

$$T = I\omega_s\omega_p$$
$$= 16.8(209)0.2 = 703 \text{ ft-lb}$$

The bearings exert the vertical forces F against the shaft, as shown in Fig. 21-5b, in order to supply this torque. Then

$$Fl = T$$

or

$$F = \frac{T}{l} = \frac{703(12)}{6} = 1{,}406 \text{ lb}$$

The forces which the shaft exerts on the bearings are F' and are equal and opposite to F. Hence $F' = 1{,}406$ lb. Since the shaft pushes downward on bearing A and upward on bearing B, the airplane tends to nose downward when making a turn to the right.

In Fig. 21-5c, the forces which the bearings exert on the shaft are shown again, and M_A is the bending moment in the shaft; thus

$$M_A = Fl$$

Thus we see that in addition to affecting the loads on the bearings, the gyroscopic effects produce bending stresses in the shaft.

Note that in this example only the gyroscopic forces due to the rotating mass of the propeller were considered. In addition, there are gyroscopic forces caused by the rotating mass of the gear on the propeller shaft. These latter forces will be much smaller than those caused by the propeller because of the smaller value of I for the gear.

EXAMPLE 21-2   A toy gyroscopic top is shown in Fig. 21-6. If the disk of weight W is given a high angular velocity ω_s about its shaft OX and one end of the shaft is placed on a pedestal, the shaft and disk will not fall but will precess around the axis OY because of the torque Wl acting on the

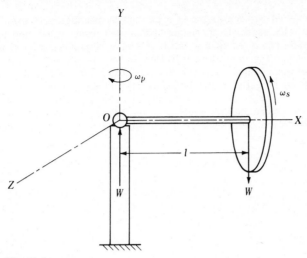

Fig. 21-6

system.   Then

$$Wl = I\omega_s\omega_p$$

or

$$\omega_p = \frac{Wl}{I\omega_s}$$

where I is the moment of inertia of the disk and ω_p is the angular velocity of precession, which is clockwise when viewed from above. If the shaft were prevented from precessing by some means such as holding one's finger in its path, it would immediately fall. Friction on the pedestal dissipates some of the energy, and the top gradually drops lower. Also, friction between the disk and its shaft causes a gradual decrease in ω_s, and because of this the angular velocity of precession increases, as can be seen from Eq. (21-4).

PROBLEMS

21-1 The rotor of a turbine on a ship has its axis of rotation parallel to the propeller shaft and is supported in bearings 15 ft apart. The rotor has a moment of inertia of 3,200 lb-ft-sec² and rotates at 1,800 rpm cw when viewed from the rear. The ship is making a turn to the right at 0.5°/sec. Make a sketch labeling the spin, precession, and torque axes, and label the front bearing A and the rear bearing B. Determine the magnitude and direction of the forces the shaft exerts on the bearings which are due to the gyroscopic effect alone.

21-2 The rotor of a jet airplane engine is supported by bearings as shown in Fig. P21-2. The rotor assembly consisting of the shaft, compressor, and turbine weighs 1,800 lb, has a radius of gyration of 9 in., and has its center of gravity lying at point G. The rotor turns at 10,000 rpm cw when viewed from the rear. The speed of the airplane is 600 mph, and it is pulling out of a dive along a path 5,000 ft in radius. (a) Determine the magnitude and direction of the gyroscopic forces which the shaft exerts against the bearings and label them F'. (b) Determine the magnitude and direction of the forces the shaft exerts on the bearings due to the centrifugal force acting on the rotor and label them R'_A and R'_B.

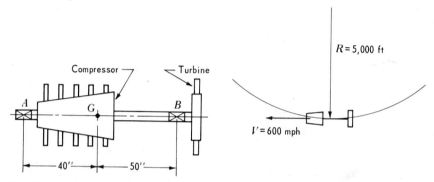

Fig. P21-2

21-3 In Fig. P21-3 a gyroscope used in instrument applications to maintain a fixed axis in space is illustrated. The rotor R is mounted in gimbals (A, B, C, D, E, and F are bearings) so that it is free to rotate about axes X, Y, and Z. The rotor has an $I = 0.002$ lb-ft-sec^2 and is driven electrically about the X axis with an angular velocity $\omega_s = 12,000$ rpm. Low-friction bearings are used to minimize precession. How much friction torque in pound-feet applied continuously is required to cause a precession of 1°/hr? Through which bearings must friction torque be applied to cause a precession about the Y axis?

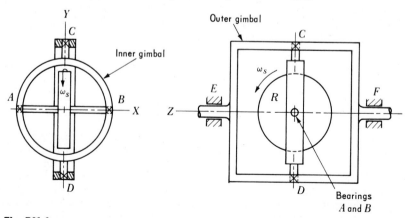

Fig. P21-3

22

CRITICAL SPEEDS
OF SHAFTS

22-1 INTRODUCTION

At certain speeds a rotating shaft or rotor has been found to exhibit excessive lateral vibrations.   The angular velocity of the shaft at which this occurs is called a *critical speed* or *critical whirling speed*.   At a critical speed the shaft deflections become excessive and may cause permanent deformation or structural damage; for example, the rotor blades of a turbine may contact the stator blades.   The large shaft deflections occurring at a critical speed cause large bearing reactions and can result in bearing failure or structural damage to the bearing supports.   This phenomenon occurs even for very accurately balanced rotors.   A machine should never be operated for any length of time at a speed close to a critical speed.   In this chapter we shall study why this phenomenon occurs and how a shaft or rotor can be designed so that the critical speeds will not be the same as the operating speed.

22-2 SHAFT WITH A SINGLE DISK

Consider the shaft in Fig. 22-1a with the disk of mass M located between the bearings.   In the following discussion we shall assume that the mass of the shaft is negligible compared with the mass of the disk.   Point O is

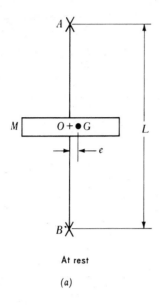

At rest

(a)

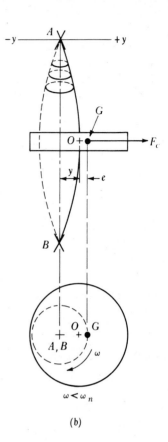

$\omega < \omega_n$

(b)

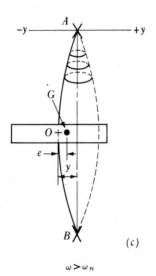

(c)

$\omega > \omega_n$

Fig. 22-1

on the shaft axis, and G is the center of gravity of the disk.   Then the distance e is the eccentricity.   In Fig. 22-1b the shaft is rotating and the centrifugal force F_c acts radially outward through G causing the shaft to bend as shown.   The centrifugal force is equal to the mass of the disk times the normal acceleration of point G; because the normal acceleration equals the radius of rotation times ω^2,

$$F_c = MA = M(y + e)\omega^2 \tag{22-1}$$

where ω is the shaft speed in radians per second and y is the deflection of the shaft where the disk is located.   The shaft behaves as a spring, and for a deflection y it exerts a resisting force ky, where k is the spring constant of the shaft in bending.   For equilibrium the resisting force equals the centrifugal force and hence

$$ky = M(y + e)\omega^2$$

from which

$$y = \frac{e\omega^2}{(k/M) - \omega^2} \tag{22-2}$$

In Fig. 22-2, the solid curve is a plot of Eq. (22-2).   From the equation we see that when ω is zero, y is zero, and when $\omega^2 = k/M$, the denominator becomes zero and y becomes infinite.   This value of ω is known as the *critical speed* ω_n and hence

$$\omega_n = \sqrt{\frac{k}{M}} = \sqrt{\frac{kg}{W}} \tag{22-3}$$

where W is the weight of the disk and g is the acceleration of gravity.   Also, we notice for values of $\omega > \omega_n$ the denominator of Eq. (22-2) becomes negative, and then y is negative as shown by the right branch of the solid curve. Further, we see from Eq. (22-2) that when ω becomes very large, y approaches $-e$.

In this analysis we have neglected the damping effect of air resistance acting on the shaft and disk.   Because of this damping the shaft deflection does not become infinite at the critical speed, and a plot of the actual deflection versus ω is indicated by the dotted curve in Fig. 22-2.   The effect of damping alters the critical speed but not appreciably.

From Fig. 22-2 we see that if $\omega < \omega_n$, y will be positive and the configuration of the shaft will be as shown in Fig. 22-1b.   If ω is constant, y will be constant, and the centerline of the shaft will remain in a fixed bent posi-

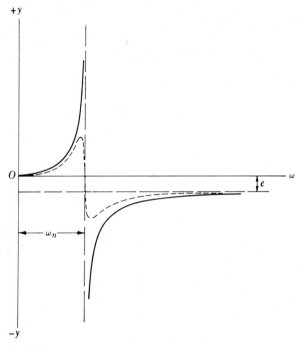

Fig. 22-2

tion whirling around axis AB, describing a surface of revolution.   From Eq. (22-2) we also note that if $\omega > \omega_n$, y will be negative, and hence the shaft bends in the opposite direction.   The configuration is then as shown in Fig. 22-1c, where y is negative.   It is known from observation that the side of the shaft which is convex below the critical speed becomes the concave side at speeds above the critical speed.   A piece of chalk brought into contact with the edge of the disk will make a mark on the heavy side.   At speeds above the critical speed a piece of chalk of another color is found to mark the disk on the light side.   Earlier we observed from Eq. (22-2) that for very large values of ω, y approaches $-e$.   Hence from Fig. 22-1c we see that the center of gravity of the disk then approaches the axis of rotation AB.   This explains why a rotor operated high above the critical speed operates smoothly.

We have noted that the shaft flexes; that is, it bends in the opposite direction as it passes through the critical speed.   Thus at the critical speed the shaft vibrates with the same frequency it would have if it were not rotating but were vibrating transversely as a beam.

From Fig. 22-2 we note that at speeds near the critical speed the shaft deflections are large, and from Eq. (22-1) we see that this makes for a large centrifugal force, and hence the forces on the bearings will be large. Because these forces are changing direction with the centrifugal force, the frame of the machine supporting the bearings will be set in vibration. Besides the objectionable noise due to vibration, the varying stresses in the bearings and the frame of the machine can result in structural damage. In addition, the large shaft deflections at speeds near the critical speed may cause permanent deformation of the shaft or contact between the rotor and its housing.   The amplitude of shaft vibration at critical speed reaches dangerous proportions only if time is allowed for the amplitude to build up; therefore if the machine is accelerated through the critical, the magnitudes of the amplitudes can be acceptable.   Machines like centrifuges and some high-speed turbines normally operate at speeds well above the critical and are brought up to their operating speeds by passing quickly through their critical speeds.

In Eq. (22-3), $k = F/y = W/y_{st}$, and thus

$$\omega_n = \sqrt{\frac{g}{y_{st}}} \tag{22-4}$$

Equation (22-4) is general and applies to any shaft of negligible weight having a single disk mounted anywhere along the shaft. y_{st} is the static deflection of the shaft at the disk produced by a force equal to the weight of the disk.

A shaft mounted in ball bearings may be assumed to be simply supported; if it is mounted in journal bearings it may be assumed to be supported as a fixed-end beam.

For a disk of weight W, mounted between bearings on a shaft of negligible weight and located distance a from the left bearing and distance b from the right bearing and where the bearings are ball bearings,

$$y_{st} = \frac{a^2 b^2 W}{3EIL}$$

and from Eq. (22-4)

$$\omega_n = \sqrt{\frac{3EILg}{Wa^2b^2}} \tag{22-5}$$

where ω_n = critical speed, rad/sec

E = modulus of elasticity for shaft, psi

I = moment of inertia of cross-sectional area of shaft = $\pi d^4/64$
  where d = shaft diameter

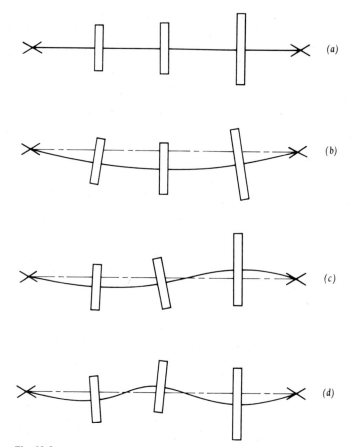

Fig. 22-3

g = acceleration of gravity = 386 ips²
W = weight of disk, lb
L = distance between bearings, in.

For the same conditions as above, except using journal bearings,

$$y = \frac{Wa^3b^3}{3EIL^3}$$

and

$$\omega_n = \sqrt{\frac{3EIL^3g}{Wa^3b^3}} \tag{22-6}$$

22-3 SHAFT WITH A NUMBER OF DISKS

The shaft shown in Fig. 22-3a, which has three disks, is representative of a shaft with any number of disks. The shaft may be considered horizontal or vertical because gravity does not affect the critical or whirling speed. The shaft could vibrate transversely in any of the three modes as shown in Fig. 22-3b to d, and the frequency would be different for each mode. Because the stiffness of the shaft increases as we go from Fig. 22-3b to c to d, the natural frequency increases.   Since the critical speeds correspond to the natural frequencies of transverse vibration, the deflection curves in the figure also represent the configuration of the shaft at its critical speeds. As the number of disks is increased, the number of modes of vibration and the number of critical speeds increases.   Usually it is only the lowest critical speed which is of importance because the operating speed of the machine is less than the higher critical speeds.   Therefore we will consider only the first order or fundamental critical speed.

The fundamental natural frequency of transverse vibration for a shaft with a number of disks can be found by Rayleigh's method. This method is based on the continuous interchange between kinetic and strain energy in the system.   Further, the method utilizes the assumption that the dynamic deflection curve for the shaft is similar in shape to the static deflection curve.

Consider the shaft in Fig. 22-4 to be of negligible weight and supporting any number of concentrated weights. To illustrate Rayleigh's method, let us suppose that there are three weights, W_1, W_2, and W_3. The solid line shows the axis of the shaft in its most deflected position as it vibrates transversely with a frequency ω_n in radians per second.   The dynamic deflections y_1, y_2, and y_3 may be assumed equal to the static deflections produced by the weights W_1, W_2, and W_3.

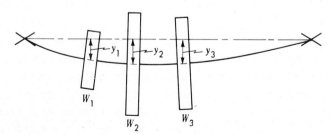

Fig. 22-4

The displacements for the vibrating shaft behave with simple harmonic motion because the restoring forces exerted by the shaft by virtue of its elasticity are proportional to the displacements. In Chap. 2, in the section on simple harmonic motion, we noted that for a particle having simple harmonic motion of frequency ω, maximum velocity is $V = y\omega$, where y is its maximum displacement, and the maximum velocity occurs when $y = 0$. Thus in Fig. 22-4 the vibrating shaft has its largest velocity and greatest kinetic energy when it reaches its unbent position. The kinetic energy for the system is then

$$KE = \sum \tfrac{1}{2}MV^2$$

and since $V = y\omega_n$

$$KE = \tfrac{1}{2}\frac{\omega_n^2}{g}(W_1y_1^2 + W_2y_2^2 + W_3y_3^2) \tag{22-7}$$

When the vibrating shaft reaches its most deflected position, all the kinetic energy is lost and is stored in the shaft as strain energy. The strain energy can be found by considering the work done on the shaft if it were allowed to deflect under the static forces W_1, W_2, and W_3. The average value of each force as it moves through its distance y is $W/2$. The work done by each force is its average value times the distance moved and is $Wy/2$. The strain or potential energy in the shaft is then the sum of the work done by the weights and is

$$PE = \tfrac{1}{2}(W_1y_1 + W_2y_2 + W_3y_3) \tag{22-8}$$

Equating the right-hand terms of Eqs. (22-7) and (22-8), we obtain the following expression for the natural frequency of transverse vibration and critical speed:

$$\omega_n = \sqrt{\frac{g \sum Wy}{\sum Wy^2}} \tag{22-9}$$

If the cross section of the shaft is uniform over the length, the static deflections in Eq. (22-9) can be most easily computed by using the equations for beam deflections given in mechanics-of-materials textbooks and by use of the *principle of superposition*, which is that the deflection at a given point along the shaft is equal to the sum of the deflections at that point as found by considering each load acting by itself. The following example illustrates the method.

EXAMPLE 22-1   Consider the shaft in Fig. 22-5a to be simply supported at A and B and to be of negligible weight compared with the two concentrated weights W_1 and W_2.   The shaft is steel and is $\frac{3}{4}$ in. in diameter.   It is desired to find the critical speed in rpm by Rayleigh's method.   The static-deflection curve is shown in Fig. 22-5b, but a better approximation to the dynamic-deflection curve will be obtained if we use the static-deflection curve shown in Fig. 22-5c, where W_2 is assumed to be acting upward rather than downward.   Thus we shall find the deflections y_1 and y_2 in Fig. 22-5c making use of equations given in mechanics-of-materials textbooks for finding deflections and slopes.

For the load W_1 acting alone on the shaft

$$y_1 = \frac{W_1 ba}{6dEI}(d^2 - b^2 - a^2) = \frac{40(4)6}{60EI}(100 - 16 - 36) = \frac{768}{EI} \downarrow$$

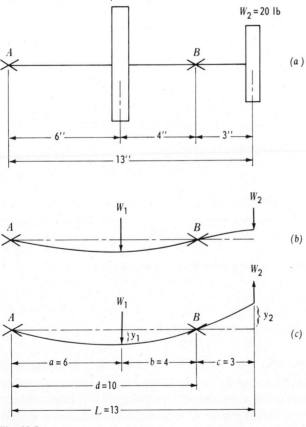

Fig. 22-5

Let θ_B be the slope of the deflection curve at B. Then

$$y_2 = c \sin \theta_B \simeq c\theta_B = \frac{cW_1ab(d + a)}{6dEI} = \frac{3(40)6(4)16}{60EI} = \frac{768}{EI} \uparrow$$

For the load W_2 acting alone on the shaft

$$y_1 = \frac{W_2ca}{6dEI}(a^2 - d^2) = \frac{20(3)6}{60EI}(36 - 100) = \frac{384}{EI} \downarrow$$

$$y_2 = \frac{W_2c^2L}{3EI} = \frac{20(9)13}{3EI} = \frac{780}{EI} \uparrow$$

By the method of superposition the resultant deflections are

$$y_1 = \frac{768}{EI} + \frac{384}{EI} = \frac{1152}{EI} \downarrow$$

$$y_2 = \frac{768}{EI} + \frac{780}{EI} = \frac{1548}{EI} \uparrow$$

$$y_1^2 = \frac{1.33 \times 10^6}{(EI)^2} \qquad y_2^2 = \frac{2.40 \times 10^6}{(EI)^2}$$

Then from Eq. (22-9),

$$\omega^2 = \frac{g \sum Wy}{\sum Wy^2} = \frac{386EI[(40 \times 1152) + (20 \times 1548)]}{(40 \times 1.33 \times 10^6) + (20 \times 2.40 \times 10^6)} = 0.294\,EI$$

and

$$\omega = 0.543 \sqrt{EI}\ \text{rad/sec} = \frac{60}{2\pi}(0.543)\sqrt{EI}\ \text{rpm}$$

For steel, $E = 30 \times 10^6$ psi, and

$$I = \frac{\pi d^4}{64} = \frac{\pi(0.75)^4}{64} = \frac{\pi(0.316)}{64} = 0.0155\ \text{in.}^4$$

$$\omega = \frac{60}{2\pi}0.543 \sqrt{30 \times 10^6 \times 0.0155} = 3{,}530\ \text{rpm}$$

22-4 CRITICAL SPEED OF A UNIFORM SHAFT

The critical speed of a shaft supporting a uniformly distributed load, or a shaft which supports no loads except its own weight, is found by considering the shaft as made up of a number of short lengths of known weight. The critical speed of a weightless shaft carrying these individual weights can then be determined by means of Eq. (22-9). By this procedure, the critical speed of a shaft of uniform diameter, simply supported at its ends and not supporting any concentrated weights, is found to be

$$\omega_n = 9.87 \sqrt{\frac{EIg}{WL^3}} \qquad (22\text{-}10)$$

where W is the total shaft weight in pounds, L is the distance between bearings in inches, and the other quantities are as defined below Eq. (22-5).

The speed obtained from Eq. (22-10) is the lowest critical speed. Other criticals occur at 4, 9, 16, 25, etc., times the lowest critical speed.

A shaft of uniform diameter, supported at its ends in rigid journal bearings so that it flexes as a beam with fixed ends, has for its lowest critical speed

$$\omega_n = 22.2 \sqrt{\frac{EIg}{WL^3}} \qquad (22\text{-}11)$$

and others at $(5/3)^2$, $(7/3)^2$, $(9/3)^2$, $(11/3)^2$, etc., times this speed.

22-5 DUNKERLEY'S EQUATION

When the critical speed of a shaft having any number of disks of weights W_1, W_2, W_3, etc., is to be determined, and the weight of the shaft is to be included, approximate results can be obtained using Dunkerley's equation:

$$\frac{1}{\omega_n{}^2} = \frac{1}{\omega_s{}^2} + \frac{1}{\omega_1{}^2} + \frac{1}{\omega_2{}^2} + \frac{1}{\omega_3{}^2} + \cdots \qquad (22\text{-}12)$$

where ω_n = critical speed of system, rad/sec

ω_s = critical speed of shaft alone, i.e., if all disks removed

ω_1 = critical speed of shaft if considered weightless and supporting only disk W_1

ω_2 = critical speed of shaft if considered weightless and supporting only disk W_2

22-6 SHAFTS WITH VARIABLE DIAMETER

Equation (22-9) can be applied to find the critical speed of any rotor shaft if the static deflections are known.   Mathematical determination of these deflections becomes very tedious if there are many loads and especially so if the shaft has different diameters along its length.   A combined numerical and graphical procedure for determining the deflections is then convenient.

Texts on mechanics of materials show that the basic differential equation which must be solved in order to find the static deflection is

$$\frac{d^2y}{dx^2} = \frac{M}{EI} \tag{22-13}$$

where y = deflection at any distance x from one end of shaft, in.

M = bending moment at distance x, in.-lb

E = modulus of elasticity of the material, psi

I = moment of inertia of cross-sectional area at distance x, in.[4]

Integrating Eq. (22-13), once we obtain the slope of the elastic curve of the shaft and integrating again, we obtain the deflection curve. Further, starting with the loading, one integration gives the shearing force and a second integration gives the bending moment. Hence, starting with the loads, four successive integrations are needed to obtain the deflections. The procedure is best shown by an illustrative example.

EXAMPLE 22-2 Consider the shaft in Fig. 22-6a to be of steel and simply supported in bearings as shown. The shaft supports a gear weighing 36 lb; it is desired to find the static-deflection curve. The shaft length is divided into a number of segments, and in order to facilitate the integration of the areas under the various curves, they are drawn to scale using any convenient scale along the x axis and any convenient scales for ordinates.

In Fig. 22-6a the loads are shown. The load at point B is the weight of the shaft from A to C and the load at D is the weight of the shaft from C to E, etc. By statics the bearing reactions were found to be as shown. The shearing-force diagram can then be drawn and appears in Fig. 22-6b. Next, the bending-moment diagram is constructed. Since bending moment is the integral of shearing force, M_B equals area A_1, which is 26.1 lb, multiplied by length AB, which is 2 in. Area A_2 is equal to 24.1 lb multiplied by length BD, which is $4\frac{1}{2}$ in. M_D equals the sum of areas A_1 and A_2. Note that throughout the entire procedure, areas above the x axis are considered positive, and those below, negative.

The moment of inertia of the cross-sectional area for each shaft diameter shown in Fig. 22-6a is computed next. Then ordinates for the M/EI diagram are obtained by dividing the ordinates in the moment diagram by the value of E, which is 30×10^6 psi for steel, and by the value of I for the corresponding cross section. Note that at section C the shaft has two values for I. The value of the moment at C can be scaled from the bending-moment diagram and is then divided by each of the values of I, in order to determine the two values of M/EI at this section. Similarly, since the shaft diameter changes at E, there will be a step in the M/EI diagram at this section.

The slope diagram in Fig. 22-6e is constructed next by laying off ordi-

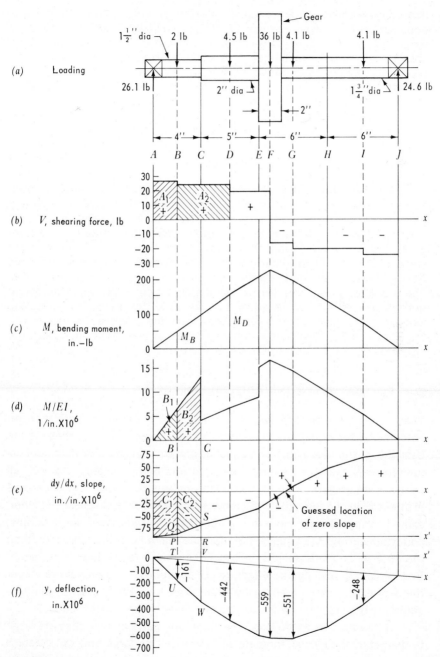

(a) Loading

(b) V, shearing force, lb

(c) M, bending moment, in.-lb

(d) M/EI, 1/in.×10⁶

(e) dy/dx, slope, in./in.×10⁶

(f) y, deflection, in.×10⁶

Fig. 22-6

nates from an axis x'.   Since slope is the integral of M/EI, ordinate PQ is equal to area B_1, which equals M_B/EI_B divided by 2 and multiplied by length AB, which is 2 in. PQ is laid off upward from the x' axis because area B_1 is positive.   Ordinate RS is equal to area B_1 plus area B_2.   We continue in this manner until the entire slope curve has been drawn.   The ordinates from the x' axis to this curve would represent the true values of slope along the deflection curve if the slope at the left end were zero. We know that actually the point of zero slope occurs somewhere between the bearings.   We can guess the location (distance x) of the point of zero slope and in Fig. 22-6e draw the x axis so that it intersects the slope diagram curve at this location.   Then areas lying below the x axis are considered negative and those above positive.

The deflection diagram in Fig. 22-6f is constructed by laying off ordinates from an axis x'.   Since deflection is the integral of slope, ordinate TU equals area C_1, and because C_1 is a negative area, TU is negative and thus is drawn downward.   Ordinate VW equals area C_1 plus area C_2.   We continue in this manner laying off ordinates from the x' axis until the entire curve is drawn.   At the bearings the deflections must be zero and thus through these points on the deflection curve the x axis is drawn as shown. The ordinates from the x axis to the deflection curve are then the actual deflections.   It is to be noted that in Fig. 22-6e, the better we guess the location of the point of zero slope, the closer the x axis will coincide with the x' axis in Fig. 22-6f.   However, regardless of how good our guess, the ordinates from the x axis to the deflection curve will be the same.

22-7   FACTORS AFFECTING CRITICAL SPEEDS

There are a number of factors which may alter the value of critical speed obtained by the equations presented in this chapter.   One of these is the flexibility of the bearings.   The usual assumption in calculating the deflection curve is that the bearings are rigid and do not deflect.   Actually all bearings deflect somewhat because of the loads upon them.   This increases the deflection of a rotor shaft and tends to lower the critical speed.   Further, some bearings deflect more in one direction than another.   For example, pedestal bearings are usually more flexible in the horizontal direction than the vertical, and this results in two critical speeds for the same shaft. Bearing length also has an effect on critical speed.   In general the longer the bearing the more it will restrain the shaft from tilting.   This results in stiffening the system and increasing the value of the critical speed.

If there are heavy disks on the shaft, and particularly if they are large in diameter, their polar moment of inertia will be large and will create a large gyroscopic effect which resists a change in direction of their axis. When the shaft starts to whirl, the disk resists the tilting of its axis.   This acts to stiffen the system and raise the value of the critical speed. This effect is greater for disks nearer the bearings where the slope of the shaft is greater.

In engineering we encounter rotors which operate in various media. Examples are grinding wheels, fans and compressors which operate in air, turbine rotors operating in gas, steam, or oil, and centrifugal pumps operating in water or oil. The operating media offer a frictional resistance called *damping*, which has little effect on the value of the critical speed and in practice is usually ignored when computing the critical. However, damping reduces the dynamic deflection of the shaft, and though this effect is slight for a rotor operating in air or some other gas, it is considerable for one operating in water or oil.

A close fit or a shrink fit for the hub of an impeller, gear, or pulley, particularly if the hub is thick and of considerable length, will increase the stiffness of the shaft and raise the critical speed.

In conclusion, experience has shown that if the operating speed of a shaft is removed at least 20 percent from any critical speed, there will be no vibration troubles.

PROBLEMS

22-1 Determine the critical speed in rpm of an electric motor designed to operate at 1,200 rpm. The steel shaft is ½ in. in diameter and the distance between bearings is 12 in. Consider the rotating element as being a single disk with its weight of 25 lb located midway between the bearings. Assume the shaft to be simply supported and neglect the weight of the shaft. Is the operating speed satisfactory?

22-2 For the shaft in Fig. P22-2 consult a mechanics-of-materials textbook for static deflection and determine the critical speed in radians per second neglecting the weight of the shaft (*a*) if the bearings are ball bearings and the shaft is assumed to be simply supported and (*b*) if the bearings are journal bearings and the shaft is assumed to be rigidly supported; i.e., the bearings do not allow the shaft to tilt at the bearings.

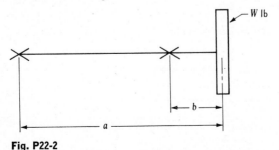

Fig. P22-2

22-3   The static-deflection curve for a shaft supported in three bearings is shown in Fig. P22-3. If the weights and the deflections are as indicated, find the lowest critical speed in rpm using the Rayleigh method.

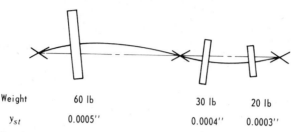

Weight	60 lb		30 lb	20 lb
y_{st}	0.0005''		0.0004''	0.0003''

Fig. P22-3

22-4   The drive shaft of an automobile is a hollow steel tube 56 in. long with an outside diameter of $2\frac{1}{2}$ in. and an inside diameter of $2\frac{1}{4}$ in. Each end of the shaft is fitted with a universal joint. Determine the critical speed in rpm. The maximum operating speed is 4,000 rpm. If it is desired to have the critical speed exceed the operating speed by at least 20 percent, is the design satisfactory?

22-5   Using Rayleigh's method, show that the critical speed for a shaft of uniform diameter simply supported is as given by Eq. (22-10). Hint: By Rayleigh's method

$$\omega_n = \sqrt{\frac{g \sum Wy}{\sum Wy^2}} \tag{22-14}$$

The static-deflection curve is shown in Fig. P22-5, and from a textbook on mechanics of materials we obtain

$$y = \frac{wx}{24EI} (L^3 - 2Lx^2 + x^3) \tag{22-15}$$

At any distance x in the figure we have a concentrated load $w\,dx$, and Eq. (22-14) becomes

$$\omega_n^2 = g \frac{\int_0^L w\,dx\,(y)}{\int_0^L w\,dx\,(y)^2} \tag{22-16}$$

Evaluate the numerator and denominator of Eq. 22-16 and solve for ω_n.

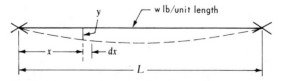

Fig. P22-5

22-6 The steel shaft in Fig. P22-6 is $1\frac{1}{2}$ in. in diameter and is simply supported in bearings as shown.   (a) Determine the critical speed in rpm of the shaft alone. (b) Using Dunkerley's equation, determine the critical speed in rpm for the shaft and the two disks.   (c) Same as (b) except neglect the weight of the shaft. (d) What percent error would there be in the critical speed if the weight of the shaft is neglected?

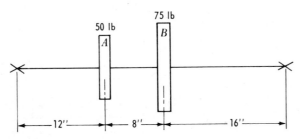

Fig. P22-6

22-7 The shaft shown in Fig. P22-7 is steel and is assumed to be simply supported at the bearings.   Determine the critical speed in rpm neglecting the weight of the shaft.   When finding the static deflections, make a drawing of the shaft using a scale of 1 in. = 2.5 in. and consider six segments having the lengths shown.   Use the following scales for ordinates: shearing force, 1 in. = 40 lb; bending moment, 1 in. = 150 in.-lb; M/EI, 1 in. = 20×10^{-6} 1/in.; slope, 1 in. = 100×10^{-6} in./in.; deflection, 1 in. = 300×10^{-6} in.

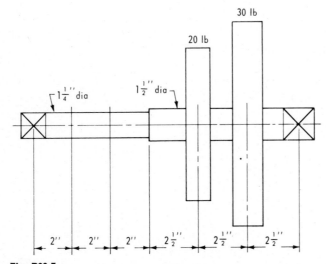

Fig. P22-7

22-8   The shaft shown in Fig. P22-8 is steel and is assumed to be simply supported at the bearings.   Determine the critical speed in rpm neglecting the weight of the shaft.   When finding the static deflections, make a drawing of the shaft using a scale of 1 in. = 6 in. and consider six segments having the lengths shown.   Use the following scales for ordinates: shearing force, 1 in. = 200 lb; bending moment, 1 in. = 1,000 in.-lb; M/EI, 1 in. = 150 × 10⁻⁶ 1/in.; slope, 1 in. = 1,000 in./in.; deflection, 1 in. = 5,000 × 10⁻⁶ in.

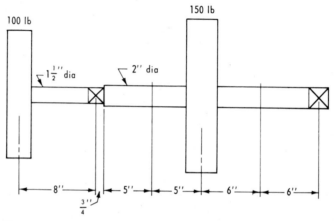

Fig. P22-8

PROOF OF
THE EQUIVALENT
FOUR-BAR LINKAGE

In Fig. A-1 a direct-contact mechanism consisting of links 1, 2, and 4 is shown. This will be referred to as the *original linkage*. An equivalent four-bar linkage is shown in dashed lines and consists of links 1, 2', 3', and 4'. Points C_2 and C_4 are the centers of curvature for the outlines of bodies 2 and 4, respectively, at the point of contact. A proof that the angular velocities and accelerations of links 2' and 4' are identical to those of links 2 and 4 at the instant is as follows:

Velocities

Let

$$\omega_{2'} = \omega_2 \tag{A-1}$$

$$\alpha_{2'} = \alpha_2 \tag{A-2}$$

where ω denotes angular velocity and α angular acceleration. Point 24 is the instant center for links 2 and 4 or for links 2' and 4'. The velocity of point 24 is

$$V_{24} = (O_2 - 24)\omega_2 = (O_4 - 24)\omega_4$$
$$= (O_2 - 24)\omega_{2'} = (O_4 - 24)\omega_{4'}$$

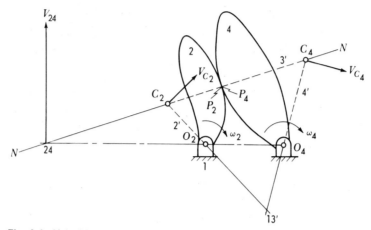

Fig. A-1 Velocities for a direct-contact mechanism and an equivalent four-bar linkage shown in dashed lines.

and

$$\frac{\omega_2}{\omega_4} = \frac{O_4 - 24}{O_2 - 24} = \frac{\omega_{2'}}{\omega_{4'}} \tag{A-3}$$

We have chosen $\omega_{2'} = \omega_2$; hence by Eq. (A-3) $\omega_{4'}$ must equal ω_4.

In Fig. 1 points P_2 and P_4 can have no relative velocity in the direction of the normal; and hence, C_2 and C_4 have no relative velocity in this direction. Thus the normal connecting points C_2 and C_4 on the original linkage behaves as a rigid link at the instant. C_2 as a point on body 2 has a radius C_2O_2, but as a point on line N-N its radius of rotation is line C_2O_2 extended.   Similarly, as a point on body 4, C_4 has a radius C_4O_4, but as a point on line N-N its radius of rotation is line C_4O_4 extended.   The extended lines intersect at 13′.   Hence, 13′ is also the instant center for line N-N on the original linkage, as well as for link 3′.   The angular velocity of line N-N is

$$\omega_{N-N} = \omega_{3'} = \frac{Vc_2}{13' - C_2} = \frac{Vc_4}{13' - C_4} \tag{A-4}$$

Accelerations

Next, it will be shown that if the end points of link 3′ are chosen at points C_2 and C_4, the angular accelerations of links 4′ and 4 will be identical. The equivalent linkage is shown again in Fig. A-2a, where A_{C2} and A_{C4} represent the linear accelerations of points C_2 and C_4.   We may write

$$A_{C_4}^n \mathbin{+\!\!\!\!+} A_{C_4}^t = A_{C_2}^n \mathbin{+\!\!\!\!+} A_{C_2}^t \mathbin{+\!\!\!\!+} A_{C_4/C_2}^n \mathbin{+\!\!\!\!+} A_{C_4/C_2}^t \tag{A-5}$$

where

$$A^n_{C_4} = (O_4C_4)\omega_4{}^2 \qquad A^t_{C_4} = (O_4C_4)\alpha_4$$

$$A^n_{C_2} = (O_2C_2)\omega_2{}^2 \qquad A^t_{C_2} = (O_2C_2)\alpha_2 \qquad\qquad \text{(A-6)}$$

$$A^n_{C_4/C_2} = (C_2C_4)\omega_{3'}{}^2 \qquad A^t_{C_4/C_2} = (C_2C_4)\alpha_{3'}$$

Equation (A-5) is represented by the vector diagram in Fig. A-2b. Since points O_2, C_2, C_4, and O_4 lie on both the original mechanism and the equivalent linkage, the directions of all the acceleration vectors in Fig. A-2b are the same for both. Further, since $\omega_{2'} = \omega_2$, $\alpha_{2'} = \alpha_2$, $\omega_{3'} = \omega_{N-N}$, and $\omega_{4'} = \omega_4$, we note from Eqs. (A-6) that $A^n_{C_2}$, $A^t_{C_2}$, $A^n_{C_4/C_2}$, and $A^n_{C_4}$ for the equivalent linkage are equal in magnitude, respectively, to those for the original mechanism. These four acceleration components are shown solid in the diagram and since the polygon must close, the magnitudes of $A^t_{C_4/C_2}$ and $A^t_{C_4}$ are consequently determined. That is, these two components, likewise, are the same for the equivalent linkage as for the original mechanism. Then

$$\alpha_{4'} = \alpha_4 = \frac{A^t_{C_4}}{O_4C_4}$$

It is known that any direct-contact mechanism has an infinite number of equivalent four-bar linkages. In Fig. A-3 point E is any point on link 4 or an extension of it, and the path which E describes on link 2 is shown. Point D is the center of curvature of this path, and the radius of curvature DE can be found by using the Euler-Savary equation. (See Sec. 7-9.) If link 4 were a point follower (shown shaded) and the path were used for the

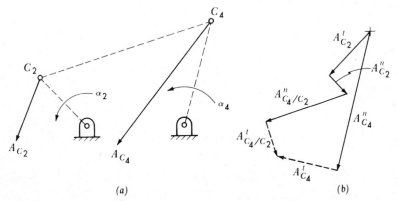

(a)   (b)

Fig. A-2 Accelerations for direct-contact mechanism or equivalent four-bar linkage.

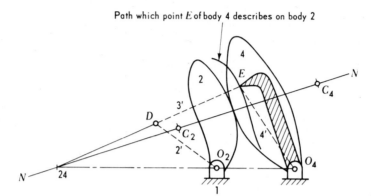

Fig. A-3 Direct-contact mechanism and another equivalent four-bar linkage shown in dashed lines.

outline of body 2, then the new body 2 and the point follower would have angular velocities and accelerations identical to those of the original mechanism. Since the center of curvature C_4 for the point follower lies at E, and the center of curvature C_2 of the path profile lies at D, the equivalent mechanism is now O_2DEO_4. Further, since E can be selected anywhere on link 4 or an extension of it, there are an infinite number of equivalent four-bar linkages. Point 24 in Fig. A-3 is the instant center for links $2'$ and $4'$ in any of the equivalent linkages, and the proof presented earlier applies for any of them.

INDEX